19.50/EM

UNIVERSITY LIBRARY
W. S. U. • STEVENS POINT

D1384762

Science and Technology in Economic Growth

Science and Technology in Economic Growth

Proceedings of a Conference held by the
International Economic Association at
St Anton, Austria

EDITED BY
B. R. WILLIAMS

A HALSTED PRESS BOOK

JOHN WILEY & SONS
New York – Toronto

© The International Economic Association 1973

Published in the United Kingdom 1973 by
THE MACMILLAN PRESS LTD

Published in the U.S.A. and
Canada by Halsted Press, a
Division of John Wiley & Sons, Inc.,
New York

Library of Congress Cataloging in Publication Data

Main entry under title:

Science and technology in economic growth.

" A Halsted Press book."
Includes bibliographical references.
    1.   Technological innovations—Congresses.
2.   Research—Economic aspects—Congresses.
I.   Williams, Bruce Rodda, ed.   II.   International
Economic Association.
HC79.T4S34   301.24′3   72–14227
ISBN 0–470–94679–2

HC
79
.T4
S34

Contents

234224

Acknowledgements

The International Economic Association wishes to express once again its indebtedness to the two bodies without which it could not have continued its general activities or held the conference that is recorded in this book – to UNESCO, its earliest sponsor, and the Ford Foundation as the major source of its income. For financial assistance in holding this conference it is greatly indebted also to the Oesterreichischer Nationalbank, and to its President, Dr. Wolfgang Schmitz. To the Austrian Nationalökonomische Gesellschaft and to Dr. A. Nussbaumer, its representative on our Council, special gratitude is due for their help in solving our problems of accommodation and in overcoming the usual minor difficulties of any conference. The Hotel Alte Post provided a most comfortable and most beautiful setting for our work. Finally we would express our gratitude to Professor Bruce Williams for all his labours as chairman of our programme committee, and as architect of the conference and of this volume.

List of Participants

Professor O. d'Alauro, University of Genova, Italy

M. Henri Aujac, Bureau d'Information et de Prévision Economique, Neuilly sur Seine, France

Professor A. L. Baumgarten Jr., Fundação Getulio Vargas/IBRE, Rio de Janeiro, Brazil

Dr. H. Choi, The Korean Economic Association

Professor D. J. Delivanis, University of Thessaloniki, Greece

Professor L. Dupriez, Université de Louvain, Belgium

Professor H. C. Eastman, University of Toronto, Canada

Mr. C. Economides, Cyprus Economic Society, Nicosia, Cyprus

Mme Michèle Fardeau, Université de Paris, France

Professor L. Fauvel, Université de Paris, France

Professor C. Freeman, University of Sussex, England

Professor J. K. Galbraith, Harvard University, Cambridge, U.S.A.

Professor L. M. Gatovskiy, Institute of Economics, Moscow, U.S.S.R.

Professor H. Giersch, Institut für Weltwirtschaft an der Universität, Kiel, Federal Republic of Germany

Professor J. C. de Graft-Johnson, Department of Economics, University of Ghana

Professor Z. Griliches, Harvard University, Cambridge, U.S.A.

Professor G. Haberler, Harvard University, Cambridge, U.S.A.

Professor Sir John Hicks, All Souls College, Oxford, England

Professor R. Hsia, University of Hong Kong

Professor E. James, Université de Paris, France

Professor Nurul Islam, Economic Growth Center, Yale University, New Haven, U.S.A.

Mr. Sukri Kaocharern, The Economic Society of Thailand, Bangkok, Thailand

Professor K. Kawano, Institute of Humanistic Studies, Kyoto University, Japan

Professor T. S. Khachaturov, Association of Soviet Economic Scientific Institutions, Moscow, U.S.S.R.

Professor E. Lundberg, The Stockholm School of Economics, Sweden

Professor F. Machlup, Princeton University, New Jersey, U.S.A.

Professor E. Mansfield, Wharton School of Finance and Commerce, University of Pennsylvania, U.S.A.

Professor J. Marchal, Université de Paris, France

Professor R. C. O. Matthews, Oxford University, England

Professor A. J. F. Misset, Waldeck Pyrmontlaan 20, Amsterdam-Zuid, The Netherlands

Professor R. Mosse, University of Social Sciences, Grenoble, France

Professor M. H. Mourad, Egyptian Association of Economy, Statistics and Legislation, Cairo, Egypt

Professor J. Mujzel, Polskie Towarzystwo Ekonomiczne, Warsaw, Poland

Professor P. Munthe, Oslo University, Norway

Dr. L. Nabseth, The Industrial Institute for Economic and Social Research, Stockholm, Sweden

Professor F. Neumark, Goethe University, Frankfurt/Main, Germany

Mr. K. P. Norris, Manchester Business School, England

Professor A. Nussbaumer, Vienna University, Austria

Professor F. Oelssner, Deutsche Akadamie der Wissenschaften, Berlin, Democratic Republic of Germany

Professor K. Oshima, Department of Nuclear Engineering, University of Tokyo, Japan

Professor G. U. Papi, Accademia Nazionale dei Lincei, Rome, Italy

Professor D. Patinkin, The Maurice Falk Institute for Economic Research in Israel, Jerusalem, Israel

Professor J. J. Paunio, University of Helsinki, Finland

Mr. A. W. Pearson, Manchester Business School, England

Professor J. O. N. Perkins, University of Melbourne, Australia

Professor F. J. Du Plessis, Economic Society of South Africa, Pretoria, South Africa

Professor P. N. Rasmussen, Institute of Economics, University of Copenhagen, Denmark

Professor E. A. G. Robinson, University of Cambridge, England

Professor J. R. Sargent, University of Warwick, Coventry, England

Dr. V. I. Sedov, Institute of World Economics and International Relations, Moscow, U.S.S.R.

Professor B. Šefer, Director, Living Standards Research Centre, Belgrade, Yugoslavia

Professor M. Simai, University of Economics, Budapest, Hungary

Professor I. Stefanov, Bulgarian Academy of Sciences, Sofia, Bulgaria

Dr. M. Teubal, The Hebrew University, Jerusalem, Israel

Professor S. G. Triantis, University of Toronto, Canada

Professor S. Tsuru, Hitotsubashi University, Tokyo, Japan

Dr. V. Wacker, Czechoslovak Economic Association, Prague, Czechoslovakia

Professor B. R. Williams, University of Sydney, New South Wales, Australia

Observers

M. M. Chapdelaine, Science Policy Division, UNESCO, Paris

Lady (Ursula) Hicks, Linacre College, Oxford, England

Mr. A. King, Department of Scientific Affairs, O.E.C.D., Paris, France

Mr. W. K. Norris, School of Social Science, Brunel University, England

Dr. H. Thomas, London Business School, England

Introduction

B. R. Williams

In the final session of the I.E.A. Conference reported in this volume, Professor Robinson asked whether there was something large in the explanation of economic growth which has defied economists: 'We can't have an amorphous residual factor for which there is no explanation, and yet this is the state in which I find myself at the end of the Conference.'

According to a much quoted assertion of Lord Kelvin's, when you can measure what you are speaking about, and express it in numbers, you know something about it; otherwise your knowledge is of a meagre and unsatisfactory kind. If Kelvin is right, our knowledge of economic growth is somewhat meagre and unsatisfactory. Econometric studies of growth still end up with amorphous residuals, though the amorphousness and the size of the residuals have been reduced substantially in recent years.

The factors that determine growth are many and varied. John Stuart Mill listed several of them over a century ago. He wrote of the causes of the degree of productiveness of the requisites of production – labour, capital, and the materials and motive forces by nature – as the second great question in political economy, but for a variety of reasons this 'second great question' was not pursued systematically over the years. Economic theorists found it convenient to treat many of the relevant factors as exogenous. Mill listed the quality of the natural resources, the energy of labour, the level of education, the skill and knowledge of the labourers and those who direct them, the security of persons and property in encouraging both production and accumulation, and the institutions of society. On science and technology he wrote that 'our knowledge of the properties and laws of physical objects . . . is advancing more rapidly, and in a greater number of directions at once, than in any previous age or generation. This increasing physical knowledge is now, too, more rapidly than at any previous period, converted by practical ingenuity, into physical power. . . . The manual part of these great scientific operations is now never wanting to the intellectual.'[1]

Marshall extended Mill's analysis and added 'organisation' to the traditional three requisites of production. His reason for so doing

[1] *Principles of Political Economy*, Book 4, Chapter 1, Section 2.

points the difficulty of putting certain numbers to the influence of inter-related factors in economic growth.

> Capital consists in great part of knowledge and organisation; and of this some part is private property and other part is not. Knowledge is our most powerful engine of production; it enables us to subdue Nature and force her to satisfy our wants. Organisation aids knowledge; it has many forms, for example, that of single business, that of various businesses in the same trade, that of various trades relatively to one another, and that of the State providing security for all and help for many. The distinction between public and private property in knowledge and organisation is of great and growing importance: in some respects of more importance than that between public and private property in material things; and partly for that reason it seems best sometimes to reckon organisation apart as a distinct agent of production.[1]

His suggestion that the Mecca of the economist lies in economic biology rather than in economic dynamics was very little appreciated until recently when the roles of organisation and education were brought back into the mainstream of economic studies.

Different types of models and analytical methods are required for an understanding of these diverse influences on economic development. There were examples of this in papers at this Conference on, for example, growth accounting, the speed of innovation, the distribution of scientific manpower, diffusion processes, and the role of organisation in the generation and use of science and technology. The extent of research in these fields has increased greatly since the war with the new emphasis on economic growth as a goal of public policy. Given the necessary variety of models and methods and the differences between countries, it is not surprising that this burst of activity has not as yet produced anything like a general theory of economic development to which the research in the different fields can be directly related. Nor is it surprising that workers in different fields often disagree about the general significance of their respective studies. Sometimes this is because of a very human tendency to overestimate the general significance of one's own work, or to forget the assumptions that had to be made, or the proxies that had to be introduced, to construct a working model. Sometimes it is because the immediate goals of research are different. There can in this respect be considerable differences between those working on time series to understand what has been, and those concerned to identify the critical or key factors which might be changed to increase the degree of effectiveness of the requisites of production. Here again, failure

[1] *Principles of Economics*, Book 4, Chapter 1, Section 1.

to be clear and explicit impedes fruitful communication. One of the purposes of conferences is to improve communications, and where researchers are concerned with different aspects of the same broad problem to see how far their different researches could be integrated.

Although the results of the burst of research activity in the last twenty years have not as yet led to a satisfactory general theory, the increase in knowledge has been impressive. It is interesting in this respect to look back at Colin Clark's *Conditions of Economic Progress*, first published in 1940 and then again in a completely rewritten form in 1951, shortly before the publication of Fabricant's *Economic Progress and Economic Change* (1954) and Solow's paper 'Technical Change and the Aggregate Production Function' (1957). On the basis of the time series available and the published studies, Clark produced an equation relating income per man-hour to capital per head of working population for various times and populations and concluded that 'the data are fairly well fitted by the equation'. 'Fairly well fitted' is a loose expression, and had economists at that time not been predisposed to believe that capital per worker was the predominant cause of the level of output per worker, the conclusion might well have been that the data 'are not very well fitted by the equation'. At any rate the conclusions drawn by statisticians at that time did little to dissuade economic theorists from elaborating models, of a kind first used explicitly by Cassel, in which growth was treated as a fairly simple function of the investment rate (or, as in pre-Keynesian economics, of the savings rate).

The publications of Fabricant and Solow constituted a major challenge to this tradition. Solow, for example, calculated that, from 1909 to 1949 for the non-farm American economy, 90 per cent of the increase in output per head was due to technical change and therefore only a minor percentage to increases in capital per worker. The dominance of technical change was a function of Solow's method. He measured increases in only two factor inputs – capital and labour – and he took into account purely quantitative changes. The size of his residual reflected this procedure. When, later, Denison included in his estimate of increased labour inputs changes in labour quality associated with increases in education, his estimate of the increase in output not explained by increases in labour and capital inputs came down to less than half Solow's figure. Nor in Solow's paper was there an attempt to identify the nature of the residual – of the relative influences of the various factors which Mill and Marshall and later writers had identified.

The first major attempt to do this was Denison's in *The Sources of Economic Growth in the United States*. In addition to allowing for the effects of education on labour inputs, he allowed also for the improved allocation of resources between industries, changes in lags in the application of knowledge, economies of scale, and 'advances of knowledge'. His later book, *Why Growth Rates Differ* (1967), extended his analysis to Western Europe. Though the statistical information was less adequate than for the United States, and at times his calculations smacked a little of 'pulling rabbits out of a hat', his results were both plausible and useful, and increased our understanding of changes in the relative importance of sources of growth from time to time and from place to place.

Although research in this field – many of the major contributions are listed in the papers by Griliches, Mansfield and Matthews – has achieved a great deal in a short period of time, it is clear from the papers and discussion at the Conference that there is still much to be done. So far the bulk of the statistical and econometric work has taken place in the United States. Owing to deficiencies in data, Denison's work on the sources of growth rates in Western Europe, for example, had to depend at times on approximations derived from United States data. It could be that this is producing certain distortions in the formulation of hypotheses in other countries. I suspect, for example, that what economists in countries with a lower level of technology expect from their own R & D – in contrast to what they should expect from better use of the existing stock of scientific and technological knowledge – has been more influenced by the results of the relatively large number of United States studies which seem to establish a positive relationship between R & D and the rate of growth of total factor productivity in different industries, than by the small number of much less positive results from data in other countries.

The precise results of statistical calculations of the sources of growth depend on the methods of measurement and on the framework of classification adopted.

The problems of measuring the inputs of labour and capital are well known. There are problems in constructing adequate price deflators, and there are many price deflators in which movements in cost are used as proxies for movements in prices. There are also important conceptual problems. Denison, for instance, in measuring inputs makes an allowance for improvements in the quality of labour inputs but makes no allowance for the improvement in the quality of capital inputs. At the moment there is no simple way of estimating changes in the quality of capital inputs, though the use of engineering data could provide an opportunity for worthwhile estimates.

It may be thought that the attempt to provide for changes in the quality of capital inputs would simply obscure the role of 'changes in knowledge'. In a sense this is true, but 'changes in knowledge' also affect the returns to education which are used to calculate quality changes in labour inputs. This is just an illustration of how the precise results are influenced by the framework of classification adopted. There is a need for much more field-study and for statistical and conceptual work in this area. Such work should yield valuable information on a question which is often relevant both to the formulation of models and to the issues of policy, namely the extent to which innovation is or is not capital-embodied.

One of the important preoccupations of growth economists has been the role of R & D. In the last ten years there has been a marked improvement in the quality of specification and measurement as the result of work by, for example, Terleckyj, Griliches, Jorgensen, Mansfield, Minasian, Brown and Conrad. But more work is needed on the extent of spurious correlations, on the length of time-lags between R & D and its application, and on the appropriate disaggregation and classification of R & D expenditures. A high proportion of useful econometric work on the influence of R & D has been conducted on an industry-by-industry basis. To go beyond this to achieve aggregated results in which we can have reasonable confidence, there are two specially important areas which call for more research. The first is the impact of defence and space research. We still know far too little about the impact of this research – much of it classified as non-economically motivated – on factor productivity. The second area, which in analytical terms is continuous with the first, is the inter-industry flow of new technology. The impact of R & D in one industry may have a substantial effect on productivity in other industries. These inter-industry flows of improved technology – and, too, intra-industry flows – are often assumed to be so important as to justify public subsidies to R & D, yet our knowledge of them is still very impressionistic.

The problem of identifying the contribution of R & D to growth in different countries is paralleled by the problem of identifying the role of education. I mentioned above the problem of classifying additional earnings associated with higher educational qualifications as improvements in the quality of labour. That education can improve the quality of labour is clear, though some forms of education have much more of this effect than others. It is also clear that a country's ability to add to knowledge through R & D is dependent on the extent and forms of education; and so also its ability to add to knowledge and to make use of new and old knowledge in adding to output. More work on the economics of education, and on the allocation of scientific and engineering manpower to the different activities of R, D, and application, is required to make possible a better understanding of the role of education in economic development.

Measures of the sources of growth, and particularly of the size of the residuals, are affected by methods used for measuring outputs. The problems of making proper allowance for changes in the quality of goods and services, and the conventions adopted for measuring Government outputs, have become more acute as the rates of expenditure on R & D have risen. Changes in quality may be positive or negative. Some changes in technology may bring very complex mixtures of positive and negative changes. Costs such as pollution of air and water and accidents are not caught in the process of measuring net output, except to the extent that Government measures to offset them are brought into output at cost. Though there can be no final answer to some of the questions relating to the conception of output, there is room for worthwhile change in the conventions adopted for the measurement of outputs.

The improved quality of work on the relations between R & D and economic development has been associated with complementary research on the extent of change that does not require formal R & D, on the richly varied nature of diffusion processes, on the inter-relations between capital expenditure decisions and production and marketing activities in successful innovation, and on the role of organisation more generally.

This Conference was concerned with the role of science and technology in economic development. This is a much more general topic than the role of R & D. Diffusion processes are, of course, important in the study of R & D, but they are of greater importance in the more general content. Over the years economic historians have provided valuable studies of the intra- and inter-national diffusion of technology. Recently, interest in contemporary diffusion processes has increased substantially. It is clear from the work of Griliches, Mansfield and colleagues, Carter and Williams, Salter, Dunning, the groups of research workers at O.E.C.D., the British National Institute, the Swedish Institute of Industrial Research and the Illinois Institute of Technology, Hall and Johnson, and others, that this is a very fruitful field in which the various inter-related factors including management, market structures, international flows of investment, capital markets and tax structures will take a good deal of unravelling.

The implications of some of these factors for public policy measures are direct and obvious. Public policy decisions relevant to economic growth do not wait for a complete, or even a reasonably full, understanding of the processes of economic growth. Nor are all economists noted for their reluctance to give advice on policy issues before they have a rounded and substantiated theory. Some of the advice given has been properly concerned to combat error at the basis of public policies or proposals; some positive advice has, regrettably, been based in error.

There are some very live issues in public policy concerning which it is to be hoped economists will shortly be in a better position to give firmly-based positive advice.

A high proportion of economists who have written on R & D have suggested that we are or may well be underinvesting in R & D. The group pressures to increase public expenditure on R & D are so strong that there is much to be said for expediting economic research on this issue. Two areas of research which seem particularly relevant are the spillover effects of R & D from one industry to others, and the effects of market structures on the type and extent of R & D. Of these two the second is likely to prove the more difficult to unravel.

In many countries inward direct investment has played and/or is playing a very important role in the diffusion of technology and growth. There are, however, strong political objections in many countries to such a process of diffusion. Economists have already contributed substantially to an understanding of conflicts between policy objectives in this area by analysing the effects of tariffs, restrictive business practices, and management weaknesses in encouraging diffusion through the establishment of foreign companies. More research on diffusion processes consistent with different policy mixes could prove very useful. It may be that our basic knowledge of diffusion processes is not yet sufficient. If so, that knowledge should provide a useful stimulus to further basic work.

There are purposes in education which go far beyond 'investment in man' to increase economic development. Yet the presumed investment effect is still extensively used in advocacy for more expenditure on education. Research in the economics of education has so far proceeded on such a level of aggregation that, with imperfections in the labour market, it is not possible to draw many firm conclusions concerning the relation between education and economic growth. The success of attempts to reconcile policy objectives referred to in the previous paragraph could be very much influenced by both the extent and forms of education. Further work on the lines of Sargent's paper in this volume is directly relevant to this issue. There is, therefore, a strong 'policy interest' in much more detailed research on the effects of education.

The early writers on economic development did not believe in unlimited growth. J. S. Mill flirted with the idea, but none the less came back to the likely coming of the stationary state, and wrote of its social advantages. Recently interest in the limits of growth has revived. In less developed countries that interest tends to concentrate on population growth. In richer countries, problems of congestion and pollution, and doubts concerning the general quality of life engendered by the modern industrial state, have resulted in a growing number of suggestions that further economic development should cease to be a policy goal. These doubts have been reinforced (where they have not been created) by forecasts that it will not be possible to use science and technology to offset the rapid depletion of natural raw materials.

We spent very little time at this Conference on the limits of growth. Most economists are very critical of the Forrester models, popularised in the 'Club of Rome' study of the *Limits of Growth* by Meadows and his colleagues. There have been criticisms of the structure of the model and of the quality of the inputs. However, it would be surprising if a significant number of economists did not apply their minds to the possibilities of improving both.

Already a greater interest in environmental issues and the quality of life has led to a greater interest in defining desirable forms of growth and the means of attaining them. This shift is likely to go further, and I guess that within growth economics there will soon be a new special field devoted to feasible or workable growth models, with the classical stationary state as a limiting case. But there is little chance that these models will have such an effect on public and private policies that the opportunity to extend our knowledge of the sources of growth will disappear.

1 The Contribution of Science and Technology to Economic Development[1]

R. C. O. Matthews
OXFORD UNIVERSITY

I. INTRODUCTION: CONCEPTS AND RELATIONSHIPS

The broad question, what is or has been or could be the contribution of science and technology to economic growth, may be asked for a number of reasons. First, we may want to know as a matter of intellectual curiosity, how important science and technology have been as a source of economic growth in the past, compared with other sources of growth. Secondly, for purposes of policy, we may want to know the likely effects on economic growth of an increase (or decrease) in the total amount of scientific and technological input. Thirdly, again for purposes of policy, we may want to know the best way of allocating a given amount of such inputs in the interests of economic growth.

These questions, as stated above and in many discussions of the subject, are not very clearly formulated, because 'science and technology' can be taken to mean a number of quite different things. It can mean: (1) the state of scientific and technological knowledge in the abstract at a point of time; (2) the extent to which scientific and technological knowledge is embodied in the work-force, as measured in particular by the number and quality of qualified scientists and engineers (Q.S.E.s); (3) the level of expenditure in the area designated as science and technology, including not only the salaries of scientists and technologists but also the cost of equipment and ancillary personnel. Moreover, the level and the rate of growth of real G.D.P. in any period are likely to be affected not only by the flow of services from science and technology (in any of the above-mentioned senses) in the current period, but also by the flow of those services in the past.

In view of these conceptual ambiguities, it will be helpful to begin by outlining a rough theoretical framework.

[1] This paper was originally written for a conference organised by UNESCO in co-operation with the International Economic Association, and held in Paris in 1968. It was published in *The Role of Science and Technology in Economic Development*, Science Policy Studies and Documents, No. 18, UNESCO 1970, and is here reproduced without change by kind permission of UNESCO.

The production process consists of using a flow of inputs to produce a flow of outputs. We may classify the inputs and the outputs in a nation's economy in different ways according to the subject under study. For the present purpose, which is to study the role of science and technology, we may use some such classification as the following.

Inputs, in the broadest sense, consist of flows of services from:

(1) Labour (L).

(2) Physical capital, including land (K).

(3) The state of knowledge (G) relevant to production – the 'state of the arts' in the terminology of the older writers. This means knowledge in the abstract – what is available to be known. We may divide it into scientific and technological knowledge relevant to production (G_s) and other knowledge relevant to production (G_n). G_n includes, for example, knowledge of the principles of accountancy, business administration and personnel management. The line between G_s and G_n is, of course, to some extent arbitrary.

(4) Human capital (H), that is to say the skills embodied in the labour force, reflecting the extent to which the labour force L is familiar with and capable of using the available stock of knowledge, G_s and G_n. Human capital may be subdivided, like knowledge, into scientific and non-scientific parts, H_s and H_n.[1]

(5) Imports (M) used as intermediate goods in the process of production.

These various inputs are used in co-operation in the different 'activities' or industries that comprise the production process. Aggregating over all activities, we may thus write $Y = f(L, K, G_s, G_n, H_s, H_n, M)$, where Y is output. The different inputs are used in different proportions in different activities, and not all activities necessarily use all the inputs. Almost all activities, however, use more than one input.[2]

The outputs consist partly of additions to the stock of inputs available for future use and partly of goods and services for current use. They are:

(1) Additions to K;

(2) additions to G_s and G_n;

(3) additions to H_s and H_n (by education and training);

[1] H can alternatively be treated as an attribute of L (the quality of the labour force) or an attribute of G (the degree of diffusion of knowledge), rather than as a separate input.

[2] About the only exception in normal national accounting classification is the activity 'occupation of dwellings', which uses only physical capital.

(4) consumption goods (C);[1]
(5) exports (X).

There are thus seven types of input, G_s, G_n, K, H_s, H_n, L, and M; and seven types of output, ΔG_s, ΔG_n, ΔK, ΔH_s, ΔH_n, C, and X. Just as each activity is likely to involve more than one input, so also each activity may produce more than one output. In the extreme case one may think of each activity as using some of each input and producing some of each output, but in differing proportions. In practice there are likely to be more zeros in the matrix relating activities to outputs than there are in the matrix relating activities to inputs. But it is important to recognise that any given output may be produced by more than one activity. For example, ΔG_s accrues not only from the activity whose primary purpose is to produce it, namely R & D; it may also accrue as a by-product in activities primarily concerned with producing C or ΔK (the process of 'learning by doing') and in the activity primarily concerned with producing ΔH_s (education). Moreover, allowance must be made for the possibility that some outputs accrue without the application of any inputs at all in the ordinary economic sense: some education (ΔH) is given by parents and friends, and some technical advances (ΔG_s) are discovered accidentally or by leisure-time inventors.

The foregoing is all very general and does not go any distance towards answering actual empirical questions. It serves, however, to indicate the complexities of the processes involved. It also helps to indicate which are sensible questions to ask and which are not.

It is not useful to ask, what is the contribution of any one input to total production. Since different inputs are used in conjunction in the production process, if the level of any one input (L or K or G_s or any of the others distinguished above) in the economy as a whole were zero, output would be zero or near to it. In this sense all of the inputs are responsible for all of the output.

It is more promising to ask, what is the contribution of the growth of each input to the growth in production. The question then is, by how much would output have increased if input i had increased as it actually did, and other inputs had remained constant; or alternatively, by how much less would output have increased, compared with what it actually did, if input i had remained constant and other inputs had increased as they actually did. Given that the changes in inputs are small and that the production function is smooth, these two questions come to the same thing.

If, on the other hand, we are considering changes over a

---

[1] Additions to knowledge not useful for the purposes of production should be included under this heading.

substantial period of time, the two questions do not come to the same thing. There is then essentially the same problem as there is in measuring the contribution of an input to the level of production. Because inputs are complementary, the increase in output resulting from simultaneous increases in several inputs exceeds the sum of the increases in output that would result from an increase in each input considered by itself. The importance of the interaction effect, for example, makes it pointless to ask, what proportion of the increase in output between 1900 and 1968 was due to capital accumulation and what proportion was due to advances in technical knowledge. What we can do is ask what proportion of the annual increase in output between each successive pair of years within that period was on average due to each of the various sources. This is quite proper, but in quoting it as an explanation of the sources of growth between 1900 and 1968 we have to remember what it means.

The appropriate procedure is thus to concentrate on the effects of marginal changes, while taking account of interaction effects where these are relevant. This applies as much to forecasting the future, in the context of policy discussions, as it does to explaining the growth experienced in the past.

The other important point which emerges from the foregoing discussion is that the contribution of science and technology to growth takes a number of quite different forms. One source of growth is the increase in G_s. This comes about partly from the expenditure of scientific skills (H_s) on R & D, but it may also come about in other ways as well. Further, H_s is used not only for the purpose of increasing G_s; much of it is used (in co-operation with other inputs) in producing current output. An increase in H_s is therefore likely to bring about an increase in output even if there is no R & D at all. Moreover, the level and/or the rate of increase of G_s and H_s may affect the rate of growth of other inputs. Finally, since R & D is both part of output and a source of increases of future output, a problem exists of identifying the direction of causation in any empirical relationship that may be found between R & D and output.

The technique of identifying the sources of growth familiar in the work of Denison [4, 5] and others is based on the assumption that the addition to national output made possible by the employment of a unit of a factor at the margin can be identified (subject to some qualification) with the payment the factor actually receives. This assumption makes it possible, for example, to use data on the rate of return on capital to infer the increase in output that has been made possible by the increase in the capital stock. This technique is not readily applicable to identifying the contribution made by science and technology, for a number of reasons. G_s is not a factor of production with a statistically identifiable rate of remuneration, like labour or capital. The return on that part of the increase in G_s that is due to R & D expenditure could in principle be measurable, but since R & D is normally performed within the firm and not by a separately paid outside contractor, the return from it cannot in practice be separated from other elements of profit. The payments to H_s can be identified more easily, as the salaries of Q.S.E.s, but the quality of the skills embodied is difficult to take into account. Finally, even if these rates of payment could be identified, there remains the problem of possible divergences between private and social marginal product, which is particularly acute in this area. It is for these reasons that the whole subject remains so unsettled. Numbers of models have been proposed on the way in which R & D, scientific training, etc., enter into the production function and on the form of the complementarities and competitiveness existing between them and other inputs [12, 15, 22, 25]. This is one that one would need to know in order to determine marginal social products. This literature remains, however, mostly at the *a priori* level.

In what follows, the contribution of science and technology will be discussed under three headings. The first and longest section (section II) deals with the growth in G_s due to R & D. Section III deals with increases in G_s due to other causes. Section IV is concerned with the effects of increases in H_s.

II. RESEARCH AND DEVELOPMENT

(1) *The rate of return on R & D and macro-comparisons*

From the point of view of the firm, expenditure on R & D is a form of investment – a current outlay incurred in the expectation of future benefits. A natural starting point is therefore to assume that except in so far as specific reasons can be shown to the contrary, rational profit-maximisation will cause the net rate of return on R & D to be the same as on other forms of investment, that is to say physical capital formation.[1] We say net rate of return, because allowance must be made for depreciation. The results of R & D are subject to depreciation in so far as they are made out of date by subsequent advances. The speed of this depreciation will vary greatly, as it does with physical capital. In some cases, for example where a whole product becomes obsolete, the depreciation may be rapid. In other cases, subsequent advances may depend on the results of the earlier R & D, so that there is in effect no depreciation.[2]

---

[1] Denison [4] argues that this may not be a bad approximation even when allowance is made for the qualifications of the type discussed below.

[2] Depreciation that results from the appropriation of the results of R & D by competitors is discussed below.

Suppose we further assume, in the absence of any firm information, that the proportion of R & D required to make good depreciation is the same as that of physical capital. Then the relative contributions of R & D and fixed capital formation to growth can be inferred from the relative magnitudes of expenditure on R & D and expenditure on gross fixed capital formation. So if, for example, R & D is one-fifth of gross capital formation, we can take one-fifth of the estimate of capital's contribution to growth (as calculated by measures such as Denison's) as an approximate indication of R & D's contribution to growth.[1]

This line of argument leads to the conclusion that the contribution of R & D to growth has increased rapidly but is still small.

Expenditure on R & D has grown very much faster than either G.N.P. or fixed capital formation in most advanced countries. In the United States, R & D expenditure amounted to 3·4 per cent of G.N.P. in 1963/4, compared with 1·4 per cent in 1953, 0·6 per cent in 1940, and 0·2 per cent in 1921 [20; 16, p. 46]. A rise almost but not quite as fast as in the United States in the proportion of R & D to G.N.P. is found in the post-war period for the other countries for which data are available (United Kingdom, Japan, Netherlands, France); Canada is exceptional in not showing an increase [17, p. 23]. The rate of increase in the U.S.S.R. appears to have been faster than in the United States [8]. Pre-war data are not generally available for countries other than the United States, but the relatively low ratios to G.N.P. found at the beginning of the post-war period indicate that, as in the United States, the spectacular growth in the ratio is chiefly a post-Second World War phenomenon. National R & D statistics do not normally cover activities of individual inventors, whose relative importance has diminished over time. Moreover, the growth of organized research laboratories may make for more complete coverage in the data in more recent years than in earlier periods, when R & D departments of firms were less frequently separated from their production departments. These considerations may lead to some overestimate of the upward trend in total R & D expenditure. It is inconceivable, however, that allowance for this could upset the conclusions that the proportion of national resources devoted to R & D in most advanced countries has increased very substantially and that this is a relatively recent phenomenon.

[1] Apart from the more general qualifications to be discussed, this is an approximation in that, if net R & D is treated as a form of net investment, the overall rate of profit on R & D and other capital taken together should be measured by $(P+R)/(K+ \int_\infty^\infty R \ dt)$, where P = profits as conventionally measured, K = (physical capital, and R = net R & D. This will not be exactly equal to the conventional measure of the profit rate, P/K.

TABLE 1.1

EXPENDITURE ON R & D COMPARED WITH OTHER ECONOMIC INDICATORS IN 12 COUNTRIES

Country	Expenditure on R & D as % of G.N.P. in 1963[a]		Gross Fixed Capital Formation as % of G.N.P.	Population in 1965 (millions)	G.N.P. per Head in 1965 (U.S.$)	Rate of Growth of G.N.P. per Employee 1960–5
	Total R & D	Economically-motivated R & D				
United States	3·4	1·0	16	193	3,560	2·9
United Kingdom	2·3	1·2	16	55	1,810	2·5
Netherlands	1·9	1·3	24	12	1,550	3·2
France	1·6	0·6	20	49	1,920	4·6
Sweden	1·5	0·8	23	8	2,500	4·1
Japan	1·4	1·0	33	98	850	8·3
Germany (Fed. Rep.)	1·4	0·9	25	59	1,900	4·1
Canada	1·1	0·5	22	20	2,460	2·6
Belgium	1·0	0·8	20	9	1,780	3·5
Norway	0·7	0·4	30	4	1,880	4·5
Italy	0·6	0·4	23	52	1,100	4·5
Austria	0·3	0·2	24	7	1,270	4·2

Sources: [19, 20, 27]
[a] 1964 for Germany (Fed. Rep.), Netherlands, Sweden; 1963–4 for the United States; 1964–5 for the United Kingdom.

Despite this rapid growth, R & D remain in all countries a small item of expenditure relatively to fixed capital formation (Table 1.1). Calculation along these lines therefore inevitably leads to the conclusion that R & D is responsible for only a small proportion of growth. (This would still be true even if we assumed that R & D has no depreciation element and therefore compared it with net rather than gross capital formation.) This is, of course, subject to what has already been said about interaction effects. But it is not possible to maintain that at the margin R & D is critically important for growth unless some of the foregoing assumptions are rejected. Possible reasons for doing this will be considered presently.

The figures in Table 1.1 show the difficulty of establishing any simple relationship between R & D expenditure in a country and its rate of growth (for indeed any other macro-economic indicator). This lack of relationship could come about because R & D is a relatively unimportant source of growth, but of course it is also consistent with the hypothesis that R & D is important but that the relationship is obscured by others.

The data shown in Table 1.1 are well known, but a few comments on them may be made at this point. It is well established that developed countries devote a higher proportion of their G.N.P. to R & D than do developing countries: for most of the latter (not shown in Table 1.1) the percentage is 0·2 or below. Apart from this, there are few general conclusions suggested by the figures. The high overall R & D figure for the United States is entirely due to the magnitude of the defence and space efforts: when comparison is confined to economically-motivated R & D (that is to say, when we exclude R & D in the defence-space-nuclear areas and also welfare and miscellaneous R & D), the United States figure is not particularly high. In fact, the ratio to G.N.P. of economically-motivated R & D does not vary much between the major industrial countries, with the exception of France and Italy where it is significantly lower. The similarities between the major countries are more striking than their differences. The table does not suggest any clear relationship between the ratio of economically-motivated R & D to G.N.P. and either G.N.P. per head or the rate of growth of productivity. The cases of Canada, Norway and Austria, and to a less extent Sweden and Belgium, suggest that countries which are small or in the orbit of a larger neighbouring country tend to have relatively low R & D; but this is not true without exception, as instanced by the case of the Netherlands.[1]

[1] Comparable data are not available on the U.S.S.R., but it has been estimated that expenditure on 'science' was 3·0 per cent of G.N.P. (factor cost, Western concept) in 1962 [8, p. 124]. 'Science 'is in some respects wider than R & D and

(2) *Influences of demand and co-operating factor supply on the return from R & D*

Using the return on investment in physical capital as a proxy for the return on R & D, even if it were valid for interpreting the past, would not give any guidance to likely trends in the return on R & D in the future. Nor does it cast any light on possible interaction effects. In considering such questions, it is useful to look at the evidence on the distribution of R & D expenditures between industries.

Industries differ enormously in their research intensiveness (measured by the ratio of R & D expenditure to sales or to value added). Aircraft, electrical engineering (including electronics), chemicals and allied, and scientific instruments come at the head of the list for research intensiveness. The relative research-intensiveness of different industries is closely similar in most countries. This does not mean that there are not some significant differences between countries,[1] but the general picture is of similarity rather than the reverse [3, 20].

The great differences in industries' research-intensiveness have as a consequence that R & D is highly concentrated in a small number of industries, out of proportion to their weight in the economy. The extreme instance is the aircraft industry which, for example, in the United Kingdom accounts for over a third of total R & D in industry and for little more than 2 per cent of industrial output. In both the United States and the United Kingdom, aircraft, electrical equipment and chemicals between them account for over two-thirds of total industrial R & D. At the other end of the scale, major industries like construction, mining, iron and steel, food and paper (to say nothing of services and distribution) have little R & D expenditure.

R & D thus operates over a narrow front. However, most of the R & D-intensive industries are producer-goods industries. In so far

---

in some respects narrower. Taking the figure as is stands as a proxy for R & D, it suggests that the Soviet R & D effort is on the same scale relatively to national income as the American; the proportion of the population engaged on R & D (for which the data are firmer) is also similar in the two countries [8, p. 72, 113]. Bearing in mind the very heavy commitment of the U.S.S.R. to defence and space R & D, and the high cost this involves in view of the much lower income per head compared with the United States, it may be conjectured that Soviet R & D in the 'economically-motivated' category is not particularly high as a proportion of national income by international standards, though it is probably higher than in most other countries with a similar income per head.

[1] There are also differences between countries in the allocation of total R & D between industries due to differences in the relative importance of the industries in the countries' economies.

as R & D leads to an improvement in the quality of these goods, it may lead to an increase in the productivity of the much wider range of industries that use them. This applies not only to capital goods but also to other intermediate goods (for example, R & D in the chemical industry leads to productivity increases in agriculture).

The similarity between countries in the relative R & D intensity of different industries points to the conclusion that the reason lies in the inherent physical characteristics of those industries – they are 'science-based industries'. It might seem natural to go on from this to conclude that the degree of amenability to science determines the rate at which technical knowledge is advanced by R & D in different industries.

This conclusion, however, is on the face of it in some conflict with the important findings of Schmookler [24]. Schmookler provided strong evidence that the state of demand for the final product is the major factor determining the level of R & D output, as measured by patents in an industry – much more important than exogenously arising scientific discoveries, which in the industries he studied had little apparent effect in the patent statistics. This result, pointing to the primacy of demand factors, is subject to certain qualifications. In the first place, it is not denied that purely technological factors determine the most efficient way for innovative activity to respond to given demand pressure. In particular, the relatively low level of R & D in many consumer industries is viewed by Schmookler as coming about because the most efficient way of increasing productivity in these industries is by product innovation in the industries producing the capital goods used by them. In the second place, it is not denied that the general level of scientific sophistication and its diffusion among those responsible for production decisions is important in determining the rate of technical advance. What Schmookler denies is that individual scientific discoveries have a major role in determining the timing and direction of inventive activity: 'the effect of the growth of science is normally felt more from generation to generation than from one issue of a scientific journal to the next' [24, p. 200]. In the third place, Schmookler's results relate to output of R & D results (patents) rather than to input of R & D resources.[1] The regularly high level of R & D in certain industries may point not so much to their amenability to R & D but to the high cost of attaining given R & D results, which are, none the less, for competitive reasons necessary.

Emphasis on the importance of demand need not therefore involve denial of the existence of technological differences between in-

---

[1] This is quite apart from the possibility that patents may not be a good measure of inventive output.

dustries. It indicates, however, that technological factors are not the only ones. In particular, Schmookler's findings appear to show that technological factors do not provide the main explanation of variations through time in the amount of useful inventive activity in a given industry. This may be compared with the commonplace finding in the histories of technology that the original inventor of a new product or process has in very many cases not been the chief financial beneficiary, if indeed he has been a beneficiary at all; the successful execution of an innovation has taken place not when the original advance in knowledge was made but when economic conditions were, and were seen to be, ripe for its use.[1]

The state of demand is not the only economic force, which, along with the state of technology, determines the profitability of R & D. Also important is the supply of the factors of production needed to carry out R & D and implement its results. This does not affect comparisons between industries so much as comparisons between countries and periods (since broadly speaking all industries may be expected to have similar access to factor supply in a country at a given time). Particularly important here is the supply of saving. R & D is a form of investment, and in so far as all saving is scarce (as it is in developing countries) R & D has to satisfy high standards of social profitability in order to be worth while. The scarcity of saving is important in another way also: innovation, as has been seen, occurs largely in capital goods industries and thus has to be 'embodied' in new capital in order to be effective. If the rate of

---

[1] The complex interaction of technological and demand factors in determining the level of R & D is illustrated by the case of defence, which is now a highly research-intensive industry but was not so to anything like the same extent before the Second World War. Demand for the product of military R & D has been high and inelastic: (a) because a cold war situation has existed; (b) because it is thought that a technological lead or lag makes all the difference between success and disastrous failure. This belief (the validity of which of course remains unproven) is perhaps the main point of difference accounting for the increased R & D-intensity of defence compared with earlier times. When weapons were less powerful, a small inferiority meant a less great absolute difference in effectiveness, and at the same time considerations of terrain absent in nuclear warfare gave a defender an advantage which permitted him to survive some degree of military inferiority. The R & D involved is extremely costly, partly because of the nature of the activity, but mainly because of the forced-march pace at which technical progress is sought; for it is well-established that a speeding-up of the pace of technical progress increases costs more than proportionally, because a greater number of problems have to be tackled simultaneously and less benefit can be had from sequential decision-making. In other sectors of the economy (except space research, which resembles defence) similar escalation is not observed, because a technological lead will not give a firm a competitive advantage if the cost of attaining it is exorbitant and because the demand for the product of the industry as a whole is not completely inelastic.

capital formation is low, R & D may go to waste, because the means to carry it into effect are lacking.[1] (This serves to lower the social profitability of R & D in so far as it has to be 'embodied'; it also increases the relative profitability of R & D leading to 'disembodied' technical progress as compared with 'embodied' technical progress.)

Scarcity of scientific manpower in developing countries may also tend to raise the cost of R & D. But the position here is not so clear cut as it is with saving. While developing countries are obviously less rich in Q.S.E.s than advanced ones, there is some evidence that the relative scarcity of highly qualified manpower is less acute than that of manpower at lower levels of skill. The 'R & D exchange rates' calculated by Freeman and Young [8] imply that R & D costs in the United States are the highest in the world – in other words that the world's most advanced country has a comparative disadvantage in R & D. These calculations rest on the highly questionable assumption that the productivity of Q.S.E.s in R & D is the same in all countries, so they cannot be interpreted at their face value; moreover, the comparison is with other developed countries, not with developing countries. But it probably remains true that the supply of qualified manpower is not the chief factor keeping down the profitability of R & D in developing countries.

(3) *Measurability and the direction of R & D*

Measurability problems have been fully discussed by Denison ([4] and [5]) and little needs to be added here.

R & D may be aimed at producing new products, at improving the quality of existing products, or at introducing new or improved processes. In the United States it has been estimated that as much as 90 per cent of industrial R & D is aimed at product innovation rather than process innovation [10]. For other countries the proportion is probably rather lower, but it is still high [7]. It is well known that product improvements are imperfectly measured in conventional national accounts statistics, so the overall rate of increase of measured G.D.P. tends to be understated. The problems of measurability apply even more strongly to R & D carried out on behalf of government in the spheres of defence and welfare.

This problem does not directly affect estimates of the overall contribution to growth made by R & D based on the assumption that it has the same rate of return as physical capital formation. It does, however, go some way to explain the lack of visible relation between

---

[1] It has been suggested that this has been the case in the United Kingdom, where the level of capital formation has been low relatively to R & D in comparison with other countries [3].

R & D and growth as between countries. More clearly still, it inter-feres with attempts to relate R & D to productivity growth by industry. Insufficient allowance for quality improvement in an industry leads to understatement of the growth rate of production in the industry and of G.D.P., if it is a consumer-good industry. If it is a producer-good industry, it may also lead to understatement of the rate of growth of G.D.P. if the effect of the quality improve-ment is to reduce the quantity of the product required per unit of output at the next stage of production. If the effect of the improve-ment is to reduce the quantity of other inputs needed per unit of output at the next stage of production, there will not be an under-statement of total G.D.P. growth, but the increase in productivity will show up in the using industry instead of where it really belongs. In view of this it is not surprising that attempts to establish a relation-ship between productivity growth and R & D by industry have been indecisive.[1] It remains a matter for further research whether the relationship would still be found to be weak if measures could be devised that made proper allowance for quality.

(4) *Appropriability*

Investment in improving knowledge differs from investment in physical capital in that its results once produced become in principle free goods, available to everyone to use unless special steps are taken to prevent it. This creates a basic policy dilemma. If exclusive property rights in new knowledge are established by law, the use of the knowledge is wastefully restricted; if no such rights are established, the benefits of advances in knowledge are competed away and no incentive exists to invest in improved knowledge.[2] The patent system is a compromise solution to this dilemma, making the benefits partly appropriable by the inventor but not wholly so. Non-appropriability or incomplete appropriability is the main reason for which economists have supposed that there is a tendency for the free operation of the price mechanism to lead to

---

[1] A relationship was found by Minasian [13] and by Mansfield [12]. See, however, the critical comments by Griliches [12] and in [10], and for a more general discussion [29].

[2] The theoretically ideal solution to the dilemma is a system of 'awards to inventors': a government agency (or a private agency subject to government regulation of monopoly powers) pays the inventor a lump sum equal to the social profitability of his invention and then makes it available free to all potential users. This solution is not feasible in practice because of another attribute of improved knowledge, uncertainty as to its usefulness. The agency cannot know the social value of the invention in advance, and even if payments were not made until the invention came into use the market would give no direct measure of what its value had been. For general discussion of the principles of patent policy, see [26].

underinvestment in R & D and for its social return to exceed its private return.

In practice, the fruits of R & D are not instantly competed away, even apart from patent protection. In fact, patent protection in most cases is less important than the temporary leadership gained by the innovating firm on account of the time taken by competing firms to follow in its footsteps. Patent royalties are a relatively small source of income to innovating firms, and the main profit comes from exploiting the innovation within the firm [1]. The value of R & D gradually depreciates as competitors follow.

The benefit of the temporary monopoly position, and hence the appropriability of the results of R & D, varies greatly between different cases. So, at least as important as any general under-allocation of resources to R & D as a whole is the likely misallocation within R & D between projects of differing degrees of appropriability. R & D will tend to be highest where the degree of appropriability is greatest.[1]

This has several applications. In general, the wider the area to which an innovation relates, the smaller the proportion of the total benefit that can be appropriated by a single firm. The degree of appropriability is likely to be less in basic research than in applied research; in perfectly competitive industries than in monopolistic or oligopolistic ones; in major innovations than in minor ones (because major ones will be imitated more quickly;[2] and, possibly), in process innovation than in product innovation. All these tendencies are apparent in the R & D pattern actually observed [16, pp. 44–88].

[1] The existence of monopoly elements makes it difficult, as always, to draw general *a priori* conclusions. However, it does not seem to be correct to argue, as is sometimes done, that the pursuit of temporary monopoly positions through R & D in general causes the private return on R & D actually to exceed the social return. Suppose that, initially, several firms are producing a similar product and the demand curve facing each is therefore highly elastic. One of them then introduces an improved product, so that its demand curve becomes higher and less elastic. It sells the improved product at a price exceeding marginal cost. This represents a departure from the Pareto optimum. However, comparing the new position with the old one, no consumer is worse off: if he does not wish to buy the improved product at the high price, he can go on buying the unimproved product from the other firms at the original price, so long as they continue to offer it. The private return from the R & D is then no greater than the social return, and the departure from optimality is merely that the social return could have been made still higher. An excess of private over social return would, however, come about if the result were to knock out the old firms altogether, so that the innovating firm could exercise a monopoly over the whole product range, or else if the loss of custom were to cause the old firms to operate at higher average costs.

[2] For a different reason (relating to indivisibilities, not appropriability) why minor innovations may be relatively more profitable, see [1].

The question of appropriability also has an important bearing on where R & D is done, and hence how efficiently it is done. Because so much of the benefit comes from temporary monopoly in use, the scales are weighted in favour of the R & D done by firms actively engaged in production and against independent inventors or specialised R & D contractors. This has become increasingly so as firms have increased in size. In industries where firms are small and competition is perfect or nearly so (agriculture, building, textiles, clothing) the appropriability rate is very low, and R & D has inevitably to be separated from production. It has typically been done by suppliers or customers, or else by government or (infrequently) by co-operative research associations. Such arrangements are not ideal, because those responsible for the R & D may be imperfectly acquainted with the situation and problems in the industry and may have difficulty in getting their innovations accepted (which comes to the same thing). The trend for R & D to take increasingly the form of in-house research by a small number of large firms is also not free from disadvantages: there is a closer link between research and production, but the histories are full of cases where conservatism by existing firms would have stifled innovation had it not been for the enterprise of outsiders and newcomers to the industry. The dilemmas created by the appropriability problem are thus deep-seated.

A word may finally be said about the international aspect of appropriability. In the case of the development part of R & D, it may be quite difficult to copy at all quickly ideas originating from abroad; even a technologically backward country may find it necessary to spend significant sums on R & D in order to adapt foreign innovations to domestic needs. This is much less true of basic research, where the benefits are difficult for the innovating country to appropriate. It is therefore in the interests of any one country to leave the basic research to someone else and concentrate on cashing in on the applications.[1] This would lead to world underallocation of resources to basic research if it were not that considerations of intellectual interest and prestige pull strongly in the opposite direction.[2]

(5) *Risk and uncertainty*

Whether a given R & D project will be profitable cannot be known for certain in advance. Many projects lead to nothing, and the cost

[1] This argument was much heard in the discussions preceding the decision of the British Government not to join in the 300 GeV Accelerator Project. It was felt that any practically useful findings that might ultimately emerge would be equally available whoever had done the basic research.

[2] As it is, it would probably be in the economic interest of most countries to be more ready than they are to acquiesce in the form of international aid that would result from basic research being done chiefly in the United States.

of these failures has to be allowed for in reckoning the overall rate of return on R & D. The risk element is not peculiar to R & D: it applies also to investment in fixed capital, if to a less degree. Risk is mitigated if a firm has numbers of separate R & D projects in hand, so that the chance of their all failing is reduced. This is a further reason making for the concentration of R & D in large firms. Where a balancing of risks within the firm is not possible (because the sums involved in an individual project are too large or because the probabilities of the outcomes of different projects are correlated) investment in R & D may be kept below the socially optimum level. It does not follow that all the tendency observed for firms to concentrate their resources on quick-yielding low-risk projects [16, p. 54] represents a misallocation. If no firms can be found willing to support some relatively small-scale project, there is a presumption that it is not worth doing. But if projects are lumpy or if the firms concerned are small, underallocation may occur.

The imperfect knowledge associated with R & D projects can in certain areas lead to overallocation [1, 28]. If the return from being first in the field with an innovation is very large, it will be worth while for a firm to work on it even though it knows that other firms are working on it too and that it is quite likely that they will get there first. The rate of return on the sum of R & D done by all the firms will not then be below the social norm, but results nearly as good from the social point of view could have been had if fewer firms had participated, so the marginal rate of return is low. This potential waste from overinvestment also exists, of course, in the case of investment in physical capital; but there is the difference that if, say, there is overinvestment in house-building, the extra houses bring some social return, though less than hoped for, whereas the social return from the R & D efforts of the firms who lose the race consists merely of the *ex ante* probability that with more firms participating the results will be reached earlier, and the benefits from this may be very small.[1]

[1] Three further points may be noted. (1) This argument points to the possibility of overinvestment in projects that are expected to be profitable per unit of R & D input. It is therefore not inconsistent with the statement made above that there may be a tendency for underinvestment in major projects, i.e. in projects expected to bring large reductions in unit costs, though possibly at the expense of very high R & D outlays. (2) Usher [28] treats this case as the result of inappropriability rather than of uncertainty. This is analogous to the argument of Buchanan that externalities in general may lead to overinvestment rather than underinvestment: e.g. in the textbook case of street-lighting the upshot may be that each householder puts a light in front of his house rather than that none of them does. However, imperfect knowledge is a necessary condition (and also a sufficient one, as illustrated by the possibility of overinvestment in fixed capital, when the question of inappropriability does not arise). (3) In certain cases, the

(6) *The role of government*

Non-appropriability creates a presumption that in a pure laissez-faire system the social rate of return on R & D will exceed the private rate of return. This presumption no longer necessarily exists when government subsidises R & D. Government is in fact heavily involved in the support of R & D in all countries – much more heavily than in the support of fixed capital formation.

This may be seen from column 1 of Table 1.2, which shows the percentage of all R & D financed by government.[1] None of the

TABLE 1.2

PERCENTAGE OF TOTAL R & D FINANCED BY GOVERNMENT, 1963–4[a]

	Total	Nuclear Space Defence	Components of Total Health, miscellaneous	Industry, Agriculture, Construction
	(1)	(2)	(3)	(4)
Austria	40	9	24	7
Belgium	24	9	7	8
Canada	55	27	23	15
France	64	44	13	6
Germany (Fed. Rep.)	41	19	22	
Italy	33	31		2
Japan	28	..	..	..
Netherlands	40	..	..	..
Norway	54	13	26	15
Sweden	48	34	12	2
United Kingdom	54	37	6	10
United States	64	58	5	1

[a] 1964 for Austria, Federal Republic of Germany, Netherlands, Sweden, United States; 1964–5 for the United Kingdom; 1963 for other countries.

Sources: [20], Vol. 1, pp. 57–8 and Vol. 2, pp. 44–5, 222–3, 226–7.

---

multiple effort may be worth while if it greatly reduces costs at a later stage; this is the famous argument in favour of parallel R & D projects [14]. This will occur if the probability distribution of the results of the different firms' efforts has a wide dispersion relatively to the cost involved; it thus applies less to development than to research. It also applies less to innovation of a relatively routine kind than it does to cases where very rapid progress is being sought in new areas (as in military and space research).

[1] This is in all countries much larger than the percentage of R & D performed by government, the difference being accounted for by government support to R & D in business, universities, and other non-profit enterprises. The difference between the percentage financed by government (64) and the percentage performed by government (18) in the United States is considerably greater than in any

countries shown has a figure below 24 per cent, and in several it is over 50; in developing countries the proportion of the (relatively small) total amount of R & D that is financed by government is probably even higher on average than in developed countries.

Defence and related R & D and welfare and miscellaneous R & D account for a large proportion of government-financed R & D. The countries with the largest government shares in the total are the ones where the defence element is the largest. Since defence is part of the function of government, government finance of R & D in this area cannot properly be considered a 'subsidy' to R & D. Government support to R & D in industries unrelated to defence (column 4) is in most countries a relatively small item.

The overall figures in column 1 therefore give a much exaggerated measure of the extent to which R & D in general is directly supported by government. Account must also be taken, however, of indirect support. In the first place, there is an element of spin-off to the rest of industry from R & D in the nuclear-space-defence area and from basic research in universities and other non-profit institutions supported by government. In the second place, and more important, the right usually allowed to treat R & D for tax purposes as a current expense rather than a capital expense gives an across-the-board subsidy, although the value of this is reduced in countries where capital expenditures themselves attract large tax privileges or subsidies.

The extent of government support to industrial R & D is thus substantial. However, it is very unevenly distributed between industries and for most industries is a good deal less than the overall figures would suggest.

(7) *Some conclusions*

If we start from the assumption that R & D has the same rate of return as other investment, we tend to be led to the conclusion that quantitatively it is a relatively small contributor to overall economic growth. The two major considerations to be set against this are interaction effects and the possible excess of social over private return from R & D.

If physical accumulation were to continue over a substantial period of time without any advances in technical knowledge, it would encounter steeply diminishing returns. R & D is not the only source of improvements in technical knowledge, but it is a major

other country, reflecting the policy of contracting out to the private sector of R & D financed by the Department of Defense and NASA. 52 per cent of the R & D performed by business enterprises in the United States is financed by government, a much higher figure than for any other country.

source, and its importance compared with other sources is increasing. It is impossible to separate the effects of R & D and the effects of capital accumulation where movements over long periods of time are concerned; and it is obvious that their joint effects have been one of the chief sources of growth in advanced countries. If the rate of capital accumulation continues to be as high as it has been in most advanced countries since the Second World War, there will be a corresponding need for a high rate of R & D. By the same token, on the other hand, countries where savings are scarce are likely to find that the rate of return on R & D is low. Moreover, the mere input of scientific and technological effort in R & D will not bring much return unless the demand for the final product has the appropriate structure and growth rate.

The degree to which private and social returns from R & D diverge is enormously variable. Over large areas of industrial R & D, the social return may be little if at all greater than the private return, especially when account is taken of the government support that it enjoys in one form or another. It is even possible that in some cases the private return is greater than the social return (even apart from the effects of government support) and that over-allocation of resources to R & D results. On the other hand an estimate of the social return on R & D at the margin must allow for the chance that the marginal project will turn out to be one of the infrequent cases where a result is produced that transforms an entire industry and brings benefits out of all proportion to the private return.

III. GROWTH IN SCIENTIFIC AND TECHNICAL KNOWLEDGE NOT DUE TO R & D

Not all advances in the state of scientific knowledge relevant to production come about from organised R & D. The other sources have to be taken into account in assessing the overall contribution of science and technology to growth. Consideration of them is relevant also to questions of the optimum deployment of scientific resources.

Three such sources may be distinguished: (1) The work of private inventors ('garage inventors'); (2) improvements that are discovered as a by-product in the process of production; (3) improvements imported from abroad.

(1) *Private inventors*

Progress due to the efforts of the typical 'garage inventor' can be called technological, but hardly scientific. In the early industrial revolution and throughout the nineteenth century, the individual inventor was the main source of technological improvements. The decline in his relative importance since then has already been referred to. His importance as measured by numbers of patents is still significant,[1] but this is probably an exaggerated measure. The question of the relative value of work by individual inventors and by large corporate R & D departments is part of the broader issue of the relative contribution of small-scale and large-scale units. The case-studies enumerated by Jewkes and his collaborators [11] provided impressive testimony to the value of small-scale research, including that of the private inventor. But as Freeman [9] has pointed out, even Jewkes' own examples show a trend through time in favour of corporate R & D.

The contribution of whole-time research not included in official R & D statistics at the present time is probably relatively small. For a view of the broad historical sweep of economic growth it is important to recollect that this has not always been so. At the same time in assessing the contribution of science in growth, it must also be remembered that most of the classic eighteenth and nineteenth century inventions were arrived at by the individual inventors concerned in an extremely empirical way and although certainly 'technological' can hardly be called scientific in the sense of depending on advances in basic knowledge of science.

(2) *Innovations as by-product*

No technical improvements occur totally spontaneously, as implied in the theorists' metaphor of 'manna'. However they may occur as a by-product (not necessarily unforeseen) of other activities. Most innovations work very imperfectly when first introduced, and are improved gradually over time as experience accrues in their use. This gradual improvement can indeed be regarded as an inherent part of the innovation process, running from research through development to production and then to improved production. More generally, scope for improvements, even radical ones, may be discovered in the course of using quite long-established methods.

We do not have any measures of the importance of this kind of improvement relatively to that resulting from formal R & D, and in some cases, as implied above, the distinction is difficult to draw even in principle. The case studies in the literature on 'learning', together with the evidence of continuous productivity improvement in industries that do little or no R & D and can point to no major innovations,[2] suggest that it may be large.

---

[1] The proportion of all patents taken out by individuals in the United States in 1960 was 40 per cent ([16], p. 45).

[2] This is not of course the only possible explanation of productivity improvements in such cases. Alternative explanations include improvements in the

Although such improvements may not result from formal research, they will not come about unless those in charge of production are alert to the possibilities revealed by experience and have the basic technical and scientific knowledge needed to perceive them. For this reason the standard of technical and scientific training of those engaged in production (as opposed to R & D) may have an important effect on the rate of growth of the stock of economically useful knowledge as well as on the efficiency with which the existing stock of knowledge is used.

(3) *Imported innovations*

Know-how may be imported directly, through the imitation by domestic firms of innovations originating abroad or through the operation of foreign-owned companies. Evidence on this has been sought from statistics of patents and international royalty payments. Know-how can also be imported as hardware rather than software, by the importation of capital equipment embodying technical advances made abroad.

The proportion of the advance in technical knowledge in a country that is due to importation in one or other of these forms is likely to be lower, the more advanced is the country and the larger is the output of its industries relative to world output. We thus find, as expected, that the 'technological balance of payments' (the balance of royalty payments on patents, etc.) is markedly favourable for the United States [8]. However, even the United States has substantial gross outward payments, indicating a significant amount of borrowing of knowledge from abroad. As a greater number of countries come to have more nearly the same level of productivity per head on the average (as has happened among advanced countries since the war), the gross amount of trade in knowledge may be expected to increase, with each country being a leader in some lines and a follower in others. It may be conjectured that widening of the front along which advances are made has contributed to the speeding up of the rate at which the frontier of knowledge has been pushed forward in the post-war period.

There is much room for dispute about the size of the contribution of imported knowledge in any one country.[1] This is important for national policy issues. However, it is irrelevant to the question of

---

quality of capital equipment used and improvements of a non-technological kind.

[1] Different ways of interpreting the patent statistics thus cause the proportion of new (scientific) knowledge derived by the United States from abroad to be estimated at a half by Denison [4, p. 234] and at around a fifth by Nelson, Peck and Kalachek [16, pp. 64-5].

how great a contribution advances due to science and technology have made to economic growth in the world as a whole. If foreign science and technology are credited with a large contribution to the advance of knowledge in any given country, the assessment of the contribution made by its own science and technology to its own economic growth may be reduced; but the assessment of the contribution to world growth made by the foreign science and technology is *pro tanto* increased. The issue here relates thus not so much to the magnitude of the overall contribution of science and technology as to the degree of externality (from the international point of view) of the benefit from each country's efforts.

The foregoing discussion of contribution to improvements in knowledge derived from sources other than a country's own organised R & D is still confined to knowledge in the general sphere of science and technology. The line between this and other kinds of economically useful knowledge is not exact, and scientific and non-scientific knowledge are in any case complementary. None the less, in a conference such as this, it is necessary in order to maintain perspective to emphasise the contribution to economic growth made by advances that cannot be called scientific or technological on any reckoning. The rapid productivity growth in retailing, in all countries one of the most important industries, owes virtually nothing to science or technology. Even in manufacturing, organisational improvements have also played a vital role; the basic ideas of interchangeable parts and the assembly line both represented essentially advances in organisation rather than technology, though they required technology to implement them. Improved techniques of accounting, personnel management and marketing have been necessary conditions for the functioning of large-scale enterprise and the economies derived therefrom. Examples could easily be multiplied.

IV. THE CONTRIBUTION TO GROWTH MADE
BY INCREASES IN SCIENTIFIC MANPOWER

The model outlined earlier in this paper involved a distinction between the contributions to growth made respectively by increases in the body of scientific and technical knowledge on the one hand and by increases in the degree to which this knowledge is diffused through the labour force on the other. We now turn to the second of these topics.

In all countries only a minority of Q.S.E.s are engaged in R & D.[1]

[1] The percentages of Q.S.E.s in R & D in 1962 have been estimated as follows: United States, 33; U.S.S.R., 20; Japan, 15; United Kingdom, 25; Federal Republic of Germany, 15; France, 17 [30, p. 5]. The percentages are naturally

The rest are engaged either in production or in education. The proportion of technicians in R & D is still smaller. Quantitatively, therefore, the economic importance of the non-R & D Q.S.E.s appears to be greater than that of the R & D ones. But this is subject to two important qualifications.

In the first place, the employment of Q.S.E.s in R & D (and also in education) is a form of investment, whereas their employment in current production is not (except in so far as they are employed in capital goods industries, in which case their contribution is allowed for as part of the contribution to growth made by capital formation). The contribution to growth over a period made by Q.S.E.s engaged in current production is therefore a function of the increase in their numbers over the period, whereas the contribution to growth of the Q.S.E.s in R & D is a function of the level of their numbers in the period. Straight comparison of the relative numbers employed in the different ways is not therefore the right procedure. Allowing for this, and using crude marginal productivity assumptions, however, the orders of magnitude still suggest that the contribution of R & D is somewhat the smaller of the two.[1]

In the second place, the partial non-appropriability of R & D output creates a presumption that factors engaged in R & D receive less than the value of their marginal social product. The same presumption does not apply to Q.S.E.s engaged in current production.

If we adhere to marginal productivity assumptions, the contribution to growth made by the increase in the number of Q.S.E.s is simply one part of the contribution made by improvements in educational standards generally. It is not a very large part either, because in most countries only a relatively small proportion of the increase over time in the average years' schooling of the labour force is accounted for by increases in the number of Q.S.E.s; and the average

---

higher for scientists than for engineers. The overall percentages would come out lower if the definition of Q.S.E.s were broadened in certain respects, e.g. to include medical practitioners.

[1] The contribution to annual growth in income (ΔY) made by the employment of Q.S.E.s in R & D is $r\,a\,w\,N$, where r = the annual rate of return on investment in R & D, a = proportion of Q.S.E.s in R & D, w = annual wage of Q.S.E.s, N = number of Q.S.E.s. The contribution to ΔY made by the increase in the number of Q.S.E.s is $g\,w\,N$, where $g = \Delta N/N$. (This last expression, gwN, includes the contribution to the increase in output of R & D, itself part of Y.) The relative magnitude of the two elements therefore depends on the ratio of ra to g. The orders of magnitude applicable to the United States are $r = 1/7$, $a = 1/3$, $g = 1/16$. This makes the growth term slightly the larger. For other countries a is generally lower and g higher than in the United States (see data in [18]).

salary of Q.S.E.s is not markedly different from that of others with an equivalent number of years' schooling. Of course the total contribution of improved education to growth turns out very substantial when calculated as Denison has calculated it, so even a relatively small proportion of it would amount to a significant contribution. The proportion would be increased, moreover, if the criterion was broadened to include not only the education of Q.S.E.s but also lower-level technical training and the scientific education given in school to those who do not go on to practise as Q.S.E.s. But even so, this line of argument does not appear to provide a basis for regarding improved scientific and technical education as a particularly key factor in growth.

We have here some inconsistency with the conventional wisdom about science policy, according to which the presence of an adequate number of scientifically trained managers, salesmen, and others engaged in current production is a vital requirement in economic growth, more important, possibly, than a high level of R & D. This view has been expressed particularly in discussions of science policy in the United Kingdom, where the proportion of Q.S.E.s engaged in production is lower than in most other countries.

What is the explanation of this conflict? The first point to be noted is the need to avoid the usual average-marginal confusion. If a country has no Q.S.E.s, obviously it cannot operate a modern industrial system at all. In this sense they are vital. But so are those with other skills, such as lorry drivers. The question rather concerns the contribution made at the margin by increases in numbers of Q.S.E.s, either in the past (for purposes of historical interpretation) or in the future (for purposes of policy making).[1] The problem then can be put bluntly by asking: if increasing the number of Q.S.E.s would have such a great effect on growth as sometimes claimed, why does the market not cause them to be paid more? Is there *a priori* or empirical evidence for a shortfall of remuneration below marginal product as there is in the case of R & D?

One possible answer is simply to discard the conventional wisdom and say that Q.S.E.s are no more valuable at the margin than their pay suggests. The second is to say that lags and market imperfections (including the major role of the government as employer) keep the rate of pay below what would equate supply and demand. The third is to say that there is no excess demand as things stand, but that there would be, if it were not that employers underrated the potential

---

[1] If we are considering growth over very long periods (past or future), moreover, we run into the same unanswerable questions, because absence of growth in any of large numbers of types of input (capital, skills of various kinds) would have been sufficient to prevent growth.

usefulness of Q.S.E.s and did not know how to use them to best advantage.[1] Yet other possibilities concern externalities that may exist in the employment of Q.S.E.s even outside R & D (because they are responsible for imitable technical improvements of the by-product kind or because they improve liaison between R & D departments and production and marketing departments), and possible deficiencies in the quality rather than the number of Q.S.E.s (which will not be well reflected in market data if much the same type of training is undergone by all Q.S.E.s).

These issues are important. But discussion of them in the literature of science policy contains more assertion than evidence – not surprisingly, in view of the intractability of the problem. For the present we have to admit that we do not know the answers.

The potential importance of Q.S.E.s in production, and more generally the potential importance of scientific and technological literacy among managers, lies largely in the effect on the diffusion of innovations. Technically under-trained production managers will be slow to perceive how innovations originating elsewhere can be used or adapted to the requirements of their own firm. The issue just discussed is thus closely related to the more general issue concerning the forces determining the speed at which innovations are diffused.

There are two opposed types of view about this. The first, which is essentially that of Schumpeter, is that the reason why some firms (or the firms in some countries) are slower to adopt an innovation than others, lies in differences in entrepreneurial ability; and one particular form of poor entrepreneurial ability is inadequate scientific or technological training or inadequate appreciation of the need to employ Q.S.E.s. The opposite view is that firms on the whole know what is good for them and do it, and that the reason why innovations are diffused only gradually is that there is a corresponding gradualness in the process by which the innovation becomes worth adopting. Some firms find it worth while to adopt it at once, whereas others, operating in different markets or natural conditions and with different factor prices or initial capital endowments, do not find it worth while until these have changed or, alternatively, until the innovation itself is available in an improved version.

A good deal of work has been done on this question. The extreme

---

[1] The issue between the second and the third views is exemplified in the contrary interpretations that have been placed on the demand and supply situation for engineers in the United Kingdom. Peck [21] maintains that a shortage exists, quoting the earlier work of Arrow and Capron [2] in support of the view that such a shortage will be reflected in rates of pay only after a substantial lag. The opposite view, that the trouble rather lies in insufficient demand from industry, has been held by other observers, and has been most forcefully stated in an unpublished paper by Christopher Layton.

Schumpeterian view is scarcely tenable. As Salter [23] showed, if innovations have to be embodied in new capital equipment, it will not normally be rational for all firms in an industry to adopt them simultaneously. Certain case studies have argued convincingly for the opposite extreme view. Another piece of evidence that can be taken as pointing in that direction is the apparent constancy in the degree of intra-firm dispersion in productivity – since if diffusion of innovations and hence dispersion in productivity resulted mainly from lack of technical sophistication, it might be expected to be less in periods or countries where technical standards are higher [6]. If the anti-Schumpeterian view were generally valid, it would make it more difficult to maintain that the demand for Q.S.E.s is kept down by the irrationality and ignorance of entrepreneurs.

But notwithstanding this evidence, it is difficult to believe that all inter-firm and international differences in industrial practice can be explained and justified in this way. It may be conceded that only extremely irrational and conservative entrepreneurs will decline to adopt a new improved method which has clear advantages and can be put into effect without any disruption. But this is not the typical case. The importance of inter-industry relations in this connection requires to be stressed. Most new products are producers' goods and most new processes require a different type of input. Irrationally slow responsiveness to technical change in a single industry casts a shadow forward or backward or both. However rational and technically sophisticated a firm is, it may be prevented from carrying out an innovation if its suppliers cannot adapt themselves or if its buyers fail to perceive the merits of the new product.[1] Under-estimation of the value of scientific and technological training in one industry may thus depress productivity and restrict the potential usefulness of Q.S.E.s in other industries.[2] This is a form of externality additional to those already mentioned.

From the practical policy point of view, the issue can be narrowed down somewhat. If a market shortage of Q.S.E.s is apparent, efforts should be made to increase their numbers. If a shortage is not apparent, this may be either because an increase in their numbers would not be very useful or because employers do not appreciate how useful it could be; in either case, a forced increase in numbers

[1] Vertical integration provides one way round this problem. But it may not be possible if differences in the scale of operations or other obstacles impede entry at certain stages of the production process. A firm producing motor car components or armaments cannot integrate forwards, and a small firm using steel cannot integrate backwards.

[2] The British reader will have no difficulty in perceiving that the mechanical engineering industry, or at least parts of it, can be cast for the role of the industry in question.

will not by itself have much immediate value, because either employers will be reluctant to take them on or else they will not know
how to use them to best advantage. The issue remaining is how far
an increase in the number of Q.S.E.s, or more generally an improvement in the standards of technical and scientific education, even if
not especially sought for by management, will in the course of time
gradually raise the quality of management itself.

There remains to be considered a rather different effect. This effect
arises from advances in technical knowledge (G_s) as much as from
increases in the extent to which it is diffused (H_s), the two being
difficult to separate in this context; but it may conveniently be
referred to here.

We have so far assumed that the rates of growth of the capital
stock and the labour force are independently given and are not
affected by science and technology. The vast effects of medical science
on mortality, and the potentially vast effects of improved methods of
birth control on fertility, show that this is quite untrue in the case
of the labour force. The issues here are too familiar for it to be
necessary to say more about them than they are extremely important.
The effects of demographic changes on the rate of growth of income
per head are, of course, debatable and may be in opposite directions
for countries at different stages of development.

The effect on the rate of capital accumulation is rather less obvious,
but may also be important.[1] The rate of growth of the capital stock
in real terms is equal to the ratio of net investment to G.D.P. at
current prices (s), multiplied by the reciprocal of the capital-output
ratio at constant prices (v), multiplied by the ratio of the price of
output in general to the price of capital goods (p) : $\Delta K/K = sp/v$. We
have seen that the science-intensive industries are largely, though
not exclusively, capital-goods industries; and it is also found that
capital-goods prices tend to be lower relative to general prices in
the richer countries [5, pp. 160–3]. It seems likely, therefore, that
advances in scientific knowledge and abundant supply of scientific
personnel will tend to lower the prices of capital goods (duly adjusted for quality) relative to other prices. In this way they will
permit a given saving-income ratio to lead to a higher rate of
increase in the capital stock. It is impossible to quantify this effect,
because the causes of trends in p are obscure and we do not know
how much of them can be explained by the differential impact of
science and technology; we do know, however, that the variation

---

[1] This is apart from the obvious point that technical advances may, by raising
the rate of return on investment or preventing it from falling, cause a higher
proportion of G.N.P. to be devoted to investment than would otherwise be the
case.

between countries and periods in p are substantial. It may be noted that this effect, in so far as it exists, has nothing to do with divergence of private and social product as normally understood; it is an effect favourable to growth as opposed to current consumption.

V. CONCLUSIONS AND POLICY IMPLICATIONS

If we are considering the phenomenon of modern economic growth in the broadest terms, it is impossible to separate the contribution made by advances in scientific and technological knowledge and its diffusion from the contributions arising from other sources, since they are all complementary.

We may instead ask a narrower question: how far are differences in growth rates between countries or between periods due to differences on the side of science and technology? Interaction problems are present here too. But the evidence is not favourable to the hypothesis that a country's rate of economic growth is closely related to its eminence in science. The United Kingdom has fairly consistently since the beginning of the twentieth century had a slower rate of growth in output per man than most other advanced countries, but its scientific standing has been high. Eminence in science seems to be rather more closely related to the level of a country's income per head; but here the direction of causation is a major problem, since a rich country is able to devote more resources to science than a poor one, just as it can to other fields of human endeavour. A rather better case can be made for the view that a country's rate of economic growth depends on its success in using scientific and technological advances for commercial purposes. The effects of this are difficult to separate from other aspects of managerial performance.

The policy problems that arise from the issues that have been discussed in this paper are various. In fact they are so various that it is doubtful whether it is really helpful to think of 'science policy' as a single field of decision-making. A number of quite different decisions about science and technology have to be made by government. These decisions are, of course, related to each other, but they are also each related, in some cases even more closely, to other questions altogether.

The decisions that the government has to take affecting science and technology in a mixed economy can be divided into those arising in the public sector itself and those concerning intervention by government in the private sector.

(i) The government has to decide what value it places on the advancement of basic scientific knowledge as an end in itself and how much it is therefore willing to spend for that purpose in universities and research institutes. This is comparable to the decision on how much the government should spend in support of the arts or sport or the preservation of scenic beauty.

(ii) The government is responsible for defence in all countries, and for road building, health, and other activities to an extent that varies between countries. It is therefore automatically responsible for R & D and for the deployment of scientific and technological resources generally in those areas. This involves problems of management and organisation not different in principle from the problems involved in the employment of other resources in the public sector. Because the government happens to be a more important buyer in the market for Q.S.E.s (or at least some classes of them) than it is in the market for most inputs into the public sector, care has to be taken to ensure that the true social cost of employing them is taken into account. Because of the government's monopsony powers, and also because of the possible divergence between the private and social product of Q.S.E.s in the private sector, it can very well happen that their proper social shadow price is different from their actual salaries. However, so long as salaries do not diverge from true shadow prices, there is no reason why spending departments of government and nationalised industries should not treat these costs on a par with other costs; there is no case for a special 'science budget' cutting across normal departmental budgeting categories. Thus, if it is thought that too many Q.S.E.s are absorbed in defence or space research, this must mean that too high a priority is being set on defence or space in general (unless there is reason to suppose that the departments concerned are not deploying their funds in the best way).

(iii) The government in most countries has the main responsibility for education. It is therefore involved in the highly important and difficult task of taking a view about future trends in the demand for personnel with scientific and technological training as compared with those with other kinds of training. Even if the choice between disciplines is left to be determined by the expressed preferences of students or their parents, this responsibility cannot be avoided, because decisions made now about the training of teachers and the provision of facilities will have an effect for a long way into the future.

The foregoing all relate to activities that fall squarely within the functions of government as normally accepted. There remains the question of intervention in the private business sector. This can be justified on the grounds of divergence between private and social marginal net product; reasons why this divergence may occur have been referred to earlier in the paper. In addition to these sources of private-social divergence in the area of science and technology itself, policies indirectly affecting science and technology may be dictated by other private-social divergences, e.g. if the balance of payments makes it necessary to give special encouragement to exports or import-saving. Finally, intervention may be justified on the grounds that firms in the private business sector are inefficient in the use of scientific and technological resources even in the pursuit of their own interests.

The most important point about this is perhaps that the need and scope for such action varies greatly between cases. The idea that non-appropriability and risk call for across-the-board subsidies to R & D is a very crude one; when allowance is made for imperfect competition and other complications, this idea has to be greatly qualified. Hence the need for selectivity in government action. This does not mean that the government should make intuitive guesses of likely technological winners and pump large resources into them even though private enterprise considers them unpromising. Rather, it means that attempts should be made to establish systematic criteria of the circumstances in which a tendency will arise for the deployment of scientific and technological resources in a particular part of the private sector to be non-optimal in a way that government intervention is capable of correcting or at least improving.

REFERENCES

[1] Arrow, K. J., 'Economic welfare and the allocation of resources for invention' and 'Comment', *The Rate and Direction of Inventive Activity: Economic and Social Factors* (Universities-National Bureau 1962).
[2] Arrow, K. J. and Capron, W. M., 'Dynamic shortages and price rises: the engineer-scientist case', *Quarterly Journal of Economics* (May 1959).
[3] *Technological Innovation in Britain* (Central Advisory Council for Science and Technology 1968).
[4] Denison, E. F., *The Sources of Economic Growth in the United States and the Alternatives before us* (1962).
[5] Denison, E. F., *Why Growth Rates Differ* (1967).
[6] Downie, J., *The Competitive Process* (1958).
[7] *Industrial Research in Manufacturing Industry, 1959-1960*. Federation of British Industries.
[8] Freeman, C. and Young, A., *The Research and Development Effort in Western Europe, North America and the Soviet Union* (1965).
[9] Freeman, C., 'Science and economy at the national level', *Problems of Science Policy* (O.E.C.D. 1968).
[10] Gustafson, E., 'Research and development, new products, and productivity change', *American Economic Review* (May 1962).
[11] Jewkes, J., Sawers, D. and Stillerman, R., *The Sources of Invention* (1958).
[12] Mansfield, E., 'Rates of return from industrial research and development' (with comment by Z. Griliches), *American Economic Review* (May 1965).

[13] Minasian, J. R., 'The economics of research and development', *The Rate and Direction of Inventive Activity: Economic and Social Factors* (Universities-National Bureau, 1962).

[14] Nelson, R. R., 'Uncertainty, learning and the economics of parallel research and development efforts', *Review of Economics and Statistics* (November 1961).

[15] Nelson, R. R. and Phelps, E. S., 'Investment in humans, technological diffusion, and economic growth', *American Economic Review* (May 1966).

[16] Nelson, R. R., Peck, M. J. and Kalachek, E. D., *Technology, Economic Growth and Public Policy* (1967).

[17] *Science, Economic Growth, and Government Policy* (O.E.C.D. 1964).

[18] *Resources of Technical and Scientific Personnel in the O.E.C.D. area.* O.E.C.D.

[19] *Economic Growth 1960–1970* (O.E.C.D. 1966).

[20] *A Study of Resources devoted to R & D in O.E.C.D. Member Countries in 1963/1964.* O.E.C.D. Vol. 1: *The Overall Level and Structure of R & D Efforts in O.E.C.D. Member Countries* (1967). Vol. 2: *Statistical tables and notes* (1968).

[21] Peck, M. J., 'Science and technology'. In *Britain's Economic Prospects*, Caves, R. E., *et al.* (1968).

[22] Phelps, E. S., 'Models of technical progress and the golden rule of research', *Review of Economic Studies* (1966).

[23] Salter, W. E. G., *Productivity and Technical Change* (1960).

[24] Schmookler, J., *Invention and Economic Growth* (1966).

[25] Shell, K., 'Towards a theory of inventive activity and capital accumulation' (with comment by K. Sato), *American Economic Review* (May 1966).

[26] Silberston, A., 'The patent system', *Lloyds Bank Review* (April 1967).

[27] United Nations. *Yearbooks of National Accounts Statistics*.

[28] Usher, D., 'The welfare economics of invention', *Economica* (August 1964).

[29] Williams, B. R., 'Research and economic growth – what should we expect?' *Minerva* (Autumn 1964).

[30] Williams, B. R., *Technology, investment and growth* (1967).

Discussion of the Paper by Professor Matthews

Professor Eastman, opening the discussion, said that he had found Professor Matthew's paper exceptionally suggestive. It raised many questions and issues relating to the role of science in contemporary advanced economic countries with large private sectors. He said that many of these questions could be rephrased so as to have universal application, and that the paper provided a survey of most of the issues to be raised in this conference. In view of this Professor Eastman said that he proposed to survey the survey. He said Professor Matthews introduced the paper with a summary or sketch of his model in which he subdivided K into physical capital, state of knowledge and the extent to which the labour force controls it, that is, human capital. He indicated firstly that only marginal analysis could lead to conclusions about the contribution of each factor to growth, and secondly he pointed out that the model was more complex than at first appeared because many of the factors were interrelated. Professor Matthews discussed the effect of various variables on growth, beginning with the role of scientific knowledge created by R & D expenditure. He found little relationship between R & D and growth, for a variety of reasons, including the improper measurement of new products in the G.N.P. Later in the paper he discussed briefly the international transfer of technology which while not explaining the mystery did suggest some possible reasons for the lack of correlation by nation. Professor Matthews pointed out that the research-intensiveness of industries varied widely, but that the pattern by industry was similar internationally. A few industries accounted for the lion's share of R & D expenditures. Why was this so? Firstly the technical characteristics of the industry; some are more amenable to science than others, and secondly the demand conditions in the industry. Both of these induced the application of knowledge. Nevertheless, although the R & D expenditures were concentrated in only a few industries these were producer goods industries and therefore their effects were widespread. International differences were due to the differences in structure between countries, and differences in demand, but were also presumably affected by the elasticity of supply of factors which would co-operate with invention and which enabled the rewards to be reaped. Professor Matthews had also turned his attention to the appropriability of returns to investment in R & D and considered the importance of this as a factor affecting decisions in industry and as a cause of distortion. R & D was a factor affecting appropriability in the form of legal provisions, patents, and of lags in ability of competitors to imitate. Inability to appropriate the full reward from R & D might lead to underinvestment and misallocation by leading to investment in industries where appropriability was easiest. Finally, in the model of imperfect competition misallocation occurred because of duplication of effort and one ended up with overinvestment not from the point of view of a particular firm, but from that of the economy as a whole. The standard solution often put forward in undergraduate courses was to make the results of research freely available, hence the government policy of free distribution of results. Indeed this would solve the first problem of underinvestment, if the bureaucrats knew what they were doing, and could also attack the problem of misallocation, but this would probably require discriminatory subsidisation. However, as regards duplication, subsidisation even on a discriminatory basis would not overcome this. It was necessary in this situation to control research activity. However, Professor Matthews warned us of the problems here and the extreme complexity because most of the restrictions on diffusion were not legal anyway.

Professor Matthews turned in his paper to other sources of invention which he said were not included in R & D. The first of these he mentioned was the garage inventor. Professor Eastman said he was not sure why this was not included under R & D – presumably it depended upon the size of the garage. He also discussed invention as a by-product of the activity of production, for example in the automotive industry, and this was affected by the skill of the labour force, a point which could not be overemphasised, for example, in the growth of the United States motor industry. Finally he discussed the role of imported innovation. This did not bear very much on innovation in terms of world growth but did affect a single country's policy. It also offered some explanation of why growth rates were not directly related to R & D expenditure. The policy issues which arise in national discussions on science policy were similar to those on the domestic side. For example, the question of distortions and of lags – the decision was made on the basis of whether importing gave a cheaper access to knowledge than producing it at home. However this tended to be overlaid with a concern for prestige and fear of an obsolescent structure if one did not keep up in a particular field, and often the lags meant the profitability from production was reduced and the receiving country got the end of the line. This last problem was analogous to the inter-firm competition. National duplication was very costly and it was better to have international specialisation if possible. On the question of human capital, Professor Eastman felt there was not much one could say conclusively. A failure of judgement in one industry might have repercussions for others. Innovation in R & D took place in producer goods industries and failure to innovate could affect the production of customers and of suppliers. This led on to the general problem of the role of government and to general guidelines on government action, in Professor Matthews' last section. There was a three-part decision – they must decide firstly on the support of basic research, secondly on the use of scientific personnel in producing public goods and services, and thirdly they must forecast the needs of science in planning future education. Finally the government had to consider the need to intervene in private firms because they were inefficient, and in this they had to distinguish between inefficiency in the use of science within the firm caused by bad management and the difference between the private and social returns from the use of science, which were two separate issues.

Professor Matthews intervened to say that he was grateful to Professor

Eastman for his lucid introduction. He would like to make some general points. The paper was a very general survey touching superficially on many things that should be dealt with more deeply during the conference. The intention of this paper was to help in indicating the issues rather than to attempt the resolution of them at this stage. However, one point Professor Matthews wanted to emphasise was that the contribution of science and technology was not coextensive with the contribution of R & D. He saw these as two overlapping sets – some parts of the contribution of science and technology were due to R & D but others lay in areas other than R & D. Science and technology on the one hand and innovation on the other were also overlapping sets, but not all innovation came from science and technology. Indeed science and technology were not synonyms – they were very different things, which one often tends to forget in discussions of science policy. Professor Matthews went on to ask what sorts of question could sensibly be asked at the macro level? Firstly the historical type of question. Professor Matthews thought it was not useful to ask how much growth over the last hundred years had been due to science and technology, but one might usefully ask the similar question about why growth in the post-war period had been *more* rapid than in earlier periods. Many areas had undergone more rapid productivity growth since 1945 than formerly and had also had a higher R & D expenditure. We ought to ask how far these were causally related. Could the faster growth, for example, be due rather to the fact that the governments were adopting Keynesian policies? The second question Professor Matthews thought we ought to ask was the policy question whether there was overall under- or over- or misallocation of expenditure in this area?

Professor Tsuru said that before making fine distinctions between science and technology on the one hand and R & D on the other it would be desirable to make clear the difference between the two basic approaches to this type of problem: one 'physical-technical' and the other 'socio-economic' or 'institutional'. He said that the former transcended the specificity of economic systems or different stages of capitalism, such as competitive and monopoly, whereas the latter had its own societal laws in distribution of returns. Science, in his view, was the application of human knowledge towards raising productivity and was essentially a free good; and that was the reason that cost was reduced as science was applied to the process of production. Of course, this cost reduction did not take place instantaneously and the institutional characteristics affected the manner in which this principle found its reflection in concrete reality. Thus, under competitive capitalism the return to the application of science might be in the form of temporary abnormal profits; under oligopolistic capitalism such returns could continue longer; in the socialist system the form of such returns again was quite different. Professor Tsuru said that Professor Matthews appeared to be aware of the difference between the two approaches referred to here and yet tried to make the 'socio-economic' approach represent faithfully the 'physical-technical' approach, which got him into difficulties.

Professor Triantis said that there appeared to be confusion in Professor

Matthews' paper. He said it was possible to look at the relationship between the level of the stock of knowledge and the level of income of a country. Both of these were high with respect to the United Kingdom. But one could also relate the yearly R & D expenditure to the level of income. Professor Matthews said this might be obscured by other factors and a two-way causality was present. One could also relate R & D expenditures to the rate of growth of income. The evidence was not clear that eminence in science was related to rate of growth. No one suggested this. But Professor Triantis said he would like to suggest that the growth rate was related to yearly expenditure on R & D as a percentage of income. The causality would be a two-way one.

Professor Simai said that he would like to emphasise the statement made by Professor Matthews about the importance of the supply of factors of production needed to carry out R & D and implement its results. He said this was especially important from the point of view of smaller countries. He felt that the distinction between the opportunities of smaller and larger countries had not been sufficiently emphasised in the paper. He also felt that the problem of the brain-drain needed more emphasis, particularly in relation to the calculation of human investment.

Professor Papi said that he had been exceedingly interested in the brilliant paper by Professor Matthews, and fascinated by his scientific imagination. The conclusions and policy implications appeared sound. He said that he was particularly interested in the statement 'if we are considering the phenomenon of modern economic growth in the broadest terms it is impossible to separate the contribution made by advances in scientific and technological knowledge and its diffusion from the contribution arising from other sources, since they are all complementary'. Professor Papi said that he would like to suggest in regard to the 'other sources' that it was important to have a knowledge of the many existing obstacles to economic development. He said that science should never neglect the necessity of removing obstacles which were no less important than the discovery of new processes. He said that we ought to collect information about these from the developed as well as the developing countries in order to arrive at a more general theory of development. Obstacles sprang from particular features of developing countries, for example dependence on agriculture, but many were common to developing and developed countries, for example the uneconomic behaviour of public authorities and the indiscriminate use of macro-economic tools. On this latter point Professor Papi cited the recent volume of essays in honour of Fritz von Hayek, in which Peter Bauer has authoritatively underlined incongruities of such utilisation. In this respect he said that one could get into the situation in which, in order to bring about increases in income, one needed an even greater growth in the costs to produce it. If this was so, the possibility of competing with other countries would weaken. Professor Papi went on to say that there was an ever greater distance between public activity and the activity of the individual. The consequence of an encroachment of public activity on that of the private individual was a progressive publicisation, nationalisation, collectivisation. This collectivisation of the

economy promoted imbalances between costs and prices, imbalances
in budgets and balance of payments disequilibria. Thus started the instability in the purchasing power of the currency – a kind of instability that
could never be attributed to shortcomings in the international monetary
system. In point of fact the instability of the purchasing power of a currency
depended to a large extent upon the scarcely economic conduct of the
public authorities of a country. This was an apparently familiar route to
communism which many governments were following. It was a route
perfectly opposite to that most favourable to the economic and social
development of a country or region. Science had to ascertain the existence
of such obstacles to economic development and to try to eliminate them
by timely and efficient action.

Professor Robinson said that he would like to make one simple point,
which was well known to all who worked professionally in this field. He
said there had already been during the conference a good deal of alter-
nation in language and description between science and technology on the
one hand and R & D on the other. He said that when he had been chairman
of a committee which made the first British R & D expenditure survey
they had found it impossible to separate R from D. If one included development as well as research, one must remember what development
might mean – for example in the aircraft industry 'development' included
not only the whole of the design but also the building of the prototype,
its testing and getting the bugs out, and anything else before reaching the
stage of quantity production. In shipbuilding, on the other hand, it was
normally assumed that ships were 'one-off' jobs, and the accounting
procedures treated the whole of design as part of the normal cost of the
production, and they were not included in the R & D expenditure. Professor Robinson said that if shipbuilding and aircraft spent the same on
fundamental research then a shift in demand from ships to aircraft would
result in an apparent big increase in R & D expenditure. He went on to
say that we must be cautious about regarding R & D figures as measures
of the contribution of science and technology.

Professor Matthews replied by saying it was inappropriate to try to
summarise or comment on the general contributions of Professors Papi
and Tsuru and Professor Simai. He said he looked forward to hearing
Professor Tsuru develop his contrast further. However he wanted to take
up some technical points which had been raised. Professor Eastman had
suggested that instead of thinking of misallocation one might look at differing degrees of appropriability and that optimum allocation might be assumed where there was 100 per cent appropriability. Professor Matthews
said he might agree with this if the government was not already subsidising R & D. The question also had to be asked, what was meant by
optimum – relative to what? Optimal allocation of savings between investment and physical capital and in knowledge, or between capital formation
and consumption? Professor Eastman had suggested that the way to deal
with the appropriability problem was by government subsidies to inventors, who would then go to work and ultimately make results freely available. Professor Matthews said the problem here was knowing in advance

what would be the ultimate outcome of research, and therefore at what level these subsidies must be fixed. On the question of the garage inventor and whether there was any difference between him and ordinary R & D, Professor Matthews said that firstly he was not included in O.E.C.D. statistics, and secondly he was not subject to the same sort of motivations as the other groups of people discussed during the session, because of the different nature of the processes in which he was involved. To Professor Triantis, on the question of the level of scientific knowledge and growth, Professor Matthews said that on the usual assumptions about the production function he was right – the rate of growth of output should be related to the rate of growth of scientific knowledge, but he did not think it would be difficult to make a good case for the opposite. Indeed, were the two very different? On the question of scientific eminence Professor Triantis appeared to think Professor Matthews meant all the Nobel prizes that a country had won; in fact he meant the average number of Nobel prizes it was currently winning, although these two might not be very different. On Professor Robinson's points, Professor Matthews said he accepted these entirely: that there was more than a nominal difference between research and development as ways of advancing knowledge. There was a great deal of preliminary work required to learn as part of the process of production. This could lead to some important differences and also to questions about how much one spent on improving knowledge. If advancement of knowledge came about as a result of solving problems in production then advance of knowledge as a separate activity was only secondary, in contrast to the case where advance in knowledge comes about only or mainly by deliberate R & D effort.

Professor Tsuru intervened to say that he had a further point he would like to make regarding what Professor Matthews had said about *Gs*, to the effect that Denison approach could not be applied to *Gs* and also that *Gs* was essentially embodied in *K*. Professor Tsuru said that he agreed with this and that Professor Matthews got into difficulties by speaking in terms of 'inputs' and 'outputs' which were classified from the standpoint of 'physical-technical' approach, which implied that his 'outputs' were not coterminous with his G.N.P. and that his attempts to find that portion of G.N.P. that was attributable to the contribution of R & D were bound to fail. Professor Tsuru suggested that it seemed to be best to consider science as 'ownership externality' – a scarce factor whose ownership could not be effectively identified – and to give up the notion of appropriability. He went on to say that especially in the era of the scientific-industrial revolution, science – even the basic sciences – becoming a central core of business activities, the contradiction heightened between the 'ownership externality' character of science and the desire of private enterprise to 'own someone's brain' and identification of specific contribution on atomistic basis became impossible and the 'societal-individual' evolved.

Professor Matthews said he had misunderstood the point made earlier and that strictly speaking pure knowledge was not a factor of production but a free good. However, he would like to take Professor Tsuru to task for the use of the word 'science' – he said this illustrated the difference in uses of the word. He said it could be used in different circumstances – for example the scholar who was interested in the pursuit of truth – but this was very different from the use of science in the development of knowledge for the increase of productivity. He said that we must be careful in how we referred to science because it did not mean the same thing to all people and in all circumstances.

2 Technology in the Developed Economy

J. K. Galbraith

HARVARD UNIVERSITY

I. CHANGING PUBLIC ATTITUDES TOWARD TECHNICAL INNOVATION

Few things have been more interesting in the last decade than the revision of public attitudes toward technical innovation in the United States and other industrial countries. Until very recent times it was, in all countries, a nearly absolute social good. The word invention was, for all practical purposes, synonymous with progress. The scientist and engineer, the sources of such innovation, were prime social benefactors. The encouragement of technical innovation was a central function of the modern state.

These attitudes are, of course, still strong. But they are no longer unchallenged; increasingly, questions are being asked about the validity or desirability of much innovation. Nor is this the esoteric reaction of intellectuals; these doubts are acquiring a substantial political base and one that is wide enough in the United States, for example, to arrest such a development as the supersonic transport. My purpose in this paper is to examine the economic foundations of modern technical innovation to see if it is possible to determine the sources and the substance of these doubts.

I am conducting this examination in a context which will be generally familiar to those who have encountered my argument in *The New Industrial State*. I assume that most technical innovation comes from the highly organised sector of the modern economy – the sector characterised by the modern large corporation. I am assuming that the directing intelligence of this corporation is collegial – that its direction unites the specialised knowledge and skills of managers and technical specialists in what I have called the technostructure. It reflects the evolution to a bureaucratic as distinct from the classical capitalist control.

The foregoing assumptions are, I believe, consistent with circumstance. Two other characteristics of this system must also be mentioned. The corporation being large and powerful, it does not react passively to its market situation. It exercises control over its prices and major costs in the commonplace manner of the firm under conditions of oligopoly. And it goes beyond the prices so controlled to seek to promote or ensure the buyer response that serves its interest. Through advertising, other forms of salesmanship, other forms of persuasion, the initiative in the formation of wants, and thus in the guidance of the economy, passes in some measure from the consumer to the producer – consumer sovereignty gives way, at least in part, to the sovereignty of the producer. This part of the economy, in other words, is loosely and informally planned. We may refer to the small number of firms that comprise it – in the United States, a couple of thousand – as the planning system. This is in contrast with the many millions of small enterprises which comprise what may still be called the market system.

In the planning system, the sovereignty of the producer is exercised on behalf of the technostructure. This interest is not served by simple profit maximisation as classical and neoclassical models assume. The interest of the technostructure requires a secure minimum of earnings. But beyond that, it has more to gain from growth than from profits. Added profits reward the owners; added growth through pay, promotion, perquisites, prestige and power rewards the technostructure itself. Earnings, including a record of improvement, are protectively necessary for the autonomy and survival of the technostructure; growth serves the affirmative interest of the technostructure.[1]

This is the context in which I examine the role of technical innovation. But first let me say a further word on how it is regarded in the conventional or neoclassical models.

II. INNOVATION IN THE CONVENTIONAL MODELS

In the neoclassical model, technical innovation does two things. It brings into existence new or improved products or services that are welcomed and bought by consumers because they better serve their needs. Or it improves the processes by which products are made or services rendered. In technical terms, innovation operates either to create or alter demand functions or to lower cost functions. Invention or improvement of products is in response to a perceived need, and if the producer or inventor can perceive need, so, obviously, can the consumer. Perhaps some persuasion is necessary to overcome the innate conservatism of the user. But the invention has

[1] It will be evident that in these assumptions as regards threshold earnings and growth, I follow the general line of argument of William J. Baumol in *Business Behaviour, Value and Growth* (New York: Macmillan, 1959), especially pp. 48–53, and Robin Marris in *The Economic Theory of 'Managerial' Capitalism* (NewYork: The Free Press of Glencoe, 1964). Cf. also Carl Kaysen in *The Corporation in Modern Society*, ed. Edward S. Mason (Cambridge: Harvard University Press, 1959), p. 90.

merit or is without merit in accordance with whether it meets an eventually recognised need. The persuasion reveals the need but does not create it.

Since technical innovation provides the user with products that better serve his needs, or lowers costs and thus improves efficiency, the neoclassical model holds it in high regard. Its disapproval of any interference with invention is especially strong. Traditionally, workers have resisted new processes out of fear they might lose their jobs. Producers have similarly sought the suppression of new processes to protect existing investment or because they feared competition. And producers have sought to suppress new products because of their fear of competition with established lines and the effect on existing investment. In all of these instances, the public interest in cheaper or better products suffered. In the traditional and neoclassical view, these were also the only purposes of interfering with invention. This being so, all such interference – any opposition to 'technological progress' – was banned and held, indeed, to be almost uniquely short-sighted and wicked.

Most economists would concede that much if not most modern innovation is now highly deliberate – that the thing to be invented or the improvement in process that is wanted is settled in advance. Then the result is pursued usually by teams rather than by individuals and often in accordance with established timetables and within approved budgets. However, the notion of spontaneous innovation by an individual deriving from a brilliant innovating thought is not quite dead. It retains its existence, in part, because such innovation, being inexpensive and independent of organisation, remains available to the small firm and thus to the market system. Without it, innovation of all kinds would be, in theory as it largely is in practice, the exclusive property of the planning system with its resources in specialised knowledge, organisation and capital.

What remains to be seen is that, as technical innovation becomes organised and deliberate, it also becomes an instrument for advancing the goals of the technostructure. It comes to serve its protective concern for its security; it becomes an instrument for promoting the growth of the firm. The motivation that lies behind it – of what is no longer an episodic or accidental development – is not the perceived need of the consumer or user but the interest of the producer. It is a response not (or not primarily) to the sovereignty of the consumer but to the sovereignty of the producer. From the viewpoint of the consumer, in consequence, it develops the ambiguity which we now observe.

This is true of innovation both in processes and in products. I take up first innovation in processes.

III. INNOVATION IN PROCESSES

Innovation in processes lowers costs and prices and thus allows, all else equal, a more rapid rate of growth. Thus it serves the affirmative purposes of the technostructure. This will not be thought socially damaging or ambiguous – no one surely can be ill-served by lower prices. But in two respects, this may be so.

All innovation in processes carries with it the possibility of new external diseconomies – of pollution of air, water or landscape. Even if this replaces older forms of environmental damage – as when thermal pollution of atomic power plants replaces the smoke of molecular plants – the new damage, to which the community is unaccustomed, will seem worse than the old damage to which it is adjusted. And the added growth will add quantitatively to the damage. It must be expected, accordingly, that increased environmental damage, or what is so regarded, will be the normal counterpart of technical innovation in processes. It will, rightly, be blamed on the innovation.

Additionally, in the modern corporation, the one factor of production that is not fully under the control of the technostructure – which, in some matters, is a challenge to its authority – is the union. This challenge is partly neutralised by the convention that excludes general union interference with what are called the prerogatives of management. It is further neutralised by control by the firm of its prices and the associated tacit understanding between firms that allows wage increases to be passed on to the public. But the power of labour is also reduced by technical innovation. All but invariably technical innovation involves the substitution of capital for labour. In the planning system, the savings that are the source of capital come extensively from the earnings of the firm – they are at its discretion and under its control. And prices of the resulting machinery and equipment are more predictable than wage costs and, once installed, do not go on strike. So innovation and the concurrent substitution of capital for labour increases the security and power of the technostructure. This means, in practical terms, that the question of labour-saving machinery in the modern corporation is not uniquely a matter of pecuniary calculation. It is quite conceivable that the substitution of capital for labour will occur at increased cost. It means that process innovation need not be of service to the consumer; it can serve the purposes of the technostructure while increasing costs to the consumer. Meanwhile, the disapproval of workers who are displaced can be assumed. And the ancient argument that acceptance of technical innovation is a sacrifice that workers make in the larger public interest is without merit.

I do not wish to argue that all process innovation is for the purpose, through capital substitution, of increasing the security and authority of the technostructure. I do want to argue that this is a possible and valid purpose from the viewpoint of the technostructure and one which justifies public doubt. However, it is with regard to innovation in products as distinct from processes that innovation, in the planning system, departs most sharply from the publicly benign role envisaged in the neoclassical model. I turn now to product innovation.

IV. INNOVATION IN PRODUCTS

As noted, once the innovation in products becomes organised and deliberate, it comes into the service of the purposes of the technostructure. The major affirmative purpose of the technostructure is growth. In pursuing this goal, it is active, not passive. The test of an invention, accordingly, ceases to be what serves a perceived need of the consumer; the test becomes instead what the consumer can be persuaded to want or buy – what, in common language, can be sold. The difference here is considerable.

To begin, newness has saleability apart from any technical improvement or enhanced usefulness. The traditional (and widely accepted) view of invention was strongly linear; it took for granted that a newly invented product was better than something that was invented a year or ten years ago. The 'latest thing' was naturally the best. This view derived from the genuine experience of the past. When inventions succeeded or failed to succeed in accordance with whether they met or failed to meet needs perceived by the user, later inventions were better than earlier ones. Those that were not, sank without a trace. The notion of such linear improvement in invention was reinforced by the pedagogy of the neoclassical model. The neoclassical model recognises no other kind of invention.

Given this social view of invention, there arises the possibility, in effect, of trading on the experience of the past. It is no longer necessary that a new product be more useful; the consumer can often be persuaded if the product is new. Newness has sales value in itself. And this value may continue even in the face of inutility or mechanical defect. Anyone who doubts the point need only notice the repetitive emphasis which all advertising, even of the most stereotyped products, places on their being new or having some feature that can be so described.

Additionally, innovation in conjunction with advertising plays a vital role in stimulating the psychic obsolescence of goods and their consequent replacement. This process has been highly successful in the past in the automobile industry although it has a wide application to other consumer products and their packages. It consists in creating a visually different version of an existing product and then, through advertising, persuading the consumer that this is the only acceptable image of the product. Although mechanical improvement, enhanced comfort or conveniences or other technical advance may be claimed this is not decisive for success. And mechanical retrogression, physical inadequacy or extreme inconvenience are not decisive handicaps. (In the automobile industry, enhanced inconvenience and discomfort have normally been associated with visual obsolescence.) The important thing is that the change succeed in making the earlier version visually eccentric and that possession and use, in consequence, reflect discredit on the person so owning or using it.

As always, exaggeration must be avoided. In addition to innovation that exploits newness for its own sake or which succeeds by rendering existing products visually obsolete, there is much that reflects mechanical or other improvement and serves a need which the consumer would, even without persuasion, come eventually to perceive. Even here, however, the role of persuasion remains important. The decision to put resources into the development of one product in preference to another turns extensively on the question of which can be sold. Minor innovations which bear on some common source of frustration or unease – which relate to inadequate sexual opportunity, a sense of social inferiority, obesity, an excessive commitment to the crypto-servant role of the modern housewife – will have attention far beyond that accorded such seemingly more important innovation as efficient surface transportation or comfortable and durable housing. As the everyday language holds, it is 'the kind of thing that can be sold'. It follows that the allocation of developmental energy is unrelated to any objective interest of the consumer. It may be frivolous and perverse.

V. INNOVATION AND THE MANAGEMENT
OF DEMAND

For the management of the demand for public goods – those purchased by the state – the role of technical innovation is crucial. It is also remarkably uncomplicated. Innovation in the most important products sold to the modern state is highly organised and wholly deliberate. The recognised effect of each innovation is to render obsolete the preceding product and thus to create a demand for the newly developed product. Work then proceeds (or in the frequent case will already be in process) on the next innovation with a view to rendering the new one obsolete and thus providing a market for the next product.

The procedure here described has reached its greatest perfection in the case of weapons and weapons systems. Here the sequence of innovation and obsolescence has been completely systematised – successive generations of aircraft and missiles are formally projected with the dates in the future when that particular item will become obsolete and will require replacement. All associated with the process recognise that the continued success of the weapons industry requires such innovation.

Payment for the innovation – for research and development – is largely assured by interaction with the planning systems of other countries and, in the case of the United States, with that of the Soviet Union. It is held that if the development that renders existing weapons obsolete does not proceed in the United States, it will proceed unilaterally in the Soviet Union. This, in turn, will accord an unsupportable military advantage. It may be assumed that the same argument as to the danger of unilateral development by the United States is made within the planning bureaucracy of the Soviet Union. The role of innovation in producing obsolescence and thus creating demand for public goods is reinforced by a competition that amounts to tacit co-operation between the two industrial powers. The result is not only a powerful role for innovation in the management of demand for public goods, but also one that is remarkably reliable. Any suggestion that innovation in weaponry is a device by which the technostructures of the weapons firms (in conjunction with the public bureaucracies) create the demand for their own products can be met with the answer that this may be true. But since development (it is held) cannot be controlled, there is no alternative that does not recklessly accord advantage to a national competitor. So it must proceed.

This use of technology to manage public demand is further protected by the fact that all intelligence on what is being undertaken abroad comes through the public bureaucracy to which the technostructure of the weapons firms are symbiotically related. Within limits, this intelligence – what other countries are doing – can be adjusted to need or wish. And, finally, military secrecy can be invoked to exclude public and legislative intrusion into the decision-making process. The use of technical innovation in the management of demand for major public goods is, in all respects, the ultimate achievement in producer sovereignty. It is not surprising that here, at least, not even the most devout defend the neoclassical model of citizen or consumer sovereignty.

VI. THE NEED TO CRITICISE INNOVATION

It is in light of the foregoing reality that popular attitudes toward innovation have been altering. The neoclassical and traditional models still powerfully affirm its merit. But the suspicion with which innovation has come to be regarded is seen now to have substance. Where processes or production functions are concerned, it substitutes unaccustomed and hence less acceptable forms of environmental damage for those to which the community has become accustomed. As an instrument of economic growth, it increases quantitatively the amount of such damage. Where products are concerned, it trades on the reputation associated with past invention. This being fraudulent, provokes an adverse reaction at least from the more sceptical consumer. So do products in which technical performance is subordinate to saleability. So does innovation that serves only to make a predecessor product visually obsolete. So also innovation which, though genuine enough, cultivates fertile but frivolous need while more important but less saleable innovation is signally lacking. And, in relation to public needs – weapons, in particular – the role of deliberate and competitive innovation and associated obsolescence becomes alarming. In consequence, the merit of innovation, both in private and public products, ceases to be something that can be assumed. Rather it becomes something the virtues of which must be assessed. It is undoubtedly in the interest of the technostructure; that it is in the interest of the public no longer follows. As in the case of supersonic transport, where the interest of the technostructure of the aerospace industry and that of the public were admirably juxtaposed, it comes to be seen that the public interest must be separately resolved.

The nature of this resolution can be readily indicated in principle; some of the practical problems are rather more difficult. It is not that the market appraisal of technical innovation is unreliable; it is rather that planning has replaced, or partly replaced, the market. And technical innovation is now tested by its service to the planning goals of the technostructure rather than to the goals of the consumer or public user. I will assume, without debating the matter further, that the market cannot be restored. To do so would mean reversing all the tendencies of economic development of the last hundred years and disestablishing the institutions, most notably the large corporations, by which it is guided and planned. The alternative is in the political process. This requires, first of all, what may be termed a new public perception. This new public perception accepts that technical innovation, in serving the goals of the technostructure, may or may not serve the public interest. The assumption that the two are identical, like the assumption that technical innovation is uniformly serviceable to the public interest, can no longer be accepted.

It follows, then, that the legislative or representative process in government must, henceforth, be brought sharply to bear on technical innovation. This is the only obvious expression of the public as opposed to the technocratic interest. It must increasingly express the public view on processes that are damaging and products that are damaging or unserviceable to the public interest. Or it must provide the administrative organs that do so. And, needless to say, it must be sharply critical of publicly sponsored technology with a view again to distinguishing between that which serves the goals of the technostructure and that which serves the goals of the public.

What is here suggested, at first glance, will seem a novel, as well as a formidable, responsibility. In fact, it is a tendency to which legislative and other public authority is already strongly committed. The problems associated with technological innovation, which I here seek to identify, have already impressed themselves strongly on the public consciousness. These, in turn, are inspiring two major modern thrusts in public policy – that which is called the ecological or environmental movement and that which passes under the general designation of consumerism. The first of these seeks to bring a public and legislative judgement to bear on technical processes and to discriminate against those that are publicly undesirable. The second seeks to bring public judgement to bear on technical innovation in products. Both movements are responsive to the failure of the market and to the conflict between the planning goals of the technostructure and the public interest. Much remains to be said about the way in which this legislative and public interest is defined, expressed and made effective. But, reflecting the underlying need as (I venture to think) here identified, it is already by way of becoming a major function of the state.

Discussion of the Paper by Professor Galbraith

Sir John Hicks, introducing the paper, said that there were several matters in Professor Galbraith's provocative paper which would no doubt be taken up later by contributors to the discussions – such as the question whether innovations were predominantly carried out by corporations. He would leave these on one side, since there were others who were better equipped to discuss them than he. He would like to concentrate on the fundamental issue. Was the activity we had to discuss, R & D – or just innovation – a good thing or not? Professor Galbraith was certainly maintaining that it was not such a good thing, as had been supposed; but he was not quite clear how far Professor Galbraith's condemnation extended. This was the theme of William Morris, of R. H. Tawney and perhaps of J. S. Mill. Probably he did not go so far as to condemn all innovation, but he was certainly saying that much of what appeared to be innovation in the sense that had been attributed to that word by economists was only pseudo-innovation. Sir John Hicks said that he would entirely agree that some innovation was pseudo, though he did not think that his estimate of the proportion would be as large as Professor Galbraith's. He hoped that some of the discussions would clear their minds as to how large the proportion was.

Sir John Hicks gathered that in Professor Galbraith's view process innovation was fairly blameless; it was product innovation that was tainted. This, one could recognise, came down from the old familiar stuff about imperfect competition and product differentiation, of which we used to hear so much in the 1930s. In terms of the early sixties and seventies it was somewhat static. Sir John Hicks did not think we needed to go into all that again, drawing the curves and so on. He would however like to suggest (also with a view to later papers) that we should be somewhat careful about this product-process division. In a completely integrated industry (if there was such a thing) it would be a clear distinction; but when the 'machines' (in the widest sense) were made in one firm and were used in another, the distinction was ambiguous. What was product for the machine-making firm was process for the using firm when the production process was considered as a whole. He supposed that Professor Galbraith's large firms were thought of as being fully integrated, but he would have thought that one source of possible disadvantage in innovation was the product innovation of the machine-making firms, not necessarily geared exactly to the true wants of the using firms, who were exposed to the same kind of Chamberlinistic monopolistic competitive pressures as was the consumer in the integrated firms. This was a matter which he hoped that we should at some stage be discussing.

So far, he was perhaps being closer to Professor Galbraith than might have been expected. There was another distinction which Professor Galbraith did not make and which must surely be brought in – the distinction between single-use goods (as he himself had called them) and durable-use goods; the meaning of the distinction was evident. He had long felt that far too much of the imperfect competition discussion ran in terms of single-use goods (such as cigarettes) when in terms of durable-use goods (even consumer durables) the whole question of want satisfaction became much more complex. It was quite clear that innovation did correspond to a considerable extent to the desire for novelty; but who were we to say that the desire for novelty was irrational? He felt that Professor Galbraith was saying that it was – perhaps not irrational, but improper. But why should we economists be committed to this extreme form of puritanism? Might he take an extreme example? What about the people who design ladies' dresses? They were clearly innovators; they spent their time designing new products, and persuading people (their customers) that these make the old products obsolete. Everything that Professor Galbraith had said about pseudo-innovation applied to them. But were we committed to an 'optimum' which implied putting the female population into a warm and comfortable uniform? Professor Galbraith had been rather clever in presenting his own ideas as 'what people are thinking', but the party he was recruiting was pretty heterogeneous. Environmentalists, Christian Socialists, Economist Presbyterians – Sir John Hicks did not see that they had much to do with one another. Here, at least, what we should be discussing was the view of Professor Galbraith, not the view of his party – if it existed.

He had said nothing about the military side, for we should learn much about that in later papers. He thought that other contributors were hoping to distinguish between military innovation and non-military innovation, as well as environment problems; Professor Galbraith was trying to convince us that they were all the same thing. He himself preferred the other view – he thought it would take us further.

Professor Khachaturov said that he had read Professor Galbraith's paper with great attention, as it had many interesting ideas, particularly with regard to the widespread but rather out-of-date idea of market economies. The situation has now been modified: free competition was very limited through the domination of corporations, and consumer sovereignty has been forced out by producers. Professor Galbraith put forward interesting ideas about obstacles to innovation, in particular the protection of existing investments but it was difficult to justify his pessimism on the negative effects of the modern development of science and technology. Technological progress was important to the progress of society. Pollution, fraudulent advertising and the like were not part of new technology as such, but were social/economic conditions. Pollution for example could be overcome by various processes enforced by legislation, as they were in the U.S.S.R.

Professor Neumark said he largely agreed with Professor Galbraith, but he had some doubts about some of the concluding remarks in the paper. He said that in the matter of public interest he saw no reason to suppose that the government could be influenced by bringing public opinion to bear on it, because there was also strong pressure from technocrats and the producers' interests might be furthered by the politicians. He said he had a pessimistic view about this.

Professor Patinkin said that on consumer sovereignty and the related question of the desire for novelty, one of the most revolutionary pieces of work on this was written by Frank Knight in 1922 – *The Ethics of Competition*. This was a critique of the view that market economies were orientated to the satisfaction of consumers, arguing instead that needs were generated by the system, and the system should be judged by the needs it generated. Knight still thought the market economy was better than the alternatives, and that social control was more frightening, bringing problems of efficiency and bureaucracy. Professor Patinkin said there had been a less inventive programme from government institutions than from private ones. Professor Patinkin went on to say that consumer sovereignty might have contracted but there had been some wonderful displays recently, for example the cutbacks in expenditure on moon probes in the United States as a result of waning public interest and willingness to pay for them. The S.S.T. experience was a further manifestation of consumer sovereignty, and there were similar developments in the automobile industry, although they were not quite as significant. While one might complain about it, the consumer did choose the new automobile model rather than the existing one at a lower price, and it must be remembered that the development of small automobiles was not producer sovereignty, but consumer. Professor Patinkin felt that many of the criticisms in Professor Galbraith's paper were exaggerated.

Professor Galbraith replied to Professor Patinkin's comments and said he was tempted to agree – Knight was the source of the original claim that consumer sovereignty had been subverted by capitalism. Professor Patinkin had said that this was not true, and quoted as examples automobiles, moon exploration, weapons, and other cases, but Professor Galbraith said there had been a resurgence of essential differences between technocratic and public interests. If this was a resurgence of consumer sovereignty he would agree, but he said he did not feel this came from the market but rather from the legislative process. The remarks on automobiles were noted, but in practice the public did not have the chance to buy the cheaper original model, as had been suggested.

Professor Griliches said that he did not agree with this. In the early fifties one company persisted with the smaller type of car (Plymouth), lost heavily on their share of the market and had to join the others in what seemed consumer preference for larger cars. In the mid 1950s the European car became available but it took a long time to diffuse. In the earlier years servicing was a problem and it has become a real alternative only within the last few years.

Professor Robinson said he was mystified because he did not think that either Professor Patinkin or Professor Griliches was addressing himself to Professor Galbraith's question. In a world in which sales pressure and advertisements had caused demand to be shifted to new products or new models their arguments did not prove Galbraith wrong or that the gain to society was positive. He thought Professor Galbraith had said you cannot take effective demands as criteria of what would have been satisfaction apart from that influence on demand. *Professor Griliches* intervened saying that there was a difference between persuasion and restriction of choice.

Professor Delivanis said there were two points he would like to raise in the context of the discussion of Professor Galbraith's paper. Firstly, Professor Galbraith believed that increased investment aiming towards the application of new technology consisted in the replacement of labour by capital and thus led to a diminution of the unfavourable repercussions of strikes. Professor Delivanis said he did not think that the latter are reduced. There might be fewer people, but they had more influence, and small groups might have great economic power, as had been shown by many recent strikes, particularly unofficial strikes. It could thus be said that our reliance on labour's goodwill had fallen. Secondly, Professor Galbraith was perfectly right in stressing that all innovation carried with it the possibility of new external diseconomies such as pollution. Professor Delivanis said that all plant and all means of transportation could create diseconomies. Accordingly in deciding to adopt an innovation, the expense of protective measures and other investments needed to neutralise these diseconomies had to be brought into the calculation.

Professor Sefer said that in underdeveloped countries new technology often meant an attack on the traditional distribution of labour, and the latter saw the process as management making the decisions for the majority. Such a distribution of labour, economic and social structure could not solve the problems created by technological progress, and society lived from one crisis to another. The trend was for new technology to be largely based on large firms, and the amalgamation of smaller units. Such integration must be freely willed. Small units must be free and responsible for themselves – if not, social conflict was inevitable. Integration should not be based on monopoly or bureaucracy – there was need for more agreement and less domination. It was difficult to say what were the best tools for doing this, social ownership was only one tool – what was really needed was full information so that people could decide about their own destiny. This was a long process and largely unknown in under-developed countries. In Yugoslavia they believed that they must think about this a long way ahead, or they would come up against the contradictions now faced by the highly developed world.

Professor Matthews said that consumption in all circumstances was a highly social phenomenon, affected by a multitude of inter-relations and pressures. There was no such thing as consumers' preferences 'in the raw' which might be compared with their preferences as moulded by the influence of producers. The work of social anthropologists had shown that the social character of consumption applied even in the most primitive societies – it was not confined to rich industrial ones. The effect of alternative patterns of consumer preferences on true ultimate measure of welfare was therefore a question that took one into very deep water. The consumer movement served a necessary and useful function by giving information on the physical properties of the goods on offer, but it could be a false as well as a dreary philosophy to suppose that the physical properties of goods and the physiological aspects of consumption were all that consumers were interested in. Professor Matthews also wondered whether Professor Galbraith's thought was best expressed by contrasting producers' interests with consumers' interests. Producers' interests in the broad sense of the conditions of work (physical and social) were not something that economists would wish to disregard, although unfortunately they often tended to do so. What Professor Galbraith feared was evidently the excessive weight of producers' interests in the very much narrower sense of the interests of the technostructure and its Napoleonic ambitions.

Professor Galbraith said he did not want to appear to disagree, in fact there might not be actual conflict. He agreed that consumption was a social phenomonen, and said that it was increasingly so at higher income levels. In all societies there were strong social influences on consumption, but this impeccable proposition was becoming a device for perpetuating neoclassical theory. There was a difference between the reaction of an individual as a product of social conditioning and the deliberate effort of a producer of goods to persuade the consumer by propaganda and devices in designs for his particular ends. If we ignored that difference we neglected an important aspect of highly developed society, particularly in the area of public goods. Professor Galbraith said that he made an important difference between the example of women's fashion and that of automobiles, a distinction which we omit to consider at cost to our understanding of modern industrial processes.

Dr. Nabseth said that Professor Galbraith had raised a number of interesting hypotheses, but that his empirical facts were not overwhelming. He said that he liked to rely on empirical information and although he was relying on special information from one country he would like to challenge some of the hypotheses. He said it might be that the old ideas of consumer sovereignty must be replaced but if so we needed to look at the meaning of producer sovereignty. Was it so that producers could plan their product in advance and fulfil their product plans? Dr. Nabseth said he had a lot of material concerning the plans and outcomes for major firms in Sweden in the 1960s. The picture did not fit Galbraith's thesis, because there were wide differences between plans and outcome. One would have expected a better outcome for Swedish firms on the home market than on the export market, but in fact the differences were larger on the home market. Some of the reasons were that in the first part of the sixties demand was more buoyant in Sweden than expected. In the second half demand was controlled and producers could not fulfil their plans. It looked as though the seventies might be just as difficult, and this did not fit the idea that the company could plan its growth. He added that for the big companies the difference between plan and outcome was a little less than for the small firms, but not much. Secondly, Dr. Nabseth said that in a competitive society, especially with heavy foreign competition, it was difficult to get far from profit maximisation just in order to have growth. This had happened with some Swedish firms, but the Swedish banks, for instance, were now saying that profit was more important than growth. The latter was not an end in itself. On consumer sovereignty Dr. Nabseth said that classical analysis assumed that consumers had full information, but for many goods they could not be expected to have the necessary technical knowledge, and for this reason we had, for example, food and building regulations. This was now being extended into other fields, and this was bound to have repercussions.

Professor Freeman said that the distinction between product and process innovation was important. He agreed with Professor Hicks that product innovations in some areas, for example machine tools, were process innovations for others. The differences between capital and consumer goods markets were highly significant, and had an important bearing on consumer sovereignty. In their inquiry they had found that on the capital goods markets buyers were on a level of technical parity with the suppliers and they spent thousands of pounds on assessing the technical performance of their proposed purchase. They had even had cases where buyers knew more about the product than the suppliers, and in these situations consumer sovereignty was very much a reality. However on the consumer goods side there was a big difference. The consumer was generally not on terms of technical and scientific parity and he agreed with Professor Galbraith that to redress this balance was an important social and political problem. There were a number of ways in which this could be done – firstly by legislation, and this was already the case in some areas, where minimum performance standards were specified for consumer goods; he disagreed with Dr. Nabseth that there was nothing new in this; the consumers' assessment was now much more difficult than in the past. Secondly, it could be done by the work of consumer associations; but this was only effective if they commanded the technical services to make the correct assessment, and if this information had the same access to the mass media as the sales information, and this was not generally the case in a market economy. However, even with increased state specification of standards and more consumer associations we could have a problem of not drying up the source of innovations. This was perhaps best illustrated in the drug industry, where there was a straight trade-off – we must save lives and yet have thalidomide-style legislation. Professor Freeman said the public bodies must protect consumers and understand the beneficial effects of innovations and also the risks involved.

Professor Griliches asked to what extent was scientific invention endogenous, and said that to answer this we must go back to Schmookler's work, which was probably the most detailed analysis. Professor Griliches said he would like to come back to defend some of Professor Galbraith's paper from the neoclassical position. He thought he had short-changed himself by not bringing in the theme of equity and income redistribution. Innovation brought changes and some groups benefited while others lost. There had been a tendency in looking at the subject to fall back on welfare economics, and say that there was a net benefit if the gainers could compensate the losers. But how was the compensation to be organised? We had not looked at the compensation problem seriously. Much of what Professor Galbraith said in terms of pollution could be discussed within a broader definition of income redistribution. The question was who gets the garbage? There had not been an effective social mechanism to compensate the losers and impose social costs on the winners. The classical example was agriculture, where technical change had been beneficial to society as a whole, but had imposed a high cost on some parts of the agricultural population, particularly on those who had to leave it. There had been attempts at social intervention, but the cure might be worse than the disease. For example, price supports had undesirable income distribution consequences – they made food dear and benefited landowners. Unless more thought was given to this problem of income distribution there were likely to be adverse consequences and growth was not necessarily good if compensation did not occur. The rise of minority groups would bring pressure for the system to provide more social compensations and insurance in face of changes. When we had that we could then look at change again through rosier spectacles.

Professor Papi said that Professor Galbraith's paper led one to a naive question – how many corporations in other countries were similar to United States companies? 80 per cent of economic activity might be undertaken by small and medium enterprises in developing countries, and one should ask whether the phenomena in advanced countries were really applicable to these less developed areas. It was the transfer of certain concepts at different stages of development that led him to be doubtful. Professor Galbraith spoke of the disappearance of market forces because of planning – as far as the experience of Italy was concerned planning remained on paper and economic life appeared to be more influenced by events. The market appeared to have won the battle with planning and would never be replaced while economic activities were in the hands of small and medium-sized enterprises. Finally, Professor Papi said he agreed with Professor Griliches that where these phenomena occurred the government must intervene on behalf of the public interest, and it was already being done. For example, the ecological movement was calling into question technological progress.

Professor James said that Professor Galbraith appeared to draw some strange conclusions. On the one hand he wanted to entrust to public opinion selection of desirable consumption and at the same time he wished the state to be entrusted with choice. But did he think that the state could make an entirely clear choice? What he said in his paper about state intervention led Professor James to doubt this. What we must ask was whether macro-decisions of the state were more rational than micro-decisions by firms. The errors in the latter tended to offset each other but the former had no similar offset. Professor Neumark had said that there were pressure groups acting upon the state, but Professor James thought they were more likely to be producers' than consumers', and doubted that protection of the consumer could be safely left to the state.

Professor Triantis said that he had found Professor Galbraith's paper enjoyable but perplexing. This was a relatively new field and he had found some of the terminology slightly confusing. He had thought that innovation was always good, but Professor Galbraith's innovation produced a worse product. He therefore wished that Professor Galbraith would define innovation more clearly. On consumer reaction Professor Triantis said he felt Professor Galbraith exaggerated his point – consumer reaction might be organised better through consumerism, but the consumer always had the final decision. He felt that the reaction of consumers to the moon exploration was not a reaction to technology but a reaction to the product mix. The consumer had reacted to mass production, but he had also favoured it, an excellent example being that of unisex clothing. With respect to product innovation it was worth referring to the essay by Francis Bacon and looking more at the rate of change. Too rapid a rate could be as bad as sticking to the old. Professor Triantis said that in economics we had neglected the effect of the rate of change in income on the formation of tastes. If it were brought in it might help explain consumers' reaction. A rate of change might be too great or not enough for consumer satisfaction.

Professor Dupriez said he would like Professor Galbraith to expand on one problem of probably necessary government intervention which he had mentioned; adverse effects of scientific development may confront us, not as a result of the nature of new processes but as the result of their extension. Professor Galbraith insisted that big enterprises had built-in concern with growth itself. On the other hand, progress brought an increase in demand in relation with decreasing costs. The conjunction of the two forces led to growth for its own sake, and there must be a limit to this, however difficult the political problems of limitation may be. At the 'logical level' neoclassical methods of measuring may not cover adequately the problems of R & D which existed and could increase seriously. However there was not necessarily a fundamental conflict. The 'quantity effects' of technological development must enter into direct conflict with the logic and the aims af growth models, and specifically with the equilibrium growth rates. Government intervention on the misuse of new processes was no new thing, but we were faced with more generalised and portentous interventions as problems arise. It was not accidental that the most crying needs still arose in the biological, not in the physical world, and more particularly where there was no authority, that is on the high seas.

Professor Nussbaumer said he wished to ask three questions relating to some of the generalisations which helped to make Professor Galbraith's paper both so convincing and so controversial. First was the advantage of large firms in R & D, so frequently referred to, primarily due to a necessary size of operation or rather to the need for substantial financial resources and the capacity to carry considerable financial risk? If the latter was true, successful R & D could also be carried on by smaller firms if their financial problems were met, for example by joint ventures or by government assistance, and industrial concentration could be avoided to some extent.

Secondly, did the influence of innovation on the labour market not depend largely on the employment situation? In contrast to the United States, labour-saving innovation was wanted in most countries of Western Europe where shortages of manpower sometimes constituted serious obstacles to economic growth.

Thirdly, could the problem of optimal allocation of resources for R & D really be improved upon by public control? If government carried on research by its own institutions and not via the capitalist industrial complex, what were the standards of profitability which could be used? Were there any? If the government entrusted private firms with research to be carried out, it generally made an *ex ante* calculation of cost including a standard amount of profit, whatever that might be, and would also reimburse unforeseen extra costs arising from the project. The cost benefit problem in planning, therefore, frequently could not be solved adequately and it always required standards of social utility which had to be set by the government rather arbitrarily.

Mr. King said that as a physical scientist he had a few general comments he would like to make. He said it was wrong to ask whether innovation was a good thing or not – it had always been man's main way of pulling himself up from subsistence, and this process had been brilliantly successful. However, we must ask whether we were now reaching the point where we needed to reassess this position. In an advanced society we could reach saturation point with material goods and this approach led to luxuries of yesterday becoming necessities of today. If this was so, the consumer began to turn from purely material goods to services, and there was a need for reorientation in innovation towards social and service sectors where the operation of the market forces was less important and the conditions for innovation different. There was also the growing problem of unwanted side-effects of technological innovation which made social acceptance difficult. This was at present chiefly manifest in environmental terms to the extent that we have created an environmental bandwagon. The public, especially the young, went well beyond the environment as the sole difficulty of our technological growth. The sterility of life in urban conditions was often wrongly blamed on technology when it was more a matter of management of technology. It was clear that we needed a reassessment of the approach to innovation in the next few years. The twenty O.E.C.D. member countries, discussing growth prospects of the next decade, decided to go for growth at about the same rate as in the recent past, with the proviso that growth was not to be regarded as an objective in itself, but as a means of getting social development to take place healthily. Changes in the next decade were likely to be similar to those in the last one, but with an increasing concern for innovation in the social and service sectors. Governments must continue to stimulate innovation as in the past and inhibit it in certain areas, where assessment warned of undesirable social consequences. This could have extensive consequences for industry, and even now multi-national corporations were discussing the reconciliation of social demand and profit maximising production.

Professor Robinson said he still remained worried, despite Sir John Hicks' and Professor Matthews' remarks. He saw many difficulties of measuring consumer satisfaction in the raw, when two things were conceptually different, and it worried him that they were swept under the carpet on the assumption that what could not be measured must be zero. He thought this should not be done. Professor Galbraith had put a big question of what we meant by progress: Is everything that is measured as progress really progress in the true sense? Professor Robinson suggested we look at the implications for public policy. For two or three generations we had recognised that big corporations could be a danger to society through their monopoly position. Professor Galbraith pointed out that another way that they can do damage to society was by rigging demand and by the use of public media. The question was what do we do about this? It was worth considering what Professor Freeman had said on this point. The grounds for fear are not great on the side of capital goods. On the consumer side, however, economists had not spent enough time thinking about the defence of the consumers. In the United Kingdom we have had two consumer associations – one private and one public. With the recent change to a government dedicated to the task of making competition more perfect, its first act was to destroy the public one. This seemed rather inconsistent, but the real question to ask is – are the consumer associations doing what is the right thing? At present they tend to measure the qualities of new products against each other. But would it not be more worth while to be told more about new products as against products of the older generation which they replace? This would help the consumer to determine whether a new product is likely to fulfil better his requirements.

Professor Galbraith, replying to the discussion, said that he had been very pleased by the fact the conference had been so kind and generous to him. He had escaped better from Sir John Hicks than he expected or indeed deserved. He said that he would still quarrel with Sir John on one or two points – he still believed that in the past the firm was essentially subordinate to the market.

On the question put by Professor Triantis, he agreed that he had probably used the term 'technical innovation' rather loosely. He said he was sure that our socialist colleagues had enjoyed a unique row between the capitalists in which they were not intimately involved themselves. But he would like to ask Professor Khachaturov whether that was really true. Any large industrial bureaucracy tended to impose its power and will on the consumer public at large, and he wondered whether this was not a problem for any highly organised society, in which case the U.S.S.R. was not exempt. Some of the points made by Dr. Nabseth were well taken, in particular that the paper was short on empirical facts. Professor Galbraith said there were more available than had been offered; however, he would suggest to Dr. Nabseth that, like definition, the demand for empirical verification was often a way of avoiding saying anything. While one must never be casual about verification one must be careful about evading reality by demanding it. He said he did not believe that the fact that planning by a firm was not perfectly achieved was any indication that there was no planning. In many socialist economies plans were not achieved, but this did not imply that the Soviet economy was not a planned economy.

Professor Galbraith said there were differences in more than degree between protecting the consumer from fraud and the endemic reaction of consumers to products. We often tended to think of common law fraud when we had a more systematic and endemic problem. There was a danger that we should go on to think that innovation was a necessary evil. D.D.T. had almost become a symbol of the wickedness of chemical innovation, but in India one saw the massive benefits it had brought, for example in the reduction of malaria. There one got a different view. Backward countries might thus have a different view on D.D.T. as an innovation than we do in the developed world. Many critics had implied that we have imperfection of market resolution, but he wondered how much worse public resolution was likely to be? 'Democracy is the worst of all systems of government except for the alternatives.' If a fully developed corporation was thought of as an extension of the arm of the state (the large weapons complex in the United States) our choice was not between imperfect market resolution, but between two macro-decisions – open and admitted planning by the state or concealed planning by an arm of the state which was not fully recognised as such. He felt sure that Professor Patinkin would agree with him in this.

Finally Professor Galbraith said that there had been many suggestions that he was speaking out of North American experience and this was a special case. He said that leaving aside the case of less developed countries he would like to warn against any temptation to take this view. He said there was a homogenising effect of big corporations and the differences between one which was American and one which was Dutch or British was minimal. Perhaps one of the great achievements of the corporations has been the denationalisation of our life and the social and economic consequences of the large corporations were just as important in many Western European countries even today. What was different about the United States was their higher state of development in this regard, and others could not ignore that this development was the way they might go. There was no advantage in being on the same planet as the United States unless one used it as an example of the horrors to which one might be exposed in the future.

3 Research Expenditures and Growth Accounting[1]

Zvi Griliches
HARVARD UNIVERSITY

I. THE EXISTING LITERATURE

Many aspects of the economics of research have been discussed ably in a number of recent books and articles and will not be resurveyed here.[2] I shall concentrate instead on a relatively limited topic, the possible contribution of public and private research expenditures to the growth in the 'residual' as conventionally measured. Sections I to III of this paper review and summarise earlier work on returns to research, explain the logic behind them, and present some additional estimates of the impact of research expenditures in United States manufacturing industries on subsequent growth in their total factor productivity. Sections IV to VI discuss how research might be treated consistently in a set of real product and input accounts and explore what traces, if any, such expenditures leave in the conventional United States productivity accounts. Section VII closes the paper with a discussion of some, only slightly related, policy implications.

Investment in research, both public and private, has been thought to be one of the major sources of growth in output per man in this century. That it has been a good investment, in the sense that it yielded a positive rate of return which has been as good and often better than the rate of return on other private and public investments, is reasonably clear. Whether it can account for a major part of the observed growth in productivity is another matter. That will be considered in the latter half of this paper.

The evidence for the statement that R & D investments yielded a high social rate of return is scattered; much of it is second-hand; but it is still quite strong. It is of three kinds: individual invention returns calculations, industry studies, and aggregate residual attribution calculations. The individual invention calculations have been done by, among others, Griliches (1958) for hybrid corn and hybrid

[1] The work embodied in this paper has been supported by a grant from the National Science Foundation. I am indebted to P. Ryan for research assistance, and to Y. Ben-Porath, R. B. Freeman, F. M. Fisher, R. J. Gordon, H. G. Johnson, D. W. Jorgenson and S. Kuznets for comments on an earlier version. The usual caveats apply.

[2] See Machlup [1962], Mansfield [1968], Nelson [1962] and Nordhaus [1969], among others.

sorghum, by Peterson (1967) for poultry breeding research, by Ardito-Barletta (1971) for agricultural research in Mexico, by Eastman (1967) for military transport aircraft, and by Weisbrod (1971) for polio vaccines. In such studies, the total social returns from a particular invention are estimated and compared with *all* the research costs in the particular area of research, not just the costs of the successful part of the research. The internal rates of return implied by these estimates are quite high (10 to 50 per cent per annum), even though they are usually based on conservative assumptions, and even at their lower end (such as the 12 per cent estimate for polio vaccines by Weisbrod) are at a level worth considering investing more if the opportunity to do so were to arise again.[1]

A major objection to such studies is that they are not 'representative', having concentrated on the calculation of rates of return only for 'successful' inventions or fields. This objection can be met by econometric studies of productivity growth in specific industries. Here all of the productivity growth is related to all of the research costs and an attempt is made to estimate by correlational techniques the part of productivity growth that is attributable to the past investments in research.[2] The list of detailed studies is longer here: the most prominent being Mansfield's (1965) study of the private rate of return to research for chemical and petroleum firms, Minasian's (1962 and 1969) study of private return to chemical and drug research, Griliches' (1964) and Evenson's (1968) studies of the social rate of return to public investments in research in agriculture, and studies of the social rate of return to research investments in manufacturing industries by Terleckyj (1960, 1967, and Kendrick 1961), Brown and Conrad (1967), and Mansfield (1965).[3] While each of these studies is subject to a variety of separate reservations, together they all point to a reasonably consistent relationship between productivity growth and research expenditures and to relatively high (30–50 per cent) rates of return on average to both public and private investments in research.

Finally, a number of studies having measured as well as they could the contribution of other sources to economic growth, attribute the rest (the 'residual') to research and compute the implied rate of return. Such calculations have been performed in the past by

[1] Private rates of return of comparable magnitude are derivable from data on returns to specific refining inventions collected by Enos [1962].

[2] That conventional measures of productivity growth may not be very good measures of the output of research is something that we shall return to below.

[3] There are a number of other studies, somewhat less rigorous in spirit, relating differences in rates of growth of output among industries to differences in the intensity of their investment in research. E.g. see Freeman (1971) and Leonard (1971).

Schultz (1953) and Griliches (1958) for total research in agriculture, and by Fellner (1970) for the total economy. Again, the implied rates of return have been very high.

Perhaps it is not surprising that different studies yield rather high estimates for the rate of return to research in the private sector. After all, research is an investment just as any other and should yield a positive return.[1] Given the great uncertainties associated with it (including the uncertainty about the degree of appropriability of its results) it is plausible that the *ex post* returns should be rather high on the average. They include a non-negligible risk premium. What is more surprising is that public investments in research, which are not constrained or guided by the profit motive, and where there is wide scope for potential bureaucratic bungling, do seem to yield not only positive but actually rather high rates of return. This, while not unexpected, is heartening.

There are two other noteworthy points to be drawn from the above mentioned studies: First, there is a significant lag between the time that the investments in research are made and the time that the products of such research activity begin to affect the average productivity of an industry or economy. For public and primarily basic research the average lag appears to be of the order of five to eight years. For the bulk of industrial research (applied and development) the lag is much shorter, of the order of two to three years, but still significant. Thus, we are unlikely to observe the effects of the current drought in research support very soon, but it may come back to plague us in the late seventies. Second, research investments both depreciate and become obsolete. They depreciate in the sense that much of knowledge would be forgotten and rendered useless without continued efforts at exercising, retrieving and transmitting it. This is what much of higher education is about. Also, some of the newer findings displace, make obsolete, large parts of previously acquired knowledge. Thus, their net contribution is smaller than would appear at first sight. In short, a non-negligible rate of investment in research may be required just to keep us where we are and to prevent us from slipping back.

II. THE MODEL IN COMMON USE

Common to most analyses of the contribution of research to productivity growth is a model that can be summarised along the following lines

$$Q = TF(C, L) \tag{1}$$

[1] That private inventive activity is subject to economic influences and calculations, just like other aspects of private investment, is documented in detail in Schmookler's [1966] important book.

$$T = G(K,O) \tag{2}$$
$$K = \Sigma w_i R_{t-i} \tag{3}$$

where Q is output, C and L are measures of capital and labour input respectively, T is the current level of (average) technological accomplishment (total factor productivity), K is a measure of the accumulated and still productive (social or private) research capital ('knowledge'), O represents other forces affecting productivity, R_t measures the real gross investment in research in period t, and the w_i's connect the levels of past research to the current state of knowledge.

An important problem arises as soon as we write down such a system, a problem that will stay with us throughout this paper. Ideally, we would like to distinguish between capital and labour used to produce current 'output' and capital and labour used in research (the production of future knowledge and the maintenance of the current stock). In fact, we are usually unable to observe these different input components and are forced to use totals for C and L in our investigations. This leads to a mis-specification of (1). Moreover, if components of L and C are weighted in proportion to their current private returns, the resulting estimates of the contribution of K (or R) represent, errors in timing apart, the excess of social returns over private. The model, as written, is strictly applicable only where all (or almost all) of the research activity is performed outside the accounting boundaries of the sector in question; as for example in the case of agriculture where most of the research is public and where the inputs used in this research are not included in the definition of agricultural capital or the agricultural labour force.

For estimation purposes, the F and G functions are usually specialized to the Cobb-Douglas form and O is approximated by an exponential trend. The whole model then simplifies to

$$Q_t = Ae^{\lambda t}K_i^{\alpha}C_i^{\beta}L_t^{1-\beta} \tag{4}$$

where A is constant, λ is the rate of disembodied 'external' technical change, and constant returns to scale have been assumed with respect to the conventional inputs (C and L). Equations like this have been estimated by Griliches (1964) from several agricultural cross sections and by Evenson (1968) and Minasian (1969) from combinations of time series and cross section data for agricultural regions and chemical firms respectively. Alternatively, if one differentiates the above expression with respect to time and assumes that conventional inputs are paid their marginal products, one can rewrite it in terms of total factor productivity

$$f = q - \beta c - (1 - \beta)1 = \lambda + \alpha k \tag{5}$$

where f is the rate of growth of total factor productivity, lower case letters represent relative rates of growth of their respective upper case counterparts $[x = \dot{X}/X = (dX/dt)/X]$, and $\hat{\beta}$ is the estimated factor share of capital input.[1] (5) is a constrained version of (4). Versions of it were run by Evenson (1968) for agriculture and by Mansfield (1965) for manufacturing industries, among others. In either form, the estimates of α have tended to cluster around 0·05 for public research investments in agriculture (Evenson and Griliches) and around 0·1 for private research investments in selected manufacturing industries (Mansfield, Minasian, and Terleckyi).[2]

Up to now I have been deliberately vague as to the operational construction of the various variables. The difficulties here are myriad. Perhaps the two most important problems are the measurement of output (Q) in a research intensive industry (where quality changes may be rampant), and the construction of the unobservable research capital measure (K). Postponing the first for later consideration, we note that $K_t = \Sigma w_i R_{t-i}$ can be thought of as a measure of the distributed lag effect of past research investments on productivity. There are at least three forces at work here: the lag between investment in research and the actual invention of a new technique, the lag between invention and complete diffusion of the new technique, and the disappearance of this technique from the currently utilised stock of knowledge due to changes in external circumstances and the appearance of superior techniques (depreciation and obsolescence).[3] These lags have been largely ignored by most of the investigators. The most common assumption has been one of no or little lag and no depreciation. Thus, Griliches and Minasian have defined $K_t = \Sigma R_{t-i}$ with the summation running over the available range of data, while Mansfield assumed that since R has been growing at a rather rapid rate, so has also K (i.e. $K/K \simeq R/R$). Evenson (1968) has been the only one to investigate this question econometrically, finding that an 'inverted-V' distributed lag form fitted his data best, with the peak influence coming with a lag of five to eight years and the total effect dying out in about ten to sixteen years. There is some scattered evidence based largely on questionnaire studies, see Wagner (1968), that in industry, where the bulk of research expenditures are spent on development and applied topics, the lag is much shorter and so also is the expected length of life of the product of this research.[4]

[1] To the extent that research inputs are included among the conventional input measures, they have already been imputed the average private rate of return.

[2] Similar results were obtained recently for agriculture by Huber (1970), using 1959 census data by the type and size of farm.

[3] See Evenson (1968) for a more detailed discussion of some of these issues.

[4] Actually much of this 'depreciation' is obsolescence, induced and hence not independent of the rate of research in the rest of the industry or economy. This

Several additional points can be made concerning the measurement of K: (1) Ideally we should be measuring the output of the research industry directly, not its inputs. Unfortunately, attempts to use direct measures such as patents as a proxy for research output have been largely unsuccessful (see, e.g., Nordhaus 1969b).[1] For inventive activity the relationship between input and output is likely to be stochastic and unstable. We shall come back to this below. (2) The time series on R have to be deflated somehow, a hard but not impossible task. (3) The level of productivity in an industry and the productivity of research expenditures in it depends also on research activities of other related industries and on the research success of other countries.[2] These cross effects are, of course, somewhat less important at more aggregative levels but may be of the essence at the firm level.[3]

Because of the difficulties in constructing an unambiguous measure of K, many studies have opted for an alternative version of equation (5), utilising the fact that:

$$\alpha = \frac{dQ}{dK} \cdot \frac{K}{Q}$$

and

$$\alpha k = \frac{dQ}{dK} \cdot \frac{K}{Q} \cdot \frac{\dot{K}}{K} = \frac{dQ}{dK} \cdot \frac{\dot{K}}{Q}$$

allowing one to rewrite (5) as

$$f = \lambda + \alpha k = \lambda + \rho I_R / Q \qquad (5')$$

where ρ is the rate of return to research expenditures (the marginal product of K) while I_R/Q is the net investment in research as ratio

---

point is emphasized by Fellner (1970). A model in which depreciation is a function of the rate of research could be constructed, but it would take us too far afield.

[1] Somewhat more positive results have been recently achieved using counts of scientific agricultural publications in a yet unpublished study of agricultural yields by Evenson and Kislev at Yale. See also Baily (1970), where the output of research in the pharmaceutical industry is measured by the number of new drugs introduced.

[2] These issues were discussed by Brown and Conrad (1967) and under the title of 'pervasiveness' by Latimer and Paarlberg (1965) and Evenson (1969).

[3] In spite of all these reservations, some estimates have been attempted. In particular, in a forthcoming N.B.E.R. study Kendrick presents an estimate of the net stock of 'Research and Development Capital' for the United States for the years 1929, 1948, 1966. He estimates it to have grown at about 9 per cent per annum between 1948 and 1966 and to have reached the level of £100 billion (in 1958 prices) in 1966. Wagner (1968) gives additional detail on the derivation of these figures. Essentially it was assumed that 'Basic Research' does not depreciate, that 'Applied Research and Development' had a life of about 10 to 15 years, and the totals were deflated by a cost index based largely on salaries of scientists.

to total output.[1] This is the kind of framework that underlies the calculation of the rate of return presented in Fellner (1970), for example. In practice, to make some connection between gross and net investment in research one needs information about the 'depreciation' of research which if available would have allowed us to construct a measure of K in the first place.

Form (5') is particularly suitable for back-of-the-envelope calculations. It requires only a 'few' guesses as to (a) the fraction of R & D expenditures that represent net investment, and (b) the rate of return (ρ) to these expenditures. For example, between 1963 and 1967 the ratio of total R & D expenditures to total G.N.P. in the United States hovered around 3 per cent. We shall see below that only about half of it is likely to have an effect on national output as currently measured. If we assume that only half of the remainder represents net investment, the rest being devoted to keeping up with where we are,[2] and that the gross rate of return was 30 per cent, we get an estimated contribution to growth of $0.3 \times 0.75 \simeq 0.22$ per cent.

III. ROUGH ESTIMATE OF THE CONTRIBUTION OF RESEARCH TO GROWTH

Equation (5') can be estimated, albeit crudely, utilising data from the National Science Foundation publication *Industrial R & D Funds* (NSF 64–25), which reports (in Table C–1) the 1958 research intensity of companies for a number of rather finely defined manufacturing industries. For 85 2-, 3- and 4-digit manufacturing industries, these data could be matched with the *Annual Survey and Census of Manufactures* data and a 'Residual' technical change measure could be computed. It was defined as follows:

$$T(\text{Residual Technical Change}) = \tfrac{1}{5}[(\ln VA_{63} - \ln VA_{58} - \ln DP) - ALSV (\ln N_{63} - \ln N_{58}) - (1 - ALSV) (\ln GBV_{63} - \ln GBV_{58})]$$

where VA is value added, DP is the 1963 price index (1958 = 1.00 from 1963 *Census of Manufactures*, Vol. 4, Table 5), N is total employment, GBV is gross book value of fixed assets, and $ALSV$ is the average share (for 1958 and 1963) of payroll to value added. Another productivity related measure is given by the 'partial price change'.

---

[1] Whether this is the social rate of return or something less than that depends on whether the research inputs are already included in the conventional input measures or not.

[2] This is a reasonable assumption for series whose growth rate is approximately equal to the depreciation rate of the stock.

$$PP = \tfrac{1}{5} \, (1n \, DP - ALSV \cdot 1nDW)$$

where PP is 'partial price change', and $1n \, DW$ is the change in the average wage rate $(1n \; DW = (1n \; \text{Payroll}_{63} - 1n \; \text{Payroll}_{58}) - (1n \; \text{Employment}_{63} - 1n \; \text{Employment}_{58})$.

Additional variables of interest are:

$R = $ R & D expenditures as a fraction of net sales.
$ROV = R \times$ Average Sales to valued added ratio $=$
$\qquad$ R & D as fraction of value added.
$D_5 = $ Dummy Variable for $R > 0.15$.

All the observations were weighted by value added in 1958. Consider first the result of estimating an approximation to (5′):

$$T = 0.0043 \; + \; 0.397 \, ROV - \; 0.107 \, D_5; \; R^2 = 0.574, \; S.E. = 0.0215 \quad (6)$$
$$\;\;\;\; (0.0033) \quad (0.038) \qquad\quad (0.013)$$

Note the highly significant coefficient of ROV and the large 'discount' (D_5) for $R > 0.15$. The latter corresponds to two industries (SIC 19 and 372), Ordnance and Aircraft and parts, where government financed R & D constitutes about 86 per cent of the total and the results of which are unlikely to show up in the residual as measured. Since the average ROV for this group is 0.305, the discount implied by D_5 coefficient is consistent with the above ratio:

$$0.107/(0.397 \times 0.305) = 0.88 \simeq 0.86.$$

The measure of T used above is based on a rather dubious estimate of the rate of growth of tangible capital. An alternative approximation to T is available, however, from the price side by the duality of the price and quantity growth accounts:

$$T = [ALSV \cdot 1n \, DW + (1 - ALSV) \, 1n \, D\Pi - 1n \, DP] \quad (7)$$

where $1n \, D\Pi$ is an index of the rate of change in the service price of tangible capital. Assuming that the latter is the same for all these manufacturing industries (which is not unreasonable), implies that:

$$PP = 1n \, DP - ALSV \cdot 1n \, DW = \; -T - ALSV \cdot \overline{1n \, D\Pi} + \overline{1n \, D\Pi} \quad (8)$$

Treating $\overline{1n \, D\Pi}$ as a constant, we can use PP as the dependent variable in an equation paralleling (6) but containing also $ALSV$ (the share of labour in value added) as an additional variable. Doing this, we get:

$$PP = 0.0207 - 0.319 \, ROV + 0.084 \, D_5 - 0.043 \, ALSV; \; R^2 = 0.635 \quad (9)$$
$$\;\; (0.0085) \; (0.027) \qquad\quad (0.010) \qquad (0.016) \qquad\quad S.E. = 0.0155$$

The coefficients of ROV and D in this equation should be similar to those of (6) but of opposite sign, while the coefficient $ALSV$ provides an estimate of the unobserved rate of change in the service price of capital. The estimates are in fact close and largely independent of each other. Moreover, those based on equation (9), do not depend on the use of questionable capital data.

They are also quite consistent with the assumption that $\rho \simeq 0\cdot 3$ made earlier.[1] Subtracting 86 per cent of the Aircrafts and Missiles R & D from the total, we have an average (weighted) *ROV* ratio of 0·04 in total manufacturing in 1958. This implies that *in manufacturing*, research investments may have been contributing as much as a full percentage point or more ($0\cdot 3 \times 0\cdot 04 = 0\cdot 012$) to the measured growth in output. Note, however, that in this period manufacturing accounted for only about a quarter of total G.N.P.

IV. THE DIFFICULTIES OF MEASUREMENT

From the point of view of 'real' national income or growth accounting, the suggestion that R & D should be treated as another investment founders on the difficulty of measuring the corresponding real output stream. There are a number of problems here and I can do no better than just catalogue them:

(1) The bulk of the R & D 'product' is sold to the public sector. It consists of research on defence and space exploration for which no adequate market valuation exists. By convention, these 'outputs' are measured by costs, resulting in zero contribution to measured productivity, by definition.[2]

(2) Much of the product of private R & D investments is in the form of new commodities or in improvement in the qualities of old commodities.[3] Whether or not this product shows up in current dollar G.N.P. depends on the extent of the short and

[1] The estimated coefficient may be too low because we used total investment in R & D rather than just net investment as is called for by formula (5'). To the extent that depreciation (or stock) levels are positively correlated across industries with the gross investment levels, the coefficient of *ROV* would be a downward biased estimate of ρ.

[2] This is true for the bulk of defence and space research which is contract research and whose end products are blueprints, formulae, or prototypes. It is also largely true for the research components of the 'hardware' sold to government because of the lack of appropriate price indexes to deflate these expenditures by.

[3] This is an important dimension of the problem. It is estimated that over three-quarters of the applied research and development expenditures in the United States are devoted to 'product innovations' as against 'process innovations'. See Wagner (1968). How this may affect the measured returns to research has been discussed by Gustavson (1962), Griliches (1962) and Millward (1964).

long run monopoly position of the firms engaging in research and on the fraction of the social gain (consumer surplus) appropriated by these firms. Whether or nor this product shows up in 'real' G.N.P. (in constant prices) depends on what happens to the price indexes with which the current net output of the particular industry is deflated. If the price indexes were to recognise these improvements in 'quality' fully, the resulting real G.N.P. measures would reflect the social product of this research. But this is unlikely.[1] If the price indexes fail to reflect these quality improvements, only the private product will show up in the real accounts, to the extent that firms succeeded in appropriating it via higher prices for the newer higher quality products.

(3) If some of these unmeasured improvements are attached to products that are used in turn as inputs in the production of other private products, their contribution will show up in the productivity measures of the industries that purchased them. Thus, for example, even though the contribution of research towards improving the performance of farm tractors may not show up in the output-input account of the tractor industry, it will have an effect on measured productivity in agriculture.

(4) There is the technical accounting problem connected with the fact that most private R & D is treated as a current expense (an intermediate good) and does not appear explicitly in the value of output. Moreover, we have no explicit income stream (return) to associate it with. Thus, any accounting scheme based on factor shares is very difficult to implement, even disregarding the fact that social returns to R & D would in any case not be reflected in these factor shares even if we could compute them. The latter fact is the basic reason for the various attempts recounted above to estimate the social contributions of research econometrically.

(5) The basic difficulty with treating research as an investment is that it is largely an 'internal' investment, without an explicit intermediate market for its product. When we talk of 'tangible' investment, meaning equipment and construction, these can be valued by what was spent on purchasing them. That is, we have a separate measure of the output of the machinery industry, independent of our measure of the resources (labour and capital) used to produce these machines. Having separate measures on two sides of the accounts is a prerequisite for the

[1] For a recent review of the status of the quality problem in price indexes see Griliches (1971).

measurement of productivity of the investment goods industry.[1]
But research, like advertising and other costs of change, is
largely internal to the firm and does not show up in its output
accounts.[2] More importantly, it may not even show up, except
with a very long and random lag, in the income account of a
firm. Consider, for example, the parallel case of a firm that
engages in drilling for oil. It will incur some costs in doing so.
The success of this activity should be measured in new wells
brought in and in new reserves discovered. Obviously, they
are unlikely to show up in the receipts of the firm in the current
or even following year. Eventually, there will be a return,
though its relationship to these costs may be by then quite
obscure. On the other hand, the finding of oil should be
reflected in the 'capital gains' column, in the appreciation of
the market value of this firm. Thus, potentially there may be a
way of measuring the private product of research, but it
cannot be derived from the current income account.[3]

Having listed all these caveats we are now ready to look at what
might be the consequences of the way we treat current research
expenditures in the national income accounts. First, let us dispose of
a few easy cases:

Research performed in the public sector is measured by input and
does not directly contribute to measured productivity, since productivity growth in the public sector is zero by national accounting
conventions (and lack of real data on the 'real' value of the output of
this sector). To the extent that it affects the productivity of resources
in the private sector, as in the case of agricultural research or the
possible 'spillovers' from the space programme, it is an 'externality'
that might be caught by the conventional total factor productivity
measures.

Research performed by the private sector but sold to the public
sector, i.e. contract research on defence and space topics, again does
not contribute directly to measured productivity since this part of
private output is deflated by cost indexes, implying zero productivity
growth in this endeavour. Only the 'externalities' of this research
could show up in the conventional accounts.

The more interesting case is that of private research investments,
spent internally on the future improvement in the productivity and

---

[1] This is of course the problem with the conventional estimates of real capital
formation by construction, where we do not have an independent measure of
output or an output price index.

[2] This point is made in a more general context by Treadway (1969).

[3] It does suggest though a line of research to be pursued further elsewhere.

profitability of the firm. To reduce the problem to manageable proportions, we shall first consider a special and perhaps trivial case: All research is private, there are no externalities, and the aggregate production function is one of constant returns to scale in C, L and K. Subject to these assumptions, we shall investigate the consequence of two 'errors':

(1) The expensing of R & D instead of the more correct treatment of it as a type of capital, and
(2) The application of the same rate of return to this investment as to the other investments while, in fact, the private rate of return to it may be higher than that for tangible capital.

For this special case a potentially correct growth accounting scheme would start from the output-input value identity:

$$p_I I + p_g G + p_K I_K = q_L L + q_c C + q_K K \qquad (10)$$

where $p_I I$, $p_g G$, $q_L L$, and $q_c C$ are the conventional measures of investment, consumption ('goods'), labour and capital returns respectively, all in current prices. The additions to the usual accounts are the $p_K I_K$ term, on the left hand side, representing current gross investment in research ('knowledge') or other intangible capital times term and the $q_K K$ term on the right hand side, standing for the current returns to the previously accumulated knowledge 'capital'. The notation used distinguishes between prices of current outputs, the p's, and the service prices (rents) of the various inputs and capital stocks, the q's.[1] Given this definition, the conventional product measure can be rewritten as:

$$p_I I + p_g G = q_L L + q_c C + q_K K - p_K I_K \qquad (10')$$

Now, it is immediately obvious that the conventional productivity measures will ignore the contribution (positive and negative) of the last two terms of this expression. In addition, there will be another source of bias to the extent that the contribution of tangible capital (its service price, or the rate of return) is estimated residually, as is usually the case. The wrong estimate of the service price of tangible capital, q^*, will be given by

$$q_c^* = \frac{q_c C + q_K K - p_R R}{C} \qquad (11)$$

where $p_R R$ are research and development expenditures which have been 'expensed' from the profit accounts and treated as if they were a cost of producing current output. Note that we are making the

[1] See Jorgenson and Griliches (1967) for a more detailed exposition of this type of accounting algebra.

distinction between $p_K I_K$, which is the current marketable product of this activity, and $p_R R$ which are the current identified costs of it. Except in stable equilibrium with perfect foresight and correct accounting schemes, the two need not be equal, a point stressed earlier. For further convenience in manipulating these expressions, I will introduce the identity:

$$p_K I_K = p_R R + p_R U \tag{12}$$

where $p_R U$ is the current excess of growth in knowledge over current costs, evaluated at the factor prices, p_R, of current research costs. With this distinction in mind, we will put it aside for a while and return to the derivation of the bias in the conventional total factor productivity measures. Given (11), the bias in the measurement of the service price of tangible capital is given by

$$q - q_c^* = \frac{p_R R - q_K K}{C} \tag{13}$$

The total (absolute) bias in the measurement of total factor productivity is given by

$$\dot{T}^* - \dot{T} = q_K \dot{K} - (q_c^* - q_c)\dot{C} - q_K \dot{I}_K \tag{14}$$

where $X = dX/dt$. Thus, measured growth in total factor productivity T^* exceeds the 'true' measure T by the contribution of the growth in the 'stock of knowledge', but falls short of it by the overestimate in the contribution of the growth in tangible capital and by the omission of the growth in the investment into this type of knowledge from the conventional output measures. Substituting (13) into (14) and expressing the various variables as rates of growth, we have

$$\dot{T}^* - \dot{T} = q_K K \frac{\dot{K}}{K} - (q_K K - p_R R) \frac{\dot{C}}{C} - p_K I_K \frac{\dot{I}_K}{I_K} \tag{15}$$

Collecting terms, and using (12) we get

$$\dot{T}^* - \dot{T} = q_K K \left(\frac{\dot{K}}{K} - \frac{\dot{C}}{C} \right) + p_R R \left(\frac{\dot{C}}{C} - \frac{\dot{I}_K}{I_K} \right) - p_R U \frac{\dot{I}_K}{I_K} \tag{16}$$

If we were willing to assume that $\dot{K}/K \simeq \dot{I}_K/II_K$, i.e. that the rates of growth in the investment in knowledge and its stock are approximately equal, which may not be too bad an assumption for a fast growing item in its early history, and that there is no discrepancy between current research inputs and outputs ($p_R U = 0$), then (12) would simplify to

$$\dot{T}^* - T = (q_K K - p_R R) \left(\frac{\dot{K}}{K} - \frac{\dot{C}}{C} \right) \tag{17}$$

Noting in addition that we can write $R = \dot{K} + \delta K$, where δ is the rate of depreciation of the knowledge stock, and assuming that the service price of this capital is given by $q_K = p_R(\rho + \delta)$, where ρ is the rate of return to this stock, (17) simplifies further to

$$\dot{T}^* - \dot{T} = p_R K \left(\rho - \frac{\dot{K}}{K} \right) \left(\frac{\dot{K}}{K} - \frac{\dot{C}}{C} \right) \tag{18}$$

That is, as long as the rate of return to this stock exceeds its rate of growth, and as long as its rate of growth exceeds the rate of growth of tangible capital, the conventional total factor productivity measure will be biased upward.[1]

We can use this formula and some of the estimates for the United States domestic private economy from the Kendrick and Wagner studies mentioned earlier to get an order of magnitude estimate for this type of bias. Their estimates imply that private K (R & D capital) grew at about 7 per cent per year during the 1948–66 period, or at about 75 per cent of the rate of growth of total R & D capital. The estimated stock of business research capital was about $35 billion (in 1958 prices), and the estimated current research investment of industries' own money was about $7 billion (in current prices). Additional numbers of use (from Kendrick) are the total private product in 1966 of $520 billion (in current prices), implicit total private product deflator 110 (1958 = 100), and rate of growth of tangible net capital stock of 3·5 per cent for the 1948–66 period.[2]

Assuming $\rho = 0\cdot3$, i.e. that the private rate of return to R & D capital is 30 per cent, we get (dividing through by the total product):

$$\frac{\dot{T}^*}{T} - \frac{\dot{T}}{T} = \frac{p_R K}{pQ} \left(\rho - \frac{\dot{K}}{K} \right) \left(\frac{\dot{K}}{K} - \frac{\dot{C}}{C} \right)$$

$$= \frac{35}{520/1\cdot1} \, (0\cdot3 - 0\cdot07) \, (0\cdot07 - 0\cdot035)$$

$$= 0\cdot074 \cdot 0\cdot23 \cdot 0\cdot035 = 0\cdot0006$$

That is, assuming a rather high private rate of return to R & D we can account for only 0·0006 in our estimates of the 'residual' by the failure to capitalise past private R & D expenditures. This is a miniscule fraction of any of the estimates of the rate of growth in total factor productivity.

---

[1] Similar expressions can be found in Gordon (1968) which deals with the same question but in a slightly different context.

[2] We are going to assume throughout that $\dot{P}/P \simeq \dot{P}_R/P_R$, i.e. the price index of research investment is equal to the implicit private product deflator. This probably underestimates $\dot{P}_R/P_R$ and overestimates the actual increase in research investment in constant prices.

This does not mean, of course, that the contribution of these investments to the actual output growth is nil. *That* is measured by $\frac{q_K K}{pQ} \cdot \frac{\dot{K}}{K}$ which given the same numbers and assuming that $\rho + \delta = 0\cdot4$ is $0\cdot0021$. This while small, is not negligible. It is not to be found in the 'residual' however, because of offsetting errors.

The above computation was made assuming that R & D investment grew at about the same rate as its associated stock, and that there is no discrepancy between the current R & D costs and output. If we go back to equation (16) we can note that the type of calculation we just performed will overestimate the bias in periods when research investment is growing very fast $(\dot{I}_K/I_K > \dot{K}/K)$ and underestimates it in periods, like the current one, when its growth is slow.[1]

We can try to extend these calculations to a more recent period. Using approximately the same methods as Kendrick yields an estimate of the private R & D stock of about \$50 (in 1958 prices) at the beginning of 1971, as against \$35 in 1966.[2] This implies an average $\dot{K}/K$ of $0\cdot06$ per annum between 1966 and 1971. At the same time, 'real' private investment in R & D was growing only at about $0\cdot04$ per cent (or less) in 1970. If we assume that net tangible capital was also growing at about 4 per cent, the second term in (12) cancels, and still assuming that the third term is zero $(p_R U = 0)$, we get

$$\frac{\dot{T}^*}{T} - \frac{\dot{T}}{T} = \frac{1\cdot3\,(0\cdot3 + 0\cdot1)\,50}{864}\,(0\cdot06 - 0\cdot04) = 0\cdot03 \cdot 0\cdot02 = 0\cdot0006,$$

which is the same as the previous estimate.

To raise this estimate by relaxing the assumption that $p_R U = 0$ we would have to assume that it is negative, i.e. that the observed research costs exceed the actual current investment in knowledge. This is unlikely in a period of decelerating growth in real research expenditures. One would assume that new returns will be still forthcoming to some of the past investments. In general, one would expect U to be negative in periods of rapid growth in R, given the lags between research investments and their actual product.

The findings in this section can be objected to because they do not allow for economies of scale, the externalities of private R & D, and the contribution of government R & D to private productivity. We shall examine these in turn in the next section.

---

[1] This assumption is about right, for the figures at hand for the year 1966.
[2] The basic data used are from N.S.F. 70–46 (1970), deflated by the implicit G.N.P. deflator and cumulated on the assumption of a 10 per cent declining balance depreciation for applied research and development expenditures.

V. THE ROLE OF INCREASING RETURNS TO SCALE

Before we proceed to an evaluation of the potential externalities we should discuss briefly the role of increasing returns to scale to private R & D investments. The discussion in the previous section concentrates on *private* R & D investments and assumes explicitly (along with the usual national income accounting conventions) constant returns to scale with respect to labour, tangible and intangible private capital. The production function we wrote down in (4) assumes, however, constant return to scale for conventional inputs and a multiplicative effect for the stock of knowledge (K), private or public. One way out of this inconsistency is to assume that private firms have enough monopoly power to 'exploit' all factors proportionately. That is, observed market shares under-estimate the true factor elasticities by a factor $1/(1 + \alpha)$ where α is the elasticity of output with respect to private K. This would lead to a rewriting of equation (15) as

$$\frac{\dot{T}^*}{T} - \frac{\dot{T}}{T} = \alpha\frac{\dot{X}}{X} + (1 + \alpha) \left[\frac{p_R R - q_K K}{pQ} \frac{\dot{C}}{C} + \frac{q_K K}{pQ} \frac{\dot{K}}{K} \right] - \frac{p_K I_K}{pQ} \frac{\dot{I}_K}{I_K} \quad (15')$$

where $\dot{X}/X$ is the rate of growth in the conventional total input index. The second part of this formula is essentially the same as before (the difference is well within the range of error of these numbers), but the first term, $\alpha\dot{X}/X$ is new. Given the numbers used earlier, private $q_K K/pQ$ and hence, α were both about 0·0 2 in 1966, while $\dot{X}/X$ is estimated (by Kendrick, for 1958–66) at 0·023. This would add only about 0·00046 to the earlier estimates. Since the effect of this is the same as that of any other source of economies of scale, I will not pursue it further here.

It is very hard to assess the claim of externalities for publicly supported research investments adequately. Two contradictory facts stand out: First are the obviously large gains from public research in agriculture, reported on in Section I. Second is the fact that over half of the total R & D spending has been on defence, space and related objectives. Whatever the social benefits of defence and going to the moon, they do not show up in the usual productivity accounts. The direct effect of these expenditures on measured productivity is nil, while the evidence for indirect effects is not particularly impressive. There is another major category of public research whose social returns are much less disputable, medical research, but they too are unlikely to show up in the productivity accounts as currently defined.

Table 3.1 represents an attempt to gauge the possible magnitude of research expenditures that could (potentially) increase measured productivity. It brings together official statistics from several,

TABLE 3.1

ROUGH DISTRIBUTION OF RESEARCH AND
DEVELOPMENT EXPENDITURES IN THE UNITED
STATES, 1970, BY SOURCE, PERFORMER, AND
POSSIBLE PURPOSE
(U.S. $ millions)

Performer, Source, and Type	Total	Defence, A.E.C. and Space	Other Expenditures with Probable Effect on Measured Private Productivity
Federal Intramural			
Basic	620	320	60[a]
Applied	1,300	850	240[b]
Development	1,740	1,610	50[c]
Total	3,660	2,780	350
Industry			
Federally financed			
Basic	400	400	
Applied	830	730	30[d]
Development	6,900	6,590	220[e]
Company Funds			
Basic	570	80[f]	480
Applied	2,480	250[g]	2,230
Development	7,700	900[h]	6,800
n.e.c.		180[i]	−180
Total	18,880	9,130	9,520
Universities, colleges, non-profit institutions, and federal research centres			
Federal Funds			
Basic	960	480	240[j]
Applied	960	310	180[k]
Development	950	830	40[l]
Industry Funds	160		160
Other Funds	1,290		650[m]
Total	4,320	1,620	1,270
Total	26,860	13,530	11,140

Notes: Totals are from N.S.F. 70–46 (1970). Breakdown of federally financed expenditures based on 'obligations' for 1970 from N.S.F. 70–38 (1970). Breakdown of industrial expenditures based on 1968 data from N.S.F. 70–29 (1970).

[a] 20 per cent of all other.

[b] All of USDA and Interior, 20 per cent of other.

[c] All of Transportation, 20 per cent of other.

[d] All of Interior, 20 per cent of other.

[e] All of Transportation, 20 per cent of other.

[f] Half of aircraft and missiles and electric equipment and communications.

[g] Half of aircraft and missiles and communication equipment.

[h] Half of aircraft and missiles, 1/10 of chemicals, machinery and electric equipment.

[i] 1/10 of motor vehicles and instruments.　　　[j] 50 per cent of rest.

[k] All of USDA and Interior financed, 20 per cent of rest.

[l] All of Transportation, 20 per cent of rest.　　　[m] 50 per cent of total.

slightly inconsistent sources and tries to allocate them between defence and space oriented research and research that could conceivably show up in real product per man as currently measured. The allocation is somewhat arbitrary and is based on 'guess-work', particularly for the figures in the last column of Table 3.1. But they should provide some impression of possible orders of magnitude and a basis for discussion. In 1970, of the $27 billion or so identified as 'research and development' expenditures, half could be attributed to defence and space activities. Of the remainder, about $11 billion could be thought as having a potential impact on aggregate productivity. These $11 billion consisted of $9·5 billion spent in industrial research, about a third of a billion in intramural federal research, and $1·3 billion in research in universities, institutes, and similar organisations.

At the same time, private for profit research in industry (as distinct from private non-profit research in institutions outside government) also equalled about $11 billion. This balancing out of about $1·5 billion in private expenditures attributed to defence (since defence is a major purchaser of the outputs of many firms, it pays to invest private resources in this field) with $1·5 billion in public expenditures with potential impact on private productivity is entirely fortuitous but instructive.

We have still not allowed for any 'spillovers' from defence and space research. An upper limit estimate might be one that took 20 per cent of these expenditures to have the same effect and the same rate of return as private expenditures devoted directly towards affecting private productivity. Table 3.2 summarises the rough allocation of various R & D investments as to their potential impact on measured productivity in the private sector, including an allowance for externalities generated by that part of R & D (such as the space effort) not deemed to have a direct effect on productivity. About half of total R & D expenditures is estimated to be 'productive' in this sense.

Before we proceed to incorporate these estimates into a revised accounting scheme, we have to decide how much of the estimated returns to private R & D are social rather than private. In the previous sections of this paper we used $\rho = 0·3$ interchangeably, both for private and social returns calculations. The econometric data which support such a number are obscure, however, as to the exact allocation of these returns. If the price indexes used to deflate output were perfect, almost all of the return estimated from total factor productivity or 'residual' studies would reflect social returns. Since these indexes are far from perfect, some significant portion of the estimated returns is in fact private. To proceed with some order of magnitude calculations I shall make the following arbitrary but perhaps not unreasonable assumptions:

(1) Private R & D expenditures earn a 'normal' gross rate of return of about 20 per cent (10 for depreciation and 10 net).
(2) They yield an externality in the private sector of 20 per cent. That is, I am assuming that of the earlier estimate $\rho = 0.3$, one-third is actually part of the private returns.[1]
(3) Public R & D investments (direct and indirect) yield a gross rate of return of 30 per cent.

Applying these assumptions and the allocations in Table 3.2 (which are based on 1970 data) to the earlier estimates of cumulated

TABLE 3.2

THE DISTRIBUTION OF RESEARCH AND
DEVELOPMENT EXPENDITURES IN THE
UNITED STATES IN 1970 BY THEIR
POTENTIAL DIRECT AND INDIRECT EFFECT
ON PRIVATE PRODUCTIVITY

	Per cent	
	Total	*Productivity Related*
Industrial R & D		
Private Funds	40	35
Public Funds	30	1
Public R & D		
Federal Intramural	14	1
Universities,	16	4
Research centres		
and related institutions		
Indirect (externalities)		12
Total	100	53

From Table 1. Indirect = 20 per cent of remainder.

stocks of R & D investments in 1966 and noting that the share of private R & D in the total was lower in 1966 and earlier years (32 per cent versus the 40 per cent in 1970), we have:

$K_{d_{p1966}}$ = \$31 billion directly productive stock of private R & D investments in 1958 prices.

$K_{d_{1966}}$ = \$7 billion directly productive stock of public R & D investments.

[1] Implicitly this assumes that these investments yield less externalities than public investments which are directly oriented towards the production of externalities.

K_{ge1966} = \$13 billion of indirectly productive stock of private and public R & D.

The appropriate accounting formula is now

$$\frac{\overset{*}{\dot{T}}}{T} - \frac{\dot{T}}{T} = \frac{(q_{Kp} + q_{Kpe})K_p}{pQ}\frac{\dot{K}_p}{K_p} + \frac{q_e(K_{dg} + K_e)}{pQ}\frac{\dot{K}_g}{K_g}$$

$$- \frac{P_R R}{pQ}\frac{\dot{R}}{R} - \frac{P_R R}{pQ}\frac{\dot{C}}{C}$$

where the first two terms represent the contribution of private and public R & D to the growth of output in the private sector while the last two terms adjust for two errors in the conventional measures of total factor productivity: the exclusion of private investments in R & D from the output measure and the over weighting of the contribution of tangible capital.

Setting $q_{Kp} + q_{Kpe} = p \times 0.4$ and $q_{Kg} = p \times 0.3$, and taking as before $K_p/K_p = 0.07$, $K_g/K_g = 0.1$, $p_R R = \$7$ billion, $pQ = \$520$ billion, $p = 1.1$, $Q = \$473$ billion, $\dot{C}/C = 0.035$, and $\dot{R}/R = 0.08$, we get

$$\frac{\dot{T}^*}{T} - \frac{\dot{T}}{T} = \frac{0.4 \times 31}{473} \times 0.07 \frac{+0.3 \times (7 + 13)}{473} \times 0.1$$

$$- \frac{7}{473}(0.08 + 0.035) = 0.0050 - 0.0016 = 0.0034$$

The total contribution of R & D to growth is now estimated at 0·005 (or half a per cent per year) of which 0·00185 is attributed to the private returns of private R & D, 0·00185 to the externalities of private R & D, 0·00046 to the contribution of 'directly productive' public R & D and 0·00084 to the 'externalities' of all other research. The net effect of all this on measured total factor productivity is somewhat less, 0·0034.

Noting that private returns and the errors in their treatment in the measurement of total factor productivity almost cancel out (0·00185 − 0·00155 = 0·0003), we can concentrate on the externalities alone. The above calculations estimate their annual contribution at about 0·0023 for 'directly productive' investments (public and private) and attribute a potential 0·0008 to the 'spillovers' from other research.

Assuming the cancellation of the private returns and the associated errors in the measurement of total factor productivity, we can use formula (9) directly for an evaluation of current contribution of R & D to growth. Total R & D with potential externalities (including spillovers) in the private sector was about \$14 billion in 1970 (see Tables 3.1 and 3.2). Assuming that about half of this gross investment was for replacement and noting that total private domestic G.N.P. was $855 billion, gives a net investment ratio of 0·0082. Our earlier assumptions and the allocations in Table 3.2, imply a weighted social rate of return of 0·234, and hence a 0·0019 contribution of R & D to the current rate of growth of private output. This is about two-thirds of the comparable number for 1966 and reflects the rather significant slowdown in the rate of growth of R & D expenditures. Given the lags involved, this reduction may not show up, however, until the mid seventies.

VI. REASONS FOR THINKING THE ABOVE ESTIMATES TO BE TOO HIGH

The estimate of the contribution of R & D investments to the rate of growth in measured total factor productivity is largely the product of three fractions: (1) the assumed excess of the social rate of return over the private rate of return; (2) the fraction of actual R & D expenditures which can be expected to have an impact on total factor productivity as currently measured; and (3) the fraction of these expenditures that represents net investment rather than replacement. In the previous sections we have provided some arguments for setting these numbers at 0·23, 0·53, and 0·5 respectively. These, when multiplied together and applied to the current (1970) ratio of total R & D expenditures to private domestic G.N.P. of 0·03, lead to the estimated effect on the residual of 0·0018[1]. The effect of these expenditures on the actual rate of growth of output is about double of that but that second half (the private returns portion) is almost cancelled out by offsetting errors in the measurement of output and the contribution of tangible capital.

These attributions are not negligible. The earlier estimates as of 1966 amount to between one-fifth and one-quarter of the 'residual' rate of growth as it has been recently estimated.[2] In my opinion, this is close to an upper-bound estimate, but the reader should feel free to insert his own 'guestimates' into the formula outlined above.[3]

[1] This is somewhat higher, but not greatly so, than Denison's estimate of a decade ago (Denison 1962, pp. 239–46). I have been more liberal in ascribing to a larger fraction of R & D the potential to effect uncovered to top factor productivity.

[2] See Jorgenson and Griliches (1971) and the literature cited here.

[3] There is an additional 'casual' contribution of R & D investments not discussed in this paper. Improvements in productivity make additional capital investments profitable and induce thereby additional capital accumulation with a consequent effect on the rate of growth of output. But according to the usual growth accounting conventions, this contribution is ascribed to tangible capital rather than R & D and rightly so in my opinion. For a contrary view, see Gordon (1969).

The reason why I believe this to be an upper-bound estimate rests on the application of a rather high social rate of return, derived from scattered evidence on actual returns in a few, largely successful, areas of research, to all research expenditures. Also, I have been lenient, I think, in drawing the boundaries as to the type of research which can have an effect on total factor productivity as measured. The major omission is the possible contribution of the scientific education process as a whole above and beyond what is counted under the label of 'research'. But that is not as large as all that. [1] Moreover, if one expands the boundaries of the relevant concept of R & D, one should probably adjust the estimated rates of return downward accordingly, since, if the productivity measures were correct, they would already contain the returns to all such R & D. Thus, for example, when we estimate the rate of return to manu-facturing R & D in Section III at about 0·3, we did not include in the definition of R & D the externalities assumed to flow from the other sectors. If these had been distributed proportionately to the indus-tries' own R & D expenditures, the rate of return to all relevant R & D would be significantly (about one-quarter) lower.

Another reason why the above estimates could be too high is our implicit assumption that the relevant stock of knowledge has grown in proportion to the growth in research expenditures. But the rapid rise in the latter could have entailed significant diminishing returns, the actual stock of knowledge not growing anywhere near the same rate. For all these reasons, the estimates presented above are probably too high perhaps by as much as 50 per cent.

VII. SOME POLICY IMPLICATIONS

A major objection to drawing policy conclusions from such findings is that they are all based on 'average' rates of return, while decisions have to be made on the margin. This objection contains an impor-tant half-truth: the next research project in the stack may turn out to be unsuccessful *ex post* and this, perhaps, could even have been clear *ex ante*. But this is only a half-truth. Any one research project may be wasteful without the same being true for other potential research endeavours in this and other fields.

Perhaps an analogy will help here. An economist can examine the accounts and experience of the wildcat oil prospecting and drilling industry and conclude that on the whole investments in this industry appear to have been profitable and that the people running it seem to know what they are doing. He could not guarantee, however, that

---

[1] In 1964, total spending at Universities on instruction in sciences and engineer-ing, exclusive of explicit R & D expenditures, was only £2.4 billion.

the next well drilled would not turn out to be dry, and much less tell them where to drill it.

At the more abstract level there is another answer to this objection. There does not seem to be any evidence that the average rate of return to research is declining and hence there is no presumption that the average and marginal rates of return are very far apart.

Most economists, if queried, would assert that there is under-investment in research by private firms because much of its product is not capturable (appropriable) by the private firm (see Arrow, 1962, and Nelson, 1959). This is the major argument for the patent system and for government support of this activity. There are only a few contrary voices. In a recent paper Hirschleifer (1971) points out that to the extent that the new knowledge may provide a firm (or an individual) with a competitive advantage, there may be a private incentive to invest in it even in excess of its social return. His point is well taken but seems to be irrelevant for most types of research supported by the government. The results of such research are unlikely to be of a form directly translatable into a market advantage. In any case, government supported research is rarely directly competitive with private research in the same area. Usually, if it is a flourishing private research area, there will be little public support for research in the same field.

Not all research contributes to growth or to the capacity of the economy to produce goods and service and to consume them wisely. Some of it, as pointed out by Johnson (1965), is a kind of public consumption like mountain climbing. We investigate certain secrets of Nature because they are there, not just and not even primarily because their solution may prove 'useful'. That some of research is of this type does not mean that it shouldn't be supported, only that it has to compete for the scarce public dollar with other public consumption activities such as the architecture of public buildings and the support of fine arts.

Both the empirical and the theoretical literature provide some arguments why public investment in science and research may be a good thing. They also provide some evidence, in the form of higher average rates of return in this sector relative to other sectors of the economy, for the argument that more investment might be even better, but almost no information on where particular investments should be made and no assurances that any particular set of investments will in fact pay off.

82 *Science and Technology in Economic Growth*

REFERENCES

Ardito-Barletta, N., *Costs and Social Benefits of Agricultural Research in Mexico.* Unpublished Ph.D. dissertation, Department of Economics, University of Chicago (1971).

Arrow, K. J., 'Comment', in *The Rate and Direction of Inventive Activity* (National Bureau of Economic Research). (Princeton: Princeton University Press, 1962) 353–8.

Baily, Martin N., *Research and Development Costs and Profits: The U.S. Pharmaceutical Industry*, M.I.T., unpublished manuscript (1970).

Brown, M. and Conrad, A. H., 'The Influence of Research and Education on CES Production Relations', in *The Theory and Empirical Analysis of Production*, ed. M. Brown, (N.B.E.R.: Studies on Income and Wealth, 1967) xxxi, 341–72.

Denison, Edward F., *The Sources of Economic Growth in the United States and the Alternatives Before Us* (Supplementary Paper No. 13) (New York: Committee for Economic Development, 1962).

Eastman, S. E., *The Influence of Variables Affecting the Worth of Expenditures on Research or Exploratory Development: An Empirical Case Study of the C-141A Aircraft Program.* Unpublished Institute for Defense Analyses memorandum (1967).

Enos, J. L., 'Invention and Innovation in the Petroleum Refining Industry', *The Rate and Direction of Inventive Activity*, ed. R. Nelson, (N.B.E.R., 1962) 299–322.

Evenson, R., *The Contribution of Agricultural Research and Extension to Agricultural Production.* Unpublished Ph.D Thesis, University of Chicago (1968).

Evenson, R., *Economic Aspects of the Organisation of Agricultural Research*, presented at the Symposium on Resource Allocation in Agricultural Research, University of Minnesota, Minneapolis, Unpublished mimeograph (February 1969).

Fellner, N. 'Trends in the Activity Generating Technological Progress', *American Economic Review* (March 1970).

Freeman, Richard B., *Engineers and Scientists in the Industrial Economy*, unpublished manuscript submitted to the National Science Foundation (1971).

Gordon, R. J., *The Disappearance of Productivity Change*, unpublished ED Report No. 105, Project for Quantitative Research in Economic Development, Harvard University (1968).

Griliches, Zvi, 'Research Costs and Social Returns: Hybrid Corn and Related Innovations', *Journal of Political Economy*, V. 66 (October 1958).

Griliches, Zvi, 'Comment', in *The Rate and Direction of Inventive Activity*, N.B.E.R., (1962) 346–51.

Griliches, Zvi, 'Discussion', *American Economic Review*, Proceedings Issue (May 1962) 186–7.

Griliches, Zvi, 'Research Expenditures, Education, and the Aggregate Agricultural Production Function', *American Economic Review*, LIV (December 1964).

Griliches, Zvi, ed. *Price Indexes and Quality Change* (Cambridge: Harvard University Press, 1971).

Gustavson, W. E., 'Research and Development, New Products and Productivity Change', *American Economic Review*, Proceedings Issue (May 1962).

Hirschliefer, J., 'The Private and Social Value of Information and the Reward to Inventive Activity', *American Economic Review*, LXI (4, 1971), 561–74.

Huber, Paul H., *Disguised Unemployment in U.S. Agriculture*, 1959, Unpublished Ph.D dissertation, Yale University (1970).

Johnson, H. G., 'Federal Support of Basic Research: Some Economic Issues', in *Basic Research and National Goals* (Washington: Government Printing Office, 1965).

Jorgenson, D. W. and Griliches, Z., 'The Explanation of Productivity Change', *Review of Economic Studies*, xxxiv (1967), 249–83.

Jorgenson, D. W. and Griliches, Z., *The Sources of Economic Growth: A Reply to Edward F. Denison*, (Cambridge: Harvard Institute of Economic Research Discussion Paper No. 131, 1970).

Kendrick, J. W., *Productivity Trends in the United States* (Princeton: Princeton University Press, 1961).

Kendrick, J. W., forthcoming NBER volume on *Postwar Capital Formation*.

Latimer, R. and Paarlberg, D., 'Geographic Distribution of Research Costs and Benefits', *Journal of Farm Economics*, 47, (1965).

Leonard, W. N., 'Research and Development in Industrial Growth', *Journal of Political Economy*, 79 (2, 1971), 232–56.

Machlup, F., *The Production and Distribution of Knowledge in the United States* (Princeton: 1962).

Mansfield, Edwin, 'Rates of Return from Industrial Research and Development', *American Economic Review*, LL, (2, May 1965).

Mansfield, Edwin, *Econometric Studies of Industrial Research and Technological Innovation* (New York: W. W. Norton and Co., 1967).

Millward, R., *Research and Development, New Products and Productivity* (Manchester, unpublished, 1964).

Minasian, Jora R., 'The Economics of Research and Development', in *The Rate and Direction of Inventive Activity* (National Bureau of Economic Research) (Princeton: Princeton University Press, 1962).

Minasian, Jora R., 'Research and Development, Production Functions, and Rates of Return', *American Economic Review*, LIX (Proceedings Issue, 2, 1969) 80–5.

National Science Foundation, *Methodology of Statistics on Research and Development*, N.S.F. 59–36 (Washington: Government Printing Office, 1959).

Minasian, Jora R., *Research and Development in Industry*, 1968, N.S.F. 70–29 Washington: Government Printing Office, 1970a).

Minasian, Jora R., *Federal Funds for Research, Development, and Other Scientific Activities*, Fiscal years 1969, 1970, and 1971, N.S.F. 70–38 (Washington: Government Printing Office, 1970b).

Minasian, Jora R., *National Patterns of R & D Resources*, 1953–71, N.S.F. 70–44 (Washington: Government Printing Office, 1970c).

Nelson, Richard R., 'The Simple Economics of Basic Scientific Research – A Theoretical Analysis', *Journal of Political Economy*, LXVII (June 1959).

Nelson, Richard R., Ed. *The Rate and Direction of Inventive Activity* (Princeton for N.B.E.R., 1962).

Discussion of the Paper by Professor Griliches

Professor Patinkin, introducing the paper, said that Professor Griliches' paper was an attempt to summarise and to put forward econometric work concerning the effect of research on productivity. It was a rather technical paper and he felt it was better in this case to concentrate on the general line of the argument rather than on the specific detail.

Part one gave a review of the empirical work on the return to investment in specific industries and projects and Professor Griliches had come out with the result that there were high internal rates of return on the research expenditures – higher in fact than on large categories of physical capital, possibly due to the risk premium. In part two Professor Griliches discussed the fitting of econometric equations of the production function type and had used one such equation to examine the unexplained part of productivity changes which were not due to capital or labour.

In examining the residual he had regressed this variable on the rate of change of R & D expenditures; Professor Patinkin said he would have preferred to hear of the results where an attempt had been made to explain total productivity in terms of capital, labour, research and development and the trend variable with all of these being free to work. He felt that this would be a more severe test of the R & D effect and suggested that Professor Griliches had increased the probability of good results for R & D by using the approach he had adopted. On the question of multicolinearity Professor Patinkin said that this should have been taken care of by fitting the rates of change of these variables and suggested that another advantage of doing it in this neutral way was that one could compare rates of return on physical capital to those on R & D.

However there were some more fundamental problems – the first of which was how did we measure R & D expenditure as the input variable for production functions? Professor Griliches said that the stock and not the flow of expenditures in any period was the correct measure and R & D expenditure was then like any other item. But how did we measure relevant R & D stock. Was it an accumulation of past R & D expenditures? This would need further qualification – there was a lag in the application of R & D to other production processes and there was also an obsolesence of new inventions and new techniques. Therefore we required some weighted total of past R & D expenditures, but the weights need not necessarily be monotonically decreasing. It was interesting to note that what we would have anticipated was shown – that is, that R & D comes in slowly and then goes out slowly, producing the inverted v-like type of structure previously indicated by Evenson. Professor Patinkin said it was always encouraging when the empirical results agreed with one's intuitive feeling.

Returning to the question of stock, he asked what was the relevant stock of R & D as far as, say, the United States was concerned. We had been speaking of ownership externality: the relevant stock of R & D for any production process was the total stock of knowledge available to the world, not just in the United States. It was accepted that there was a longer lag in the adoption of knowledge from abroad, but fundamentally all this knowledge was available. This raised more questions: Is this stock available to all countries of the world, equal quantities to all, no matter what their size or stage of development? Professor Patinkin said that R & D must operate in connection with other factors of production and with certain systems of entrepreneurs; it was tied up with a social structure which was receptive to change and to competitive atmosphere. Everybody could make use of techniques but not all of them did so. There was an essential complementarity of factors of production and a broad framework of capital and labour was not sharp enough. In this sense Professor Patinkin said he preferred Professor Matthews' model in which a distinction was made between different kinds of labour and in which the way in which R & D influenced productive processes was brought out.

On the contribution of R & D Professor Griliches first presented this as an overestimate, subsequently modified it and concluded by saying that in the 1960s, 0.45 per cent of the growth in productivity could be attributed to R & D expenditures. Professor Patinkin asked was this large or small, and said that he thought this could not be answered without a basis of comparison. A half of 1 per cent would be a fantastic contribution given that the contribution from labour was 1 per cent and from physical capital was 1 per cent over the same period. Professor Griliches explained that this was an overstatement. R & D expenditures included government expenditures and these by definition were not designed to increase productivity. Therefore half of the R & D was not relevant. He also added that much of R & D was a consumption item – for example moon exploration, from which there was considerable public satisfaction to be extracted out of the race, stimulated perhaps by the media.

A further reduction in the contribution of R & D as a growth factor arose because of technical points. As Griliches pointed out, if R & D was a factor of production it was also a component of the output, and yet it was not included in measures of output; in the work to date, these considerations produced opposite forces, some tending to overestimate and some to underestimate. He also suggested that in general we would anticipate that these would reduce the contribution of R & D by a half – this is further reduced by half because of the government, leaving a figure of 0.1 per cent as a contribution by R & D. Professor Patinkin said that he did not think that one-tenth of the contribution of growth as compared to that of physical capital was small, and suggested that another comparison might be worthwhile – the physical investment in the United States was approximately 16 per cent of the G.N.P., whereas the investment in R & D was 1 per cent of the G.N.P. This made a ratio of 16 to 1, but the contribution to output was only 10 to 1. This might be a rather rough comparison, but it provided an indication that the return of R & D was better than the return on capital and was certainly worth thinking about.

Professor Patinkin ended with the comment that he felt the tone in the paper was one he welcomed – it was one of humility with reference to econometric studies which was all too often missing from this field. He said that Professor Griliches was willing to conclude that, if we got a lower estimate of productivity of R & D, then we intuitively felt that maybe something was wrong with our measures of productivity. If we were to get at measures of contribution of R & D to specific economies, then we must look wider and at new measures, and give consideration to models giving greater complementarity between R & D and other factors.

Professor Patinkin concluded by saying that he had benefited considerably from being able to discuss the paper with the author.

At this point *Professor Griliches* intervened to say that he would like to make some minor corrections to his paper. He said that the paper as presented was not consistent on the question of whether there are or are not economies of scale with respect to R & D and that he hoped the revised version would clear this up. He also said that there was some language confusion in that the paper did not distinguish between the the contribution of R & D to growth and to the explanation of the residual[1]. He said that these were two entirely different things and made an order of magnitude difference. In trying to explain away part of the residual by capitalising R & D expenditures and imputing a rate of return to them we must also treat R & D as an investment and therefore include it in the definition of output and allow for the fact that the share of capital was over estimated previously due to the attribution of all the R & D effects to fixed capital. When one subtracted these the net effect on the measured residual was small. This did not say, however, that the contribution of R & D to growth of output was necessarily small. In the private sector this was estimated to be of the order of $\frac{1}{4}$ per cent, and Professor Griliches said this might or might not be small depending upon one's view of the world. The paper tried to bring out that much of what was measured as R & D did not show in the conventional productivity measure since a lot of R & D was spent on defence, health, and similar research topics in or for the public sector. Such expenditures were clearly important and socially productive but their product was not reflected in the current G.N.P. accounts. There was thus an underestimate of the true contribution of R & D. The exercise performed in this paper was limited to R & D as defined by official statistics and one could also argue that much research in science and invention went on outside the framework of the official statistics. Professor Griliches said he was not sure how far one should go, because it might be necessary to subtract from these statistics things which were not really research and invention and also one might have to add in the cost of teaching graduate science which was not generally counted as research. However these were not likely to change the conclusions significantly.

Professor Rasmussen said that he was greatly attracted by having the stock of R & D knowledge introduced into the production function rather than the flow, in the same way as the stock of capital was considered rather than a corresponding flow of investment. This raised, however, the question of deciding when an increase in the stock should enter and this

[1] This has been made clearer in the revised version of the paper printed in this volume.

was the old problem of the production period discussed ever since Böhm-Bawerk. Professor Rasmussen asked how this could be tackled; in this connection he wondered about the sort of time lag, five to eight years, which had been indicated in the paper. He said he thought this was a very low figure if one included basic research and in the same line of thought the question of depreciation of knowledge came to the surface. On this question he felt Professor Griliches had been somewhat inconclusive, perhaps for good reasons.

Professor Griliches said that most of the R & D they were talking about was applied R & D, where the lag was not so long. He admitted that depreciation was a hard problem, and said that Evenson had had a long time series on agricultural productivity and agricultural research expenditures and had tried various combinations of lags in an attempt to find which depreciation formula fitted best. Professor Griliches said that he was not convinced that one got much mileage at this level with fancier production functions. These concentrated their questions on curvature (elasticity of substitution) when there was even difficulty in getting answers about the slope. On Professor Patinkin's point about knowledge being freely available but used by only a few, Professor Griliches said he thought it was misleading to assume that all such knowledge was a free good. He felt that it took a lot of effort to transmit it. It might all be available but we could not always use it easily. There was a real cost of transmitting and acquiring knowledge. He said that all of physics was in theory available to him, but it would certainly cost him a lot to learn.

Professor Patinkin agreed with this and said he was including this in complementary factors and perhaps, where Professor Griliches was thinking of difficulties of learning the subject, Professor Patinkin was thinking of physical capital rotting in developing countries because no labour force was available to use it.

Professor Triantis wanted to make a small point on obsolescence, which was important whether one was measuring the stock of knowledge or the flow of knowledge. The rate of obsolescence would be related to the rate of R & D expenditures and innovation. He suggested it would be useful to distinguish between additive and destructive innovation. The former might raise rates of return on previous investment and this makes it difficult to separate the returns on R & D from those on existing physical capital. As to destructive innovation Professor Triantis suggested that much of military technology was in this category. He suggested that the rate of diffusion of new technology varied between additive and destructive innovations, since it depended upon the degree of market control, and on who benefited and lost through externalities and other such factors. Unless one had more knowledge on these matters we had difficulty in estimating the rate of obsolescence. This was also relevant to the question whether the rate of obsolescence of imported technology was greater or smaller than that of domestically developed technology – do we import more additive or destructive innovation? Professor Triantis suggested this might depend partly upon how it was imported, for example through patent rights or through capital investment.

Professor Matthews said that some presumption existed that the marginal rate of return on R & D was less than the average rate of return, simply on the grounds that firms knew what they were doing and gave priority to the most promising projects. This annoying problem of having to use the average as a proxy for the marginal also arose, of course, with the rate of return on investment in physical capital. In the simplest type of neoclassical analysis one said that current investment made only a small difference to the slope of the production function. But this might not in reality be true either for physical capital or for R & D because of embodiment effects. A major new breakthrough might mean that there was a high return on the R & D that was needed to translate it into production in any given period but a considerably lower rate of return on R & D in areas where there had been no recent major breakthrough.

Professor Lundberg said that Professor Triantis had raised many questions. He had been puzzled by Professor Griliches' answer to Professor Rasmussen on the Böhm-Bawerk point and on the time interval between discovery and application to the last firm. In any branch of industry techniques might vary between those that were thirty years old and the most modern plants. The oldest plants might be applying techniques with low profitability which were, however, rational in their context. Professor Lundberg asked for clarification on the point made by Professor Griliches when he said he had no reliance on the production function apparatus, and yet he devoted much of his paper to its application. Were we paying lip service to the Cobb-Douglas production function although we knew that new techniques were embodied in both labour and capital?

Professor Robinson said he would like to add a footnote to the exchange between Professors Patinkin and Griliches on the extent to which R & D expenditure was expenditure on adding to knowledge or expenditure on communications. In the United Kingdom when there had been increased expenditure on research institutes, firms had to hire scientists to understand the results. When they had looked at this, it was found that research units of less than ten people were almost purely channels of communication and were not adding to knowledge. On the question of scale effects in research about which nothing has yet been said, Professor Robinson suggested that we could neglect work which was going on in research units of much less than 100, if we were concerned with substantial additions of new knowledge.

On the question of the timing of research effects *Professor Griliches* said that econometrics was a very blunt tool and the lag structures which were used allowed for the possibility of a long tail. The lag structure had a mean of eight years but a long thin tail. On the question of embodied versus non-embodied technical change, he said this was not a question of Cobb-Douglas or not – one could have technical change embodied in the Cobb-Douglas framework – his problem was that the data were hardly available to disentangle it, and it was not reasonable with the available data to try to ask whether the effect of research done in 1957–8 was different from that done in 1962–3 – the data did not have this power of resolution. Professor

Griliches thought this might be better investigated with the help of micro data. He also said that the Cobb-Douglas form had several lives and was difficult to kill off – he had tried! In replying to Professor Triantis on obsolescence, Professor Griliches said that it was not just a question of arguing how many angels were standing on a pinhead but also what was the colour of their eyes? On marginal versus average returns he said that again we had no good answers. He could think of two things going on simultaneously – fundamental research was like sampling from some urn, but was the sampling with replacement or not? In sampling from the same urn we were likely to get diminishing returns, but a lot of research had to do with the finding of new urns or changing the standard deviation of the old urns, and in this case there was less expectation of diminishing returns. He thought that there was little chance of diminishing returns because one kept discovering new urns, but it was surely preposterous to assume that we would never run out of new urns. There was some evidence that this might be happening in particular areas or fields, where a particular line of research had become temporarily exhausted, but there was a surprising mobility of scientists from problem to problem. In the short run pumping money into the field with only a limited number of people in it would produce diminishing returns, but that was a different question.

Professor Sir John Hicks said that in view of his old connections with the elasticity of substitution he had better begin by explaining that he was now a member of the anti-production function party. He was not an extreme member of that party, and he did not believe that any employment of the production function was sinful. He was quite prepared to believe that one could get interesting results by employing a poor theory even when it was applied to poor or unlikely facts, but all the same he had the greatest doubts about the physical capital which appeared in the production function. He did not think that there were any measures which would do anything like all that was required for that formulation to make sense. Despite that he was prepared to be tolerant and believe that if one worked in the conventional way followed by Professor Griliches one might get interesting results.

The production function was being continually shifted by technical progress, so in the analysis of time series what we really wanted was a production function which contained technical progress, e.g. $P = F(L, C; T)$. The semi-colon was important as T was quite different from the other variables. All sorts of things might go into T. It seemed fair to look for proxies and for some measurable magnitudes which might account for a fair amount of T. R & D expenditure even when narrowly defined as Professor Robinson would define it (groups greater than 100) was a plausible candidate, but no one would expect it to account for the whole of T; in fact he said he would be surprised if it accounted for more than a small part. Anyhow, it was surely by no means the only possible indicator. Number of Q.S.E.s was clearly another, in which case we might split L into two components – that referring to Q.S.E.s and that referring to non-Q.S.E.s, and leave T to be a sweep-up for everything else, including the accumulation of capital, and no doubt changes in the standard of qualifications.

Professor Mansfield said that he would like to make some comments on the comments as well as on the paper. He said that economists often thought of R & D and basic science as the same thing, and that this could be quite misleading. The length and extent and types of uncertainty were quite different in industrial R & D than in basic research and the latter formed only a very small proportion of the total effect. Secondly Professor Mansfield suggested that economists often tended to think of R & D in terms of major innovations like the transistor or nylon. These innovations had great impact and were well worth studying but it should be recognised that the bulk of the work carried out in industry was not directed to such advances. Instead it was aimed at marginal improvements involving small advances in the state of the art and a quick payout. He said there was a good deal of information on this score in the United States. For example, according to one survey of all manufacturing, about 90 per cent of the firms expected their R & D expenditures to pay-out in five years or less and 55 per cent in three years or less. Because of this it did not seem surprising that the peak effects of R & D on productivity might be about five years later. This did not mean of course that the results cited by Professor Griliches were necessarily right, but they were not perhaps as unlikely as some had stated. Thirdly Professor Mansfield suggested that economists often tended to underestimate the costs of technology transfer, that is, the costs of transferring technology from one organisation to another. There was some tendency to view technology as being a stock of readily available blueprints that were usable at a nominal cost to all. The truth was that this was far from the case. For example the United States decided some years back to copy a British jet engine and when the process was complete it cost more to transfer the technology than it would have to develop a new engine. Even when the problem was to transfer technology from one plant to another of the same firm the costs and headaches involved in transferring technology could be quite high. Fourthly Professor Mansfield pointed out that economists tended to work with models that were so highly aggregated that it was a bit difficult to put much stock in the results; in particular it seemed to him to be extremely important to break down R & D into various components, since it contained work of vastly different kinds with different lag structures and entirely different sorts of externalities. He thought that many of the most realistic and important science policy questions would never be touched upon, let alone resolved, until R & D was broken down and its various components treated separately. Further, he felt somewhat more comfortable with models which looked at the effects of R & D within a particular industry or sector of the economy, rather than attempting to determine the return from R & D in the entire economy. Finally, even if R & D expenditures were broken down and other refinements were made, Professor Mansfield shared Professor Griliches' scepticism concerning the extent to which the returns to R & D could be measured at all accurately through models of this sort. For example, the problems due to inadequate treatment of quality improvement in the G.N.P. or in output measurement figure were particularly difficult to overcome.

Professor Patinkin commented that what came out of Professor Mansfield's remarks was that R & D was so tied up with specific programmes and processes that one should not expect to get much out of the macro type analysis. We should also be aware that R & D was likely to be already included in labour and capital and one explanation of what we believed to be low results was due to this.

Professor Griliches said that specific empirical results he referred to were at the industry or firm level. The econometric work he had tried to summarise, of which there were only a few scattered studies, tried to grapple with the problem at some intermediate aggregation level where the questions and results were not *a priori* nonsense. The aggregate part of his paper really dealt with national income accounting conventions and what one would expect if one tried to treat R & D similarly to other investments. Returning to an earlier question Professor Griliches said that it was unlikely that R & D components were captured in capital measures because capital measures did not allow for improvement in quality. Labour measures also did not allow for differential quality of the labour force, and to this extent R & D was not caught in either labour or capital. Even when the distinction was made, the scientist and engineer categories were very heterogeneous in the United States. The data information was very poor and there was also often substitution within firms between resources and people. Therefore one expected that better results would be obtained using measures of R & D expenditures rather than measures of the number of people involved in the R & D process. Professor Griliches said that Professor Patinkin had suggested that the measures of capital might reflect also the role of R & D – was not the rate of change of stock of R & D similar to the rate of change of stock of capital?

Professor Mansfield interjected to ask why we should expect this relationship between the rate of change of stock of R & D and the rate of increase of stock of capital. *Professor Patinkin* said this might be inferred from Professor Mansfield's description of the nature of it, i.e. most R & D was short term, implying a more direct relationship than Professor Mansfield had in mind. *Professor Mansfield* asked how small improvements in, say, a toaster fitted in – this was not necessarily equipment?

Professor Patinkin then asked whether there was a relationship between short-run R & D and capital embodied R & D?, and *Professor Griliches* and *Professor Mansfield* both said 'No'. If one worked at the individual firm level, R & D and capital investment were different things both in time-scale and the speed with which they might affect output.

Dr. Nabseth said he would like to draw on some information on a study of diffusion of technology in different countries which had particular relevance to Professor Mansfield's points. Information about new technology moved slowly and it was the diffusion of technology which was important. Regarding introduction of new technology, it was not always R & D people who were involved in this; people on production lines were involved, and those were not usually included in R & D, so that R & D figures might be too low when considering transfer of new technology.

Professor Tsuru commented on the point made in the paper that 'any accounting scheme based on factor shares is very difficult to implement, even disregarding the fact that social returns to R & D would in any case not be reflected in these factor shares, even if we could compute them'. Professor Griliches then went on to say that that was the reason why he was going to use the econometric approach. But did this overcome the difficulties which had just been mentioned? If one began with the equation of Q not Y, Q is what Adam Smith called 'value in use' and Y is 'value in exchange'. Q is a physical measure and the production function is a physical relationship. Professor Tsuru said he would make a distinction between the two, although the marginal revolution allowed a synthesis. The role of science and technology drew attention again to Adam Smith's distinction. Were we trying to obtain a relationship in terms of physical Q or Y, or did we think they could be synthesised in a harmonious relation?

Professor Griliches said he had used econometrics because a significant part of R & D resulted in externalities and this did not show up in the standard accounts of companies and there was no hope of getting at it from accounting schemes alone. The rest of the paper was devoted to what one could get from national income accounting – that part for which there was some hope. He was not sure he understood the question about value in use versus quantity; was it asking whether we take adequate account of consumer surplus, as he had tried to do, in hybrid corn? Standard figures on output were bad and he had a mixture of badly deflated value-added figures. Unless one believed in changing levels of monopoly power the distinction between quantity and value was not an important problem, because all the rest was so large in comparison with it. *Professor Tsuru* intervened to say the difference between the competitive and monopoly period worried him. Consumer surplus was a relative and a dangerous concept. *Professor Rasmussen* asked whether Professor Tsuru assumed that there had been substantial changes in degree of monopoly over the period studied. If not, Professor Tsuru's point was not important. *Professor Griliches* said that for the United States in the mid-fifties to late sixties it was hard to find evidence for a change in monopoly power in either direction. *Professor Tsuru* said he agreed within those decades, but what about the eighteen-fifties? *Professor Griliches* thought this was not the point – he burrowed in a small neighbourhood.

Professor Oshima said the importance of military and defence R & D was typical of the United States but not of Japan, and asked how one assessed the impact of this part of R & D. Did one assume that the impact was large or small or nothing? *Professor Griliches* said that unless we knew how to evaluate the worth of military expenditures we could not evaluate the worth of military R & D. In his paper, he said, as far as the national accounts were concerned military R & D did not contribute to aggregate productivity. He did not say there was no contribution, only that it was not measured. On the question of spillovers to the rest of the economy, the proponents of military R & D or space budgets had made claims about spillovers, but provided few facts. The evidence of actual spillovers was meagre and anecdotal. In the end Professor Griliches said he was forced to make assumptions and took one-fifth of military and other such public R & D to be as productive as private R & D, but this was a number out of a hat and he did not necessarily want to defend it. *Professor Oshima* asked whether with this analysis it was possible to evaluate the impact on the economic sector if one-fifth was spinoff? *Professor Griliches* replied by saying it was a guess – merely an assumption for back-of-the-envelope calculations – to add one-fifth of military R & D to the R & D in the private sector.

Professor Oshima said that if one took the stock of R & D as the basis, in the case of Japan most of the innovations contributing to growth were transferred. It might be possible in countries such as the United States to think of a stock of R & D. But in a country like Japan one had to look at the infrastructure. Most of the R & D in Japan was oriented towards the transfer and import of innovations. At the extreme one could think of a stock, but the transfer cost was the important factor. *Professor Robinson* said he had been puzzled by this point in an earlier session. Thinking in terms of stocks of R & D made sense. But for Japan or India what was the stock? Was it the world stock? Or was it the national stock? Was there a cost of transfer, particularly for a country which was not generating research? *Professor Griliches* suggested that we needed to think of it as a number of effective blueprints transferred – once transferred there would be a stock, and when accumulated it might be that this stock had a high return, but the transfer was cost intensive. In the end one had to value it at the resources used to acquire it. If the cost of acquiring it was low it would have a high return.

Professor Lundberg, following Professor Robinson's point, said that he would like to stress that the capital stock of R & D from an international or a national point of view could not be separated. It was a question of utilisation of (a more or less common) stock of capital where there were problems of rate of utilisation. *Professor Robinson* intervened to ask what was the stock of knowledge available to India. Did it include knowledge in the United States of how a rich country could solve problems at high capital cost. *Professor Hicks* added on the transfer of technology and its costs, that this was relevant to questions raised previously on Professor Galbraith's paper, particularly regarding the division between machine making and machine using firms. Bugs and teething troubles were really symptoms of the problem of transfer.

Professor Freeman commented on the inadequacy of R & D statistics as a proxy measure and shared the doubts of Professors Hicks and Mansfield on this question. Anyone who was familiar with R & D statistics at the enterprise level realised the difficulties of using them. R & D was only part of a wider scientific and technical activity and it would be better if we all had the type of statistics which were collected in the Soviet Union, and which included information about scientific and technical services. With this wider rubric we would have more accurate measures of R & D. However, we also need better breakdowns of R & D. Further complications arose because of the fact that science and technological activities were changing very quickly over time and at different rates in different societies. The relative contribution of new knowledge won on the production line and the new knowledge won by R & D was different. On the nineteenth century accounting systems we would have come out with an absurdly low figure for R & D because few firms measured this activity separately from their other activities. In the twentieth century R & D had become increasingly departmentalised and specialised, was a separate activity in companies and government, and this process was continuing. It had occurred because of the complexity of science and technology itself and of the nature of the processes used in manufacturing industry. With batch processes it was easier for the people on production lines to experiment and to learn by doing, and to win new knowledge this way. With flow production or mass production people on the line could make experimental adjustments, and many improvements in flow production could only be made in laboratories or in prototypes. Because of this the proportion of new knowledge won by R & D to that on the production line was changing and it had to be remembered that it was very different in different countries and we must be very careful about using R & D figures in any long range sense in production functions we are discussing. Professor Freeman continued by saying that he agreed with Professor Robinson about the scale factor being important in R & D, but disagreed that only those employing over a hundred people contributed to new knowledge. He said there was plenty of empirical evidence to suggest that the contribution could be large from R & D establishments with less than 100 people. *Professor Robinson* said that he was comforted by what Professor Freeman said. This was a different impression from what a D.S.I.R. study had revealed fifteen years earlier.

Professor Freeman said that the difference in measured R & D between say the aircraft industry and shipbuilding was not merely due to accounting procedure but rather to the type of R & D being undertaken, and also in the amount of science and technological change in the design process. Fashion change or variations in routine processes were excluded from the concept of experimental development and might be more appropriately called design engineering, but building a prototype and learning from it was different from design changes in shipbuilding.

Professor Mujzel said that two years earlier the Polish planning authorities had tried to apply a Cobb-Douglas production function to measure the efficiency of R & D but he maintained that a crucial role was played by the management system. They had recently introduced an incentive system based on bonuses and promotion, and they hoped this approach would overcome the sluggishness of innovation. It was assumed that innovation was efficient for society, and the efficiency of labour was growing faster than R & D per unit of labour – this difference should be maximised by most progressive firms. Preliminary calculations in 1966–8 were astonishing. Some innovating industries got low or negative indices; others with poor general reputations got good indices, clearly showing that the economic and social process of progressive innovation was very complex.

Professor Griliches, replying to the discussion, thanked the contributors to it, many of whom fell into the category 'other things should have been taken care of'. He agreed with this, but he preferred to take small steps at a time. On the question of the production function he thought many of the comments were not important at the aggregate level but probably would be important at the micro level. It would be easy to get into an elaborate mathematical mess, and at this aggregate level there was not much mileage to be gained from using weighted harmonic rather than weighted geometric or arithmetic means of the various inputs. He agreed that the treatment of capital and labour left much to be desired. He had done some work on the measurement of capital and labour in other contexts and on the complementarity of capital and skilled labour and referred to a paper in Hansen's *Human Capital* volume. Public R & D should be treated as a stock of government capital available to everybody who wanted to use it. Professor Griliches said he had tried this for agriculture. It was not conceptually different from the contribution of government road building to the efficiency of the transportation industry. On the question of transfer of technology he said the diffusion studies focussed more directly on this problem. He suggested that some ideas were clearly not embodied and were easily transferred, but that most ideas had to be adapted for use in a particular country. For example, the development of hybrid corn was the invention of a method of inventing but it was up to a particular area to do something with it, to invent something appropriate to the ecological conditions of the area. While some ideas might be free floating it might still require considerable individual effort to exploit them. On the poor quality of the figures we worked with, Professor Griliches said this affected not only R & D but output, capital and labour statistics as well. The figures might be bad, he said, but all one could do was to try to improve them and then use them. The only alternative was to despair of inquiring into such questions.

4 A New Approach to Technological Forecasting in French National Planning[1]

H. Aujac

SCIENTIFIC ADVISER
BUREAU D'INFORMATIONS ET DE PRÉVISIONS ÉCONOMIQUES
NEUILLY S/SEINE, FRANCE

I. THE PROBLEMS OF TECHNOLOGICAL FORECASTING

There seems no doubt that the role played by technological fore-casting in French planning is undergoing a radical change. Until quite recently, the preparatory work underlying the National Plan basically involved reconciling assumptions about trends in the individual components of final demand with the projected level of output of the various sectors for the last year of the five-year period covered in the Plan. The reconciliation had to be made in the light of the general aims pursued by the public authorities – full employ-ment, a degree of price stability, and rising living standards – as well as a certain number of specific objectives, for example in the fields of defence, education, infrastructural investment.

As regards methodology, the reconciliation was sought by means of an industry input-output table. This meant that a five-year fore-cast of technical coefficients, capital/output ratios and labour productivity had to be made. The coefficients were calculated to match the classifications of some 70 component cells in the table.

In this approach to the Plan, 'technological forecasting' was concerned with these coefficients, and it was an intermediate stage enabling the transition to be made from final demand to output. It had little intrinsic interest otherwise, and the underlying assumptions were not even published.

Certain comments may be made about the process of forecasting technological coefficients. Firstly, the quality achieved was at best only fair. Admittedly the reports filed by the 'Commissions d'Indus-tries du Plan' contained many useful indications as to trends in consumption of raw materials, energy, capital and labour per unit of output, but the information tended mainly to be provided in terms of physical quantities, using an extremely fine product breakdown. In the majority of cases, a satisfactory linkage between this information

[1] Special thanks are due to Madame Guiriec, Director of Studies, B.I.P.E., for her help in the preparation of the technical sections of this paper.

and the coefficients in the input-output table could not be established. Thus, as a practical matter, when the table was projected, the coefficients were forecast by extrapolating past trends; special studies were made only for the substitution of various forms of energy between each other.

Secondly, the forecast, although rudimentary, proved to be sufficiently accurate for the purpose served, namely the transfer from final demand to output. In terms of the classifications of the items in the table, the errors it induced were of secondary magnitude compared with the uncertainties attending the forecast of final demand, especially as regards the foreign sector.

Lastly, it did not seem useful to publish this technological forecast. Only national accounts specialists could interpret it. Business firms were unable to connect up technical forecasts at this level of aggregation to their own production functions. The public authorities, despite their praiseworthy efforts to raise productivity, were far from having it in mind at the time to become active in the area of R & D, or to develop policies regarding new products and processes.

For all these reasons, technological forecasting was necessarily a very background feature of the French planning effort.

It seems safe to affirm that the role played by technological forecasting will henceforth be more important. But it will be technological forecasting of a different kind. The shift corresponds to a quite radical but apparently inevitable modification of the content of national planning, and will reflect the necessary adaptation of forecasting methods to the new requirements so generated. Several factors make this development unavoidable. They are outlined below.

The opening of borders and the progressive integration of France into the Common Market will markedly reduce the interest of business circles in the results of a market forecast which, however broad its product coverage, is limited to the French market alone. There are admittedly forecasts of foreign trade, but they are independent estimates, since foreign trade is considered as an exogenous variable. At the same time, the domestic production forecasts which are closely dependent on trends in foreign trade, necessarily become even more arbitrary, subject accordingly to greater margins of error, and therefore difficult to project five years forward in the kind of breakdown which is useful for business and industrial purposes.

Again, competition is waged nationally, and even more so internationally, between enterprises, rather than between industries as defined in the national accounts, and between products and processes which are defined by an incomparably more detailed classification than any currently in use.

It is also becoming steadily more essential for the preparatory forecasting in the Plan to extend well beyond a five year span: ten, twenty, thirty or more years are necessary if a valid appraisal is to be made of the consequences of current decisions, or those to be taken during the years covered by the Plan proper. Some of these decisions will be taken by the public authorities in such areas as education, health, defence, and transport facilities; other lie in the competence of enterprises, and concern for instance R & D policy, new products and processes, investment, or industrial reorganisation and restructuring through co-operative arrangements or mergers with domestic or foreign firms.

In addition, a feature of the kind of technical progress being made today, especially in its most advanced and revolutionary forms, is that its real importance is out of all proportion to the value of its quantifiable parameters: economic flows produced and consumed, number of establishments, labour force size and even value added. The electronics sector is a good example of that class of new industries whose real importance is their role as growth catalysts.

Two consequences flow from all this. In the first place, the content of the information provided to industrialists by the Plan needs to be changed. The main purpose of the French Plan, apart from the attempt to co-ordinate the activities of government departments with economic functions, has always been to educate business and the public, whereas in practice, only the largest firms were in fact able to benefit from the information put at their disposal. To remain or become useful to enterprises, the Plan's guidelines must henceforth be formulated in terms of trends in European and world markets, target costs of production, competitive industrial and enterprise organisational structures, targets for the rate of technical innovation, and so on.

Secondly, the input-output table as at present designed cannot help in formulating this type of information forecast, or at best must be used with extreme prudence, and its value in forecasting for periods beyond five or ten years is, to say the least of it, dubious. It fails in these requirements because it identifies industrial branches but not firms, average costs but not the cost structures of competing firms, and so on. Further, it has to treat foreign trade as an exogenous variable, and has only an elementary vocabulary available to express the content and growth of technical progress whose qualitative aspects and role as a catalyst of economic growth are therefore given altogether inadequate emphasis. Thus it is incapable of contributing to the formulation of this kind of predictive information, or at best, must be employed with extreme finesse. The scope for using it to forecast beyond a five or ten year time horizon is practically negligible.

The table must therefore be replaced, or at least supplemented, in the function in which it has been used hitherto as the touchstone of the French Planning process. Among the new instruments, technological forecasting with certain well-defined specifications will certainly fulfil a major, and possibly central role.

Technical progress has always occurred in one of two ways. At times, there has been a continuous process of improvement of existing techniques and products, both at the technical and the economic level. At others, there has been an abrupt shift rendering some techniques and products obsolete, to be replaced by new ones. The technical history of the production process might well be described as a succession of 'eras' during which contemporary products and techniques were improved more or less rapidly, separated by 'periods' of technological leap forward, with eras and periods succeeding one another at variable speeds in relation to individual industrial sectors, but overall, linked together by technical and economic interdependence.

Technical progress today, appraised in these historical terms, appears to have certain original features:

the shift from one phase of the process to the other occurs at even shorter intervals;
the technical interdependence of economic activities is growing, and with it, economic interdependence. An innovation in a given sector is liable to revolutionise products, sectors and techniques which might earlier have been thought totally unrelated.

These characteristics of current technological trends augment the insecurity of enterprises. Each enterprise must take the greatest care to avoid being taken by surprise by a technological shift which may occur in its own sector, but may equally easily arise anywhere in the production process at large. In either case, its growth – or its existence – could be jeopardised. The steps taken by firms to control this risk are well known. They range from attempts to manipulate private demand through advertising to attempts to control public sector demand through the 'military-industrial complex' denounced by Eisenhower in his farewell Presidential address to the American people.

Technical risk, in other words, is and is felt to be becoming more dangerous. Fortunately it now seems possible to determine in advance (or is this presumption?) how much time is required following a decision to develop a product or a technique for the industrial prototype to be built, and then for large scale marketing to be undertaken. In other words, forecasting technical progress and therefore its probable consequences now seems to be feasible.

Let us imagine for a moment that for each major industrial or other economic activity a document is available giving two items of information: the probable list of R & D results having some chance of being transformed into important innovations, and for each innovation, a schedule indicating the approximate dates of preparation of the industrial prototype, of initial marketing, and of full-scale market growth. Such a document would be of enormous value to the public authorities and to industry, subject of course to the limits set by the accuracy of the information it conveys.

The public authorities would be advised of the direction, in terms of types of products, equipment and techniques, of trends in world technology. A firmer basis would be available for government policies in such important and varied areas as R & D, technical assistance, education, labour force training, conversion or revitalisation of stagnating industries or regions, and the rest. There can be no doubt that the document would contribute to better management of taxpayers' funds, and help in taking timely measures to safeguard groups of the population threatened by economic progress, as influenced in particular by technological progress.

Likewise, as part of its essential educational function, the Plan would now be able to provide French firms wishing to protect their position in the international competitive race with a blueprint of likely long-term and even ultra-long-term technological developments. Is it not reasonable to expect that the winning management team will be the one which fosters progress during technological 'eras', and prepares to take the corner in to the next 'leap forward', not too early, but also not too late? Poor timing, by contrast, may even lead to the ruin of the enterprise.

Enterprises would have an 'early warning system' of technological shifts or innovations from which they could draw conclusions as varied and useful as: 'What is the most reasonable amortisation period for this product, or that machine?'; 'Is our product or process development policy dynamic enough to keep us abreast of international competition?'; 'With what other firms whose technical know-how will prove indispensable should we be organising co-operation?

Without doubt, such a dated list of probable innovations is one of the new management tools called for by present-day changes in the international environment. It would be capable of use in the context of the new problems facing both the public sector and private industry.

It is precisely this kind of technological forecast which the Commissariat Général au Plan has asked the B.I.P.E., to prepare, as an experimental project. The main initial problem was to specify and perfect the method to be used. The results achieved are described briefly below.

II. A CASE STUDY OF TECHNOLOGICAL FORECASTING: THE MAIN RESULTS

The textile sector has been selected to illustrate the construction of the method of long and ultra-long-term technological forecasting.

The main results obtained are presented in the attached forecast timetable,[1] which is simple to interpret: on the vertical scale, the various stages of production of textile products are classified in sequence, starting from the final product;[2] the horizontal scale refers to the development, initially technical and then economic, of a new process or product, using a five-year span as the unit of time. Four phases of the R & D stage have been distinguished, from fundamental research to the development of an industrial prototype.[3] The results presented are in principle valid internationally. Three phases of the marketing stage are identified. Here the results relate to French firms only.

As can be seen, the timetable does two things. It lists the inventions which may turn into important innovations, and it gives the schedule of the main technical and economic stages of their maturation.

The timetable is well worth examining, for it indicates the presence of a remarkable and rather exceptional phenomenon. Towards the period 1980–85, the current 'era' of continuous improvement of existing techniques and products will draw to a close, and be succeeded by a 'period' of technological mutation. Labour-intensive today, the textile industries will become heavy industries based on electronic and chemical methods. This will inevitably create difficult problems of employment, financing of new equipment and industrial reorganisation, and they will emerge soon. If the French textile sector fails to carry through its transformation in time, large segments of it are doomed to disappear.

The value of the results obtained lies in the fact that they could only be derived by the methodical synthesis of specific individual forecasts. Before this study was made, managements and industry

[1] Taken from the B.I.P.E. study *Essai de définition à long et à très long terme de l'évolution des structures techniques et économiques de la filière des industries textiles* (B.I.P.E., 1970). This study was commissioned by the Commissariat Général au Plan and carried out under the direction of Madame Guiriec, Director of Studies, B.I.P.E.

[2] Represented here by hosiery alone. Unfortunately it was impossible within the limits of this study to analyse the products of the clothing industry.

[3] In practice, for technical reasons, fundamental research is only rarely taken into account.

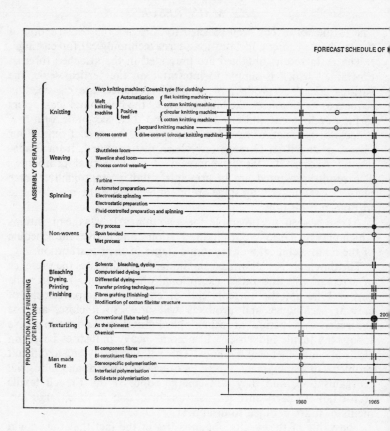

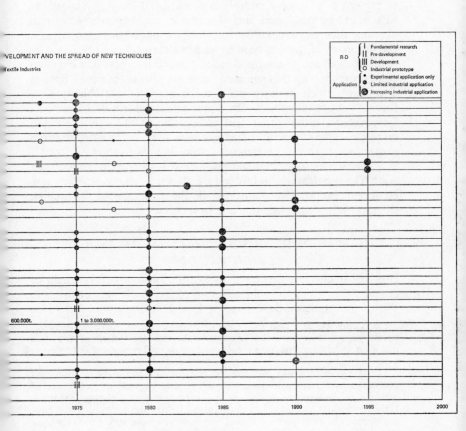

VELOPMENT AND THE SPREAD OF NEW TECHNIQUES

Textile Industries

R-D	{	I	Fundamental research
		II	Pre-development
		III	Development
		○	Industrial prototype
Application	{	•	Experimental application only
		●	Limited industrial application
		◉	Increasing industrial application

600.000t.　　　1 to 3.000.000t.

1975　　1980　　1985　　1990　　1995　　2000

federations alike were totally unaware of the scale and imminence of the forthcoming technological mutation.[1]

The document in other words, may well be one of the very management tools which we know will become more and more indispensable in present conditions. Certainly it would be if its prognostications turn out to be well-based. But of course it would be utterly naive to presume that they are, unless and until the underlying method, and the use made of it, have been explained clearly.

Space does not permit the detailed presentation here of a long and complex study;[2] all that can be done is to indicate the broad lines of a method which appears suitable for carrying out this type of forecasting. In so doing, the approaches actually used will be systematised, although at the outset, they were developed and employed quite empirically: this was after all an exploratory research project whose specific purpose was to define the method.

The best way to explain what was done is to go right back to the start. The problem was to forecast significant innovations which may come to play an important role in the textile sector in coming decades. What approach was one to try?

III. HOW SHOULD ONE FORECAST SIGNIFICANT INNOVATIONS?

It is best to consider first the nature of the knowledge it is sought to acquire; this will help in the selection of a suitable type of method. But of the whole range of ongoing R & D projects, to select those

[1] It is only too clear that the textile industry both in the developed and the developing countries has failed to perceive the true size and closeness of this mutation. Behavioural patterns provide immediate confirmation of this sobering fact:

The textile industries in industrialised countries are still invoking social and financial considerations to claim state aid and protection – protection of domestic manpower and capital against competition from low-wage countries. If the conclusions reached in this study are valid, these arguments no longer have any force. The anticipated technical progress will more surely and radically eliminate certain skills and investments than anything that developing countries' competition could possibly bring about. If a case is to be made out for temporary protection of the textile sector in an advanced country, it must rely on the future that technology will bring, not on social or historic arguments, and the competition most to be feared will be that levied by other advanced countries, not by the less developed world.

The developing countries should beware of the poisoned gift constituted by the introduction of current textile techniques into their economic fabric. These techniques will be outmoded in a not very distant future, and the host countries are poorly placed to make a successful transition from the traditional to the new technical structures of the textile sector.

[2] Nevertheless, some illustrative details will be given in Section VII below, using the techniques of the spinning industry as an example.

liable to result in major innovations which will respond to the needs of society between fifteen and thirty years from now, and then to go on to attach dates to the stages of development of each, is by no means the kind of research whose results can be presented with the irrefutable conviction of a mathematical demonstration. The real world that we must observe, and if possible forecast, is one in which new research projects and inventions are being born all the time and open up prospects of development in various, often unforeseeable, directions. Some new inventions will compete with each other. Some will fall by the wayside; others will thrive. Behind their struggle we perceive human competition – management teams and enterprises seeking each to secure the future of their own technical advances, with the life of their firm and their own careers as the stake. The results obtained in this struggle are governed in large part by the extent to which the socio-economic characteristics of the markets concerned are matched by the technical and economic specifications of the stream of products and new techniques launched by firms, and their timing. Admittedly, success in this (or failure) is conditioned by human calculation and foresight. But these are not the only factors. Luck and the goddess Chance also have their word to say.

To understand this reality, we can perhaps do no better than to follow Cournot, who viewed the unfolding of history as consisting of causal sequences, interweaving and interfering with one another at random. Each encounter gives birth to a new causal sequence, so that the motive force of the historical process lies in these incessant meetings between determinism and chance.

Our research therefore has no place in the world of cause and effect and evidence. It can only belong to the domain of verisimilitude, plausibility and probability. For our views to be acceptable and carry conviction, we must contend: no proof is possible.

Thus the nature of the problem makes it necessary to adopt the framework of an open model. The reality to be described by the results we are seeking to achieve is governed by too many factors which are absolutely unforeseeable today, and which will progressively take on clearer form only as the mists of the future dissipate. The essential purpose of our research must accordingly be to provide a framework for reflection whose contents are explicit, and to try to synthesise all the information – of whatever sort – that can humanly be gathered pertaining to the probable evolution of technical progress. This framework should be capable of absorbing a continuous supply of new facts, providing an ever-updated synthesis as it 'digests' data on changes in the economic and social environment, new achievements, and the discovery of paths of technical advance that cannot even be imagined today.

Thus the method must be characterised by the 'open' character of the model used, the explicitness with which its contentions are formulated, and the light it casts on the procedures used for synthesising.

IV. THE LIMITS OF WHAT IS TO BE FORECAST

It is now time to specify the limits of what is to be forecast. Should attention be focussed on a certain textile innovation, or should the textile and clothing industries in general be considered? or again, these industries along with final demand, whose main component is the demand of the household sector?

In our view, the last option is the proper one. The structure of final demand to a large extent determines the type of technical progress which will occur in the manufacturing chain which services it. In textiles, a 'rich' market will demand certain of the sector's products which will in turn foster the development of materials and techniques having specific features, while a 'poor' market will mobilise a different set of products and techniques. Technology is not an end in itself, but a means of adjusting supply to demand. In the last analysis, of the set of techniques that progress makes possible, it is the market that sorts out those which it will promote. This is the dominant criterion which we can use to select the inventions and research projects currently on record to write into the league table of future innovations. Obviously, to use this table we must predict quantifiable characteristics, but a forecast of the qualitative characteristics of long-term demand trends is equally and perhaps even more necessary.

In the specific case of the textile industries and their market, the problem is that the specific demand pattern of the 'rich' market of the industrialised countries (we assumed that the French textile sector would orient to this market) will move steadily to an accelerated rate of product changeover and wider production ranges. This runs counter to present production constraints. Until recently, it was impossible to reach acceptable cost levels without embarking on mass production, thus making use of relatively specialised processes and raw materials, with a correspondingly inflexible production process.[1]

Concern to keep costs reasonably low will inevitably work in favour of techniques which contribute to resolving the conflict between product diversification, which raises costs, and series mass production, which lowers them. Electronic techniques now exist which abate the impact of this conflict. Management-type computers,

[1] For example, using present-day techniques, production of fashion clothing involves a 15 to 20 per cent decline in overall productivity.

originally employed for business processing, can be used to harmonise production conditions and marketing requirements. The scheduling of manufacture can be continuously and immediately adjusted as a function of the inflow of new orders. Production can therefore be aligned to respond to the market from day to day. This is essential for the future of the textile industries. Further, the adaptation of process computers will enormously speed up production and so raise productivity levels, lower costs, and at the same time guarantee that the due quality of output is maintained. Certain chemical production techniques which are currently well advanced in the R & D process will sooner or later provide similar results.

Such, schematically presented, was the type of reasoning that was used to analyse the types of technological progress that would correspond to a given forecast trend of demand.

The analysis itself cannot, of course, be developed at this level of generality, but must be in terms of raw materials and equipment and use the same classifications as the research and engineering departments of textile producers and the firms which supply them with their material inputs and machinery. The main burden of supplying information for our forecasting machinery accordingly lies on the executives of these departments and the technical service units of the industry. To communicate efficiently with them, we must speak their language, and the probable 'innovations' we are trying to identify must be expressed using their classifications for describing their research projects and prototypes and the products and techniques stemming from them. Some examples of the analytical process will be presented below.

V. THE IDENTIFICATION OF MAJOR INNOVATIONS

The criterion of correspondence with the trend of final demand is fundamental to the selection operation, but it clearly cannot be used in isolation. It is neither possible nor desirable to catalogue every expected innovation, large or small, major or insignificant. The purpose being to identify research products with a significant potential impact in terms of national economic growth, we must limit ourselves to an appraisal of only those projects which may lead, sooner or later, to major innovations. A second criterion is therefore necessary.

For this purpose, a major innovation can be defined in terms of the disturbance it would be capable of causing in three areas:

redundancy of machinery presently in wide use, which would face firms, and doubtless also the public authorities, with financing problems and problems of industrial conversion;

irrelevancy of the technical skills of a sizeable number of employees, with the corresponding problem for firms and the government of technological unemployment and the need for retraining schemes;

modified structural organisation of the industry, to the extent that (and the case is met frequently) some kinds of technological progress are associated with such high costs or technically sophisticated processes that they can only find a home in sufficiently large firms.

These criteria were applied to select the major innovations likely to emerge in the textile sector over the next few decades; their effect, incidentally, will be to make it less and less 'textile', and more 'chemical' and 'electronic'. For a project to appear in the 'honours list', it has necessarily to satisfy the trend-of-final-demand criterion, and at least one of the three size-of-innovation criteria described above.

Once the list is drawn up, the next step should be to collect a considerable volume of information from sources all over the world. Even more important, the closest collaboration with the research departments of enterprises and industry associations is necessary to compile a timetable such as that presented in this paper. This is easier to say than to do.

We have now reached the heart of the problem, namely, how to assemble and organise the information so as to acquire the conviction that research project X, for example, will lead to results; that these results will be exactly in line with demand requirements; and that the product or process that has been developed will involve a substantial effect in terms of rendering machinery, labour force skills or even industrial structures obsolescent. Similarly, we must decide what kind of reasoning should be applied to turn an opinion into a conviction that research project Y will lead to construction of a prototype five years hence, but that the project will experience a rapid expansion of its market fifteen years from now, rather than ten – or twenty.

The answers to these questions determine the degree of credence and interest attaching to the results in the over-all summary table.

VI. THE AVAILABLE SOURCES OF INFORMATION

The information needed to underpin the technological forecasts to be developed, and which it is feasible to collect, is of many different types. Some of it is related to general economic and social trends (e.g. predictions of quantitative and qualitative changes in demand). In other areas, technical information must be obtained about products and the production equipment that will be promoted by the research projects or prototype developments presently in hand (e.g. forecast production rates, delivery periods, product quality, adaptability of equipment for alternative uses, etc.). We must be aware of the economic characteristics of products and techniques (cost of machinery, manpower skill requirements, etc.); trends in the relationship between a firm's size, its ability to procure finance, and the type of technical progress it can harness; or again, anticipated changes in the length of the production chain, inter-process or inter-product substitutability and so on. Information is also required on the waiting periods to be accounted for in the technological forecast: how long it takes for a process to be brought to technical maturity, for the industry to gain experience with it, for manufacturers of new equipment to tool up for regular production, and over what period equipment installed should be amortised. Allowance must be made for the decisive effect on technological trends in certain industries of public sector intervention, so that government policy in this area must be appraised and assumptions developed about the possible direction (or directions) it might take.

The length of the list of items of information which it seems possible to collect should not cause any illusions. Of course there are snippets of documentation available on almost any of these points, but really firm information is rare. Above all, the available facts are only too often inconsistent with one another. This emerges clearly as soon as an attempt is made to weigh up the chances of a new product or process by confronting the arguments for and against it, using a dialectical procedure similar to the courtroom arguments leading to acquittal or conviction.

The unsuitability of much available information for use in technological forecasting is in some cases due to accidental causes which may be remedied, up to a point. Unfortunately, there are also a number of fundamental causes.

One of them is that data of a technical nature are generally of better quality when they concern a prototype already in operation than they are when they concern a new research project just being launched. Similarly, if a product or machine has already been marketed, more accurate knowledge can be conveyed about costs, labour force qualifications and so on than is possible for a prototype still under development. It is obvious why this should be so. The technical characteristics of a product, and later its economic characteristics only reveal themselves gradually as a project advances through the various stages of R & D to enter the marketing phase and ultimately, becomes an accepted innovation.[1]

[1] Even allowing for this phenomenon, the information that can actually be collected is often limited in volume and quality by comparison with what ought

Another fundamental difficulty in exploiting the available information is that individual items are as often as not inconsistent with one another, or else leave obvious gaps. There are intrinsic reasons for this. Materials are gathered from a wide variety of sources – official and business economists, researchers, engineers, management, trade union leaders, etc. These people have different specialities, work in different organisations, and they have different objectives and therefore different concerns, outlook and scope for action. The division of labour and the differentiation of functions and institutions are responsible for this.

These difficulties exist along with others. We should be aware of them, but they are not reason for despair. When all is said and done, the data collected in this way are not so much worse than those underlying reputedly well-based forecasts in other areas.

VII. AN EXAMPLE FROM THE SPINNING SECTOR

This can be illustrated by a rapid review of the materials available concerning the techniques which are or may be used in the spinning sector (see the diagram), which will shed light on the nature of the information that led to classification of some innovations as important, and correspondingly, to establishment of their development timetable.

Two of these techniques, the turbine and automated preparation, are well beyond the R & D stage. The former is being tested operationally; the latter is coming into widespread use. The problem is to assess how long it is likely to take for French industry to apply them on a routine basis.

A substantial volume of reasonably accurate technical and economic data on these techniques is available. For example, we know that fully automatic spinning, a chain process in which machines transform cotton lint into ribbons of fibre ready to be spun, was perfected in Japan during the period 1960–65; it was also being developed in the United States about this time. It involves investment outlays 20 to 50 per cent higher than normal continuous process spinning, and is inflexible in production terms: a narrow range of products only can be spun from necessarily large and homogeneous input batches. Against this, it is enormously labour saving. In present market conditions, given the relative cost of

to be available. In recent years, large firms in France, and indeed the public sector have in practice lost interest in economic forecasts going beyond a five-year horizon, and they first embarked on technological forecasting only a short while ago. It is therefore not very surprising that the collection of information should have proved a disappointing area.

capital equipment and labour, it would not be economic to introduce into France just yet. But by 1980, it may well offer worth-while returns.

By contrast, partial automation, which integrates the process from raw fibre preparation as far as combing, is already an economic proposition for output of 100 tons per month, and is spreading rapidly. It lowers spinning costs by about 10 per cent, reduces labour requirements by about 50 per cent and raises the unit yield of machines by 20 to 35 per cent. It also reduces waste considerably and improves product quality by eliminating joints.

Towards 1980, the French industrial and economic setting will justify introduction of full automation. But by that time, technical competition will be intensive between conventional continuous spinning on spools and turbine spinning which, available only experimentally today, replaces the sequence of roving frames, continuous spinning and possibly also winding machines by a single open-end spinning installation. It remains to be seen which of the two techniques ultimately takes the dominant place of industrial processing.

To choose, we must list the factors that may work in favour of the turbine process, and those that will militate against it, and confront this list with the corresponding list for fully automated spinning. We can then make a plausible assessment which, hopefully, will not be invalidated by History ten years from now.

For the time being, the turbine's uses are too limited. Only an excellent quality of cotton or short staple chemical fibres can be processed, and the only yarn that can be produced at reasonable cost is of average fineness with below average resistance characteristics and light and fluffy appearance. Any change in these operating conditions makes it too costly by comparison with conventional spinning.

Nevertheless, it already has considerable advantages, and these will augment steadily. The production chain is shortened by cutting out some machines. Spinning is twice as fast and the quality of yarn more regular. Most importantly, there is a massive reduction in labour requirements; this is of the order of 50 per cent for machine minders at the yarn spinning stage, which is the most labour-intensive of all the stages in the spinning process generally.

Thus two techniques may well be in effective competition by 1980. Fully automated spinning will in any case be incomparably in advance of the processes in current use, but most of its technical and economic potential will have been reaped. The turbine may already be competitive by this time, having been adapted to produce wider and more flexible ranges of output, but it will still have considerable potential for further development. Rightly or wrongly, we concluded the the future industrial organisation of the spinning sector would be centred on the turbine.

In order to form a view as to how quickly this technique might spread, we then took account of the age structure of the 'park' of existing machinery. It was assumed that modern continuous spinning units now being installed would be amortised during the period 1980–85 and replaced by spinning turbines, which would account for some 15 per cent of production capacity in 1980, and 50 per cent in 1985. This trend would be supported by the probable lowering of unit costs resulting from series fabrication, associated with probable difficulties of labour recruitment.

Two other techniques presently at the R & D stage were also deemed promising candidates for future development, namely electrostatic preparation (originally studied in Czechoslovakia, and now being developed by a number of machinery suppliers), and 'fluid' spinning. Little technical or economic information is available about either of the two, but that little is enough to suggest a plausible development timetable for them.

Electrostatic preparation would cut both capital investment and manpower requirements very heavily. A prototype electrostatic spinner should be demonstrated in 1972, but the first operational versions will reach the market in 1985 at the earliest, and be sold in significant quantity only towards 1990.

The use of 'fluid 'techniques for spinning calls for no machinery, and produces a very fine yarn about thirty times faster than present day continuous production machinery. The technical difficulty to be overcome is how to obtain a sufficiently great reduction in the consumption of energy used in imparting movement to the fluid. It is unlikely that this technique will have been perfected before the period 1980–85, so that it will only be marketed in quantity towards the year 2000.

The foregoing will have served to illustrate the information and the type of reasoning applied to select between new techniques and specify the likely timetable for the development of each.

VIII. THE RELATION OF INFORMATION TO METHODOLOGY

The impact of the state of the corpus of available information on the method to be adopted for technological forecasting is clear on the face of it. The best method is that which draws most advantage from the volume of information that can be assembled at a given stage of the technical, economic and political development of society, despite its sparseness, poor quality and internal inconsistencies.

One of three approaches may be adopted to deal with the problem. They may be described in terms of the relationship of the forecaster with his sources of information:

(a) distribute a questionnaire, collect the answers and start collating them. A selection process is necessary, since only those elements which appear compatible with one another can be used. This is a delicate and thankless job. If one is too demanding, only a few odd shreds of the materials originally gathered will remain. This creates a temptation to work up these shreds by methods whose sophistication must be in proportion to the reduction in the volume of information available for processing. However, the sophistication of a method is no proof of its efficiency. What counts in this field is balance between the volume and quality of information and the procedures used to mobilise it.

(b) A second, more active approach goes beyond the mere despatch and summarisation of questionnaire results by the forecaster. This is the method used in the study described above, and it consists of seeking to exchange views with correspondents with the specific aim of translating the replies into economic terms. The essential, and the most difficult question in technological forecasting is, after all, to predict whether the market will accept a certain new product or technique, and it obviously cannot be put in terms of technology alone. But there are practical difficulties in expressing it in economic terms, i.e. in terms of raw material, machinery and labour costs, quantities marketed and product price. As a general rule, the engineers and researchers working to perfect a product have more or less firm ideas about its expected technical performance. For example, in the case of textile machinery, they can indicate its expected production rates, quality of output, etc. In general, however, they are reluctant to hazard an estimate of its capital or operating costs. It is possible to understand their refusal to enter the economic field: this is not their business (although at this stage of R & D it is sometimes difficult to find the people in the firm whose business it is), and in any case, to make an economic appraisal, they would have to be aware of the development of the economic and social environment the product or technique they are working on will have to enter. This gap in the knowledge of R & D research workers and engineers regarding the economic aspects of their activities is doubtless a temporary result of insufficient co-ordination between the R & D, production, and sales departments: it may be expected to be remedied as time passes. For the time being, however, we must consider it as one of the data of the problem.

In these circumstances, the most the technological forecaster can hope to achieve is to get answers from the set of people contacted which can be expressed by the forecaster himself in economic terms, and which tally well enough for them to be integrated into the general forecasting frame. He must therefore try to describe to each respondent one or more plausible scenarios of the general long-term trend of the economic and social environment, together with the main provisional results already attained as regards the projected timetable for the development and marketing of new products and techniques. These descriptions should help the forecaster to standardise the general reference frame in which the technical replies regarding a specific product or process are incorporated. Interrogation is then conducted so as to facilitate as far as possible the translation into economic terms of the technical information provided by R & D specialists.

This approach places a considerable responsibility on the forecaster, who alone is able in this case to translate his interlocutor's experience into terms that are efficient for technological forecasting. Thus informants only play a passive role in working out the forecast, and they may well get the impression that their statements are being milked to yield more substance than they put into them.[1]

Nevertheless, this method proves markedly superior to the first. Its defect is its failure fully to mobilise the experience of all who have participated in the forecasting process: only very imperfectly does it wield a language common to all, nor does it achieve efficient communication between the experiences acquired in the different universes of action, thought and prediction of the participants.

(c) It is now possible better to perceive the approach which would still more efficiently mobilise the varied experience of society, reflecting the division of labour and the different types of social organisation, especially at the various stages of the production marketing chain. Men, placed in different environments and undergoing different experiences, have become as strangers to each other. Yet it is this very diversity of experience which, properly used, can give technological forecasting a reasonably solid foundation.

It is easy to formulate the method needed to unite the practical conditions in which these varied experiences may be mobilised: the participants should be integrated into a system of forecasting-oriented collective self-education, intended progressively to develop a common language and modes of thought acceptable to all. None

---

[1] It may also be noted that this approach to forecasting does no more than duplicate what we have seen happen only too often in some enterprises, namely, serious difficulties of communication due to differences in the language used by research workers, financial executives, economists and engineers.

need sacrifice the originality of his own experience, but he would now be able to convey it in a language all understand. The group would provide him with an evolving frame, constantly adjusted through the collaboration of all participants, which he himself has contributed to formulating, and within which his own predictions find their place.

Carried out in this way, technological forecasting would become truly collective, mobilising the whole range of the scattered experience of society as efficiently as is humanly possible.

This is the most general method that can be designed.[1] Further, it is valid for any forecast, whatever its purpose. But it is particularly important in technological forecasting, whose results are especially difficult to justify, and which therefore calls for mobilisation of the greatest number and widest range of experiences. There is another reason too. Technological forecasting, by reason of the diversity of experience to be mobilised, is the area which raises the most problems of inter-participant communication.

As has been seen, the underlying principle of the method is very simple, but requiring as it does the collaboration of a fairly large number of different kinds of people, its implementation must be done with great care. It lends itself well to description by Napoleon's remark about warfare: 'It is a simple art: all lies in its execution.'

---

[1] A forthcoming paper will show that the methods presently used are particular cases of this general approach, distinguishable from it only by the nature of the information collected and the degree of iteration involved in its processing.

Discussion of the Paper by M. Aujac

Professor Giersch said that in introducing M. Aujac's paper he hoped to be very brief, since not being a specialist in the subject of this conference, he must be careful not to lower the level of discussion. The shortage of time available for reading the paper had led him to have a personal conversation with M. Aujac which had helped to clarify some points which otherwise might have led to lengthy cross-examination. The first few pages explained why there was to be a change in the role of technological forecasting within the French National planning. M. Aujac favoured a radical change. He, and Professor Giersch supposed also the Planning Office, were no longer satisfied with the method of merely extrapolating past trends into the future, and notably the trends exhibited by the technical coefficients in the input-output table underlying the five-year plan. Instead M. Aujac proposed, for each industry, a concise prognosis of the technological advance to be expected over the next fifteen to thirty years. An illustration of the type of prognosis was given in a table for the French textile industry which listed for each stage of production seven phases of techno-economic development. They extended from fundamental research to mass application in industrial production. As one could immediately perceive, there was to be expected a kind of technological revolution in French textiles in about fifteen years, indicated by the numerous black spots on the vertical axis for 1985. Around that time M. Aujac expected to find in France a textile industry based on the electronics industry and the chemical industry. If the transformation were not carried out in time, large segments of the French textile industry were doomed to disappear. It was suggested that no other industry was faced with such a technological revolution.

Professor Giersch said that the interesting point was the type of research and the methods which had led to these interesting conclusions. Although there seemed to have been much trial and error, the method in retrospect looked like this: First: you find somebody who is familiar with the textile industry and has studied it for more than a decade. If it is a bright female economist with good judgement you are in a particularly good position, as M. Aujac was. Second: you conduct interviews in order to prepare a survey of the whole range of ongoing R & D projects within the industry. Third: on the basis of some personal judgement you make a selection of those projects which are likely to be economically sound in fifteen or even thirty years. Judgement in these matters probably meant that there was a fairly clear view of how production conditions and demand would look after 1984. To do that it was necessary to know a certain amount not only about the size and structure of domestic demand, but also about the size and structure of foreign demand, and as regards supply about whether or not, and to what extent, domestic firms would have shifted some or all stages of production into less developed countries with lower wage costs.

Professor Giersch thought that this was the right place to make a point of reservation. The European Commission some time previously had asked Professor de Bandt to prepare a paper on the future of the European textile industries. This report, which had drawn attention to the possibility or necessity for the textile industries to move towards countries with lower wages, immediately induced the West German textile industries to commission an anti-de Bandt report from one of the leading research institutes. The institute's report stressed the good chances for survival by pointing to past technological trends. However, bygones were bygones. How many times happier the West German textile industries would have been if they had not merely been persuaded to be more proud of their past achievements and to have more courage for the future, but if instead they had been supplied with M. Aujac's technological forecast. It would have been easy for them to draw a conclusion which would have suited their interest much better rather than a policy of telling the government how badly off they were and how badly they needed more protection against imports from Asia. More protection and a few subsidies of this and that kind would have been fully justified as a purely temporary device, if one could be fairly certain of a technological revolution around 1985. Professor Giersch said he was tempted to ask a question which carried this line of thought a step further. Would it not be in the interest of the textile industry and any other industry facing competition from developing countries to participate very actively in a technological forecast with a view to influencing thought in a way that would provide strong support for temporary protection? If such a thing should happen, the consequences for the division of labour between developed and developing countries might be far-reaching indeed.

Happily enough for the study on the technological future of the French textile industry, all such considerations did not seem to have played any role. The only – or at least the decisive – market criterion applied in the selection process seems to have been the notion of the rich market and the assumption that the French textile industry should be its principal supplier. The rich market made selection easy. Such a market, with an accelerated product changeover and a wide diversification, was likely to favour computerised production processes – or as M. Aujac said 'processes that can be continuously and immediately adjusted as a function of new orders'. The example of fashion clothing was of course very plausible, and the prospect that one could have diversification and a quick change from fashion to fashion together with a reduction of cost between 15 and 20 per cent would be warmly welcomed by women and their husbands alike. But would the mass of consumers after 1984 really have the same taste for exclusiveness or diversification and for quick change from one fashion to another as did today's members of the upper classes? However, he was sure that in any case supply would create its own demand, quite in line with Professor Galbraith's proposition.

Professor Giersch went on to say that, among the technological developments that fitted the market criteria set out, M. Aujac and his collaborators had concentrated on major innovations. This required judgement on various quantitative issues: First: how much obsolescence would be caused? Second: how many people would become unemployed, and would have to be retrained? Third: how far would the new techniques favour the large firms as compared with their smaller competitors? An innovation which had sufficiently large effects in any direction would qualify as a major innovation and would remain in the list. M. Aujac was well aware that the whole exercise of drawing up a list of major innovations for a specific industry with seven phases clearly dated from each specific innovation process amounted to a tremendous exercise in collecting information of all sorts. It was not necessary to draw attention to all the questions to which one needed to know the answers; anybody would have sympathy for M. Aujac's complaint that the scanty information which was available was very often unsuitable for his purposes.

M. Aujac seemed, however, to be optimistic in pointing out that the largest French firms and the public authorities which formerly had not looked beyond the five-year horizon of the French plan were now showing more interest in long-term technological forecasting. Given the large gap between the need and the availability of information, M. Aujac was looking for a method that would maximise the flow of information even if it was poor information and inconsistent information. The method of interviewing correspondents and translating the answers into economic terms was the method adopted. It was, however, with moderate success only. The engineers who were contacted seemed to have been very reluctant to make an economic appraisal, but M. Aujac was confident that this difficulty would be overcome when R & D research workers had been induced to maintain closer contacts with the production and sales departments. At present the interviewer or forecaster himself had to judge the economic significance of the technical information received, and Professor Giersch said that M. Aujac had been doing it by confronting the engineers with one or possibly various alternative scenarios of long-term economic development.

The method which M. Aujac considered better and which he was going to try out and develop in the future was briefly described in the last two pages as a system of forecasting-orientated self-education. This resulted from bringing together for forecasting purposes groups of some ten people from the best French firms in the industry at each stage of production together with one or two full-time specialists in techno-economic forecasting. The latter would provide the group with estimates of the trend of final demand and with an account of the expected economic trends at the national and institutional level, including relevant changes in legislation. As a result of the exchange of information the members of the group were expected to arrive at a common forecast of the technological development by iteration and by means of the conceptual framework used in the table presented to the conference. Included in the group would be R & D researchers, production experts and salesmen, including, perhaps, representatives from large department stores and mail order houses. They were expected to develop a common language, a kind of techno-economic Esperanto. Whether or not they would talk if they came from competing firms is an open question. But M. Aujac in a private conversation with

Professor Giersch had said that competitive attitudes would not be a great obstacle under certain conditions: Firstly you would have to shift people's attention and concern from the short term to the long term and secondly you would have to confront them with a hypothesis of your own in order to evoke their objections and counter-arguments. In conclusion Professor Giersch said that he thought the task of technological forecasting was an important one and that M. Aujac's approach was well worth trying.

M. Aujac thanked Professor Giersch for his clear introduction of his paper and was pleased to learn that Professor Giersch might become a consultant for West German firms. M. Aujac said he had prepared his study by way of an exploration and in order to examine whether there were likely to be sudden changes in one industry, with major national repercussions, involving machinery becoming obsolete, problems arising of finances of firms and public authorities, and also problems of employment. He had been particularly interested in changes in structure because these might be necessary. M. Aujac said they had just happened to choose this branch because they had a well-informed assistant at the time. Their first problem had been to find how to approach the problem. In considering long-term studies they had found that even detailed input-output information was not enough. Technical progress brought improved productivity and changes of techniques and they had not found it possible to quantify them and draw up a useful table. They had come to the conclusion that what they needed were global hypotheses of various trends of the economy. Initially they thought they could have used inter-industry exchange tables coupled with a list of new products and processes, but they had found that analysis on the basis of new products was too vague. They had therefore attempted to look at the quality of demand that French industry would be satisfying, and it had seemed that if the textile industry were to remain alive it would have to supply a wealthy market with increasing diversification of production. Considering what particular technology would allow such diversification with low cost and rapid change they reached the conclusion that electronics could make it possible and that chemistry could produce almost any answer. They had tried to identify all innovations that seemed important to replace the range of raw material and produced the list found in the table. This was not exhaustive, but the problem was to achieve a realistic table. M. Aujac said they had then gone on to establish a dossier for each product and considered where technologies were available for each of these areas. For example, automatic spinning already existed. There had been a number of fully-automated plants in Japan since 1965. One thus had knowledge of the economic and technical features of these, and in placing them in the French context one was asking whether they would now be economic. In this case it was decided they would not. They had broken the process, therefore, into two parts: to automate the preparation of the fibre but not the actual spinning, and this had many advantages. Semi-automation meant 8–10 per cent reduction in costs. The equipment could be used for longer hours and there was a 15 per cent cut in manpower. This could be fully economic by about

1980. But by then another process might be developing, for example turbine spinning. In this field there was a Czechoslovak patent and it had been studied in many countries. If successful it would cut time and provide a 50 per cent manpower reduction, and cost could be cut by 25 per cent. But it was expensive equipment, approximately three times the cost of classical equipment at present. It also might have limited use, because one needed a high quality input. Hence it made it an advantage to use chemical fibres. The power costs were high and the capital requirement was large, but if the quality improved the costs worked out at a fairly economic level.

M. Aujac said they estimated that by 1980 turbine spinning would be ready to replace the amortised equipment of automated spinning. By 1980 turbines would be used for 10–20 per cent of total production, by 1985 about 50 per cent of total production, subject, of course, to the large margins of error inevitable in this sort of projection. They concluded, therefore, that if they were not mistaken there would be a considerable change in the textile industry which might be characteristic of changes in the economy as a whole. There was a lot more that could be said about this. There were other methods of operation, for example electrostatic equipment might be coming along, although this would take a long time and might not be in operation until about the year 2000.

M. Aujac said that they had tried to produce similar tables for other French industries. These had shown large and continuous changes, but no such sudden changes as in textiles. He concluded by saying that the work was worth what it was worth. They had submitted the table to all who helped them and there were numerous people involved. They hoped that a continuing dialogue would help to get over the inevitable difficulties of forecasting. He said that all concerned in technological forecasting began by speaking in different languages. But in the end he felt that they understood each other, and that forecasting depended on drawing experience from all parties and then turning the forecasts back to them for discussion.

Professor Oshima said that he had been very interested in the paper and impressed by the changes which were being introduced into French national planning, particularly in the areas of forecasting. He said the points concerned the Japanese economy as well, and that in future labour-intensive industries would have to become fully automated. They would also need to meet the requirements of markets not just by mass production but by versatility, and would have to go for higher added value. Professor Oshima asked why the textile industry was selected. Was it because the French textile industry was weak in relation to future competition and that the government was willing to give assistance to it? He also asked whether the forecast was a guideline for future government assistance.

He felt this was an attractive field for the future. Profits were declining in the chemical industry, and this group might be interested in moving into other fields. At the same time other groups would be prepared to join in with technology which was entirely new to this industry but which was highly developed in other applications. Finally he asked whether they were looking at the whole restructuring of industry in France or just the textile industry.

Mme Fardeau said that she agreed entirely on the importance of long-term forecasting. M. Aujac was very right to stress the need for forecasting where there was national or international competition. She asked how one could hope to get the necessary documentation together for technological change on the international scene, and also how one could mobilise experts in different industries who might be competing with each other? She said that the main difficulty was in finding a common language and also in achieving collaboration between different sources of information with competing interests. One was asking for their most precious and secret form of capital, their research and innovation plans. This was necessary, but she felt it was hardly likely to be obtained unless we had some form of social change, and she thought it might be expecting too much to expect French planning to make this attitude possible.

Professor Dupriez said that his one regret regarding the paper was its title. He would have much preferred the problem to have been posed in the abstract because the research which has been carried out had a fundamental application for the world at large and emphasised the fundamental nature of the methodologies and of the problems they were trying to overcome. He would like to congratulate M. Aujac on choosing textiles, as it was a very interesting area. He pointed out that in the earlier industrial revolution innovations in textiles had been first in the field, for example the flying shuttle was a major innovation. Today the textile industry was still developing its traditional tools. Yet future progress by new methods might well be very much greater. One might measure accomplished progress in hundreds of units of productivity of labour. No other sector had done this and yet was still faced with profound changes which were no longer speculation but ascertained facts. Industrialists were already involved. For example, turbine spinning had produced good results already on a semi-industrial plane. Professor Dupriez said that on account of capital costs there was a strong tendency to lower capital cost in industry by the widespread introduction of three or four shifts a day, and that socially one had to take account of these processes of change. We had also to consider the effects on the siting of industries. Professor Giersch had quoted the de Bandt report saying that the textile industry would have to move to less developed countries. But the discussions on progress in the textile industry show this might be the wrong decision. We should not give a cup of poison gift to developing countries. We could now represent textiles as an easy industry with low capitalisation. In this field, as in others, the siting would be where the consumption centres were, and not where raw materials existed. We might ask whether this was to condemn the less developed countries to not developing textiles. *Professor Papi* intervened at this point to ask whether account had been taken of the growth of population in the developing centres as well as in the developed countries.

Professor Robinson called in question whether, in the European Economic Community with increased openness of the economy, the indicative planning of France could or could not work. He said he would not want to kill French planning because of the E.E.C. but only to look at the change of degree. Professor Robinson wanted to distinguish between separate questions, both of which he felt were important. Firstly, how many million square metres of textiles were the French going to produce? Secondly, for each million square metres of textiles, what were the future inputs going to be? He pointed out that the answer to the second did not depend on the answer to the first question. One must not think that one should not try to answer the second question if one could not answer the first. He had tried to use input-output techniques in developing countries and had been horrified by the difficulties of projecting them. Undynamised input-output tables were more dangerous than nothing. But the question was how should one dynamise. He said that he had seen many examples of statisticians doing this by extrapolating technical progress. This he felt was hopeless, and he was therefore grateful to M. Aujac for dynamising input-output tables by rational technical methods. But were the results realistic? Mme Fardeau had said that we could not get access to technical information. Professor Robinson believed there was plenty of information available which was public. It was often more important to know what the machine producers were doing than the users of machines, and they would often talk more freely because they wanted people to get information about the machines in advance of them becoming available. Finally, Professor Robinson asked whether the results were likely to be made nonsense because of competition. He said competition could reduce the millions of square metres made by one firm and increase those made by another firm and alter the relative strengths of firms which were in competition. But he thought the results were still useful. One could use them as indicators of the problems facing the machinery making industry and also of the problems which the country would have with manpower adjustment. And one could see what R & D activity was needed to maintain an industry in the situation that we foresee for it in the future.

Professor Delivanis said that he believed M. Aujac had been over-optimistic in his estimate of the possibilities of forecasting. He particularly believed M. Aujac should give us more information, firstly on the possibility of forecasting for more than five years ahead when a well-known British economist had rightly said that such forecasts were not worth the paper on which they are written; secondly, on the uncertain position of the firms which were assumed to try to maximise their profits, to increase their reserves, to reduce their dividends and to influence demand; thirdly, on the testing by the market of new products, inasmuch as the latter could occur only when the new product would be available; fourthly, on the possibility of inducing the various firms to part with the necessary information; fifthly, on the possibility of linking economic and technical aspects of forecasting in the way that he was expecting to do.

Professor Marchal said that he was most interested by M. Aujac's forecasting and would like to ask a few questions. Firstly it would seem that the development he contemplated presupposed that financing would be available for the technical transformation, either by self-financing, or by credits, or by subsidies from the state budget. According to the rate at which they could be obtained the dates could be accelerated or slowed down. People would not necessarily wait until machinery became completely obsolete to replace it if credit was available, but if finance was not available change would be delayed. There were not only the problems of the means of financing, but the social problem of transferring the excess manpower into other branches of industry and also re-training it in new skills. Again the policy followed by central government might slow down the process of transformation or speed it up, and of course in these respects we could not study French industry in isolation from foreign industry. Professor Marchal thought that perhaps we should really consider forecasting as described in this paper as conditional – subject, that is, to finance being assumed to be available, social conditions being changed in a certain way, and so on. But this raised the big question of the basis on which M. Aujac had concluded that these changes would be possible on the time-scales envisaged.

Professor Simai said that he had read the paper with great interest as they had a very similar position in the Hungarian textile industry. He therefore asked to what extent in the French technological forecasting socio-economic considerations had been included and also to what extent in other areas technological forecasting had achieved results similar or otherwise to those in the paper.

Professor Williams intervened to say that M. Aujac in a different institutional context and for one small sector had been talking about the same issues as were raised on other papers by Professors Khachaturov and Gatovskiy. We must not think that we were in different worlds.

M. Aujac thanked all those who had contributed to the discussion. He said that the role of this forecasting in French planning should not be over-estimated. If it was regarded as worth while the plan would take account of it. The study had been prepared at the request of the planners and the planning department might or might not agree with it. In the present context they could only cover five years with input-output tables and even five years was hardly credible. The question was what should take their place? They had tried this alternative approach which they had found useful. On the question of how many million square metres would be produced M. Aujac said they did not know and they had not in fact asked. They had been thinking only of the wealthy markets. There would be other producers but they did not know what their share would be. The aim of the planning was to determine what might be needed and to guide firms in the country. Industries needed to keep up with technical progress in other countries and we should provide them with warnings about what things might happen. The choice was then left up to the firms and the managers to make for themselves. We had to prepare them as well as we could and hope that they could stand up to the competition. All reports would be published and if, for example, the Belgians found them to be any use this was all to the good. M. Aujac said that the forecast consisted essentially of two parts. The first part, the technical part, was valid for all centres in the world and was available to all countries anyway in the form of publications and widely disseminated scientific information. They had

had the assistance of the Institut Textile de France which had a number of highly qualified engineers who had gone through the documentation from a number of major sources. He said the producers of equipment were only too keen to talk, and again this was generally available to everyone in the world. This provided information about processes which might be contemplated in the future and also those which might die out, and therefore this part of the report was international. The second part of the report related to specific conditions applicable to their own country, and was aimed at providing information which would help such problems as those of financing to be sorted out, whether the industry was aiming to be self-financing or whether there was to be support from outside. M. Aujac said that it was interesting to ask whether the plan would change the structure. The planning department has no real executive powers in France. But it was clear that the chemical industry, for instance, would very probably integrate large parts of the textile industry, and it would probably form the basis of many of the changes which would take place in the future. The textile industry and the government must be aware of this. M. Aujac said that secrecy did exist at the level of know-how, but we did not need to know that this or that tool will be used in such and such a way. We needed to know roughly the costs and lifetime of the materials. Secrecy did not exist as soon as you could show people that you knew the secrets. If you put the information in a general framework and gave it to them in that form they would tend to co-operate more. Finally M. Aujac said that the government might make decisions on the basis of the evidence brought out in such a study and push forward certain technological developments. But without progress in the market this would be of no use. One needed to be one, not two steps in advance. This was an open model, in the sense that changes, for example in the E.E.C., might bring dates forward or back and the tables had to be kept up to date. He maintained that people from different sources could pool their resources, and that was in fact what they were doing. Otherwise we were in the land of crystal-gazing. We had to mobilise the experience of people speaking different languages and this was at the heart of the whole question of planning. The great point was that we should all be working together towards the same objectives, allowing the maximum freedom and diversity, and this was possible within this framework.

5   Estimating the National Economic Effect of a New Technology

L. M. Gatovskiy

INSTITUTE OF ECONOMICS, MOSCOW

I. THE PROBLEMS

As in the case of the efficiency of production taken as a whole, the efficiency of any technology with given targets of production takes the form of progressive reduction of costs. Essentially this means reduction of specific costs (the total of current and capital costs appropriately defined) achieved by the application of the new technology and measured by cost per unit of production using the new technology. Any raising of the quality of the product is assumed to be reflected in its price.

Since the definition of the effect of the new technology is the total reduction of specific current and related capital costs, it follows inevitably that existing costs serve as a starting point for planning any particular type of new and more efficient technology. The economic effect of the new technology as compared with the existing one can be expressed in the first place by means of the formula of costs (relating to a certain specific year) used to compare alternative capital investments:

$$NZ = C_j + E_n K_J = \min \tag{1}$$

where NZ = resulting costs
C = current costs
K = capital investment
E_n = planned norm of effectiveness of capital investment

Preference is given to the variant with the minimal value of existing cost. The efficiency of the new technology as compared with the existing one can be expressed as $PZ^n \leqslant PZ^s$. That is, it must yield such a condition that the value of costs with the new technology is equal to or lower than the costs with the old technology.

The ultimate saving derived from the new technology represents the lowering of costs per unit of production made with the help of that technology. This is the effect contributed to the national economy by the utilisation (the productive consumption) of technology.

The reduction of costs (even if some incentive has been required) in an enterprise producing technological equipment – assuming it to have normal profitability – is reflected in a reduction in the price of the new technology, which in turn will reduce the capital costs of its user. Thus it follows that cost saving by the producer is taken into account in the final national economic effect of employing the new technology.

The advantage derived by the consumers of the new technology already takes account of the advantage obtained by the producers, as reflected in costs and results. The volume of production costs of the new technology producers is reflected (on the assumption of the normal profitability level) in the price for this technology when bought by enterprise-consumers, and thus in their capital costs.

It goes without saying that such calculation of the final national economic effect does not imply any underestimation of the importance of planning or of measures to encourage the cost-reduction and to improve the economic performance simultaneously with the adoption of a new technology of production by the relevant industries, combinations and enterprises. The quality of management, the structure of production costs, the capital investment and capital output ratio, the material input per unit of output, the labour productivity, the profits, the return from producing the new technology are all of immense and quite independent importance. But a quite different problem arises in this case – to evaluate the economic performance of the relevant specific industries rather than to define the over-all national economic effect of the new technology applied in these particular industries.

II. THE DISTINCTION BETWEEN PROFITS AND TOTAL BENEFIT

It is necessary to distinguish the total of profits created as a result of the output and consumption of the new technology in the national economy as a whole from the total of the final national economic effect. By the latter we mean the complete effect attributable to the whole range of the use of new technology, whereas the effect on its producers is only reflected (through the price of the technical equipment) in the effect for them of the utilisation of this technology. The repercussions on the final national economic effect are also of great importance. Apart from this progressive reaction through costs, three other basic elements contribute to the total of the technological effect. Firstly, the total of the national economic effect can be so calculated that the estimate of the effect brought by a new technology covers not one single industry only but includes all industries (as well as all enterprises in the given industry) in which this technology is used. Thus, having estimated the total planned effect of the technology in this way for all industries, we can then use it for purposes of comparison in the selection of the most effective technology to be used in the national economy. And on the basis of such selection we can plan, estimate, compare and provide incentives for the effective utilisation of the selected technology in each enterprise in each particular industry.

Secondly, the total of the national economic effect can be so defined that it is measured not by one embodiment of the technology (related to one particular job), but by the total of the technology developed in a given industry, and finally by the whole trend of technological development, namely by all technology developed in the whole national economy.

Thirdly, the total of the national economic effect is a concept that requires to be measured not by the technology produced and applied during a single year but extending over a whole period of time, defined either as the whole period of service of the given technological asset (its yearly output) or as the whole period of production and employment of all units of a certain model (all years of production of such models and their whole period of use); as the whole period of production and employment of models based on the given technological principle; as the whole fixed period over which technological concepts embodying a given technology remain in use.

An attempt to measure the effect of a new technology over a period of time implies that the national economic effect is defined neither exclusively nor primarily as the current effect, but rather as the perspective effect corresponding to the character of the technological development. It is of particular importance in the sphere of scientific and technological progress to ensure the decisive role of perspective planning (including long-term plans). This makes obligatory the use of long-term forecasting.

III. THE TOTAL EFFECT OF A GIVEN INNOVATION

The ultimate effect of a new technology (∃), defined as the difference between the values (measured as the reduction of costs) between its consumer effect and the additional capital investment required during the whole planned service period of this technology can be expressed by the following formula (specific values being measured per unit of production with the new technology):

$$\exists y' = \sum_{t'=1}^{t'=T'} (\exists p t' - \Delta K t') \alpha t' \ldots, \tag{2}$$

where $\exists y'$ = the complete economic effect of the given technology, as estimated for the whole sphere of its application during the whole period of its use, the effect be related to year t;

T' = the planned period of use of the new technology $t' = 1, 2, 3, \ldots T'$;

$\exists pt'$ = the consumer effect of the new technology in year t of its use;

$\Delta Kt'$ = the additional capital investment required for the use of the new technology in year t of its use.

The rate of discount to bring costs occurring at different dates to the initial year (the first year of utilisation of the new technology) can be shown as

$$\alpha_{t'} = \frac{1}{(1 + E_H)^{T'}} \tag{3}$$

The value of the consumer effect in the form of reduction of current operational costs as the result of utilisation of the new technology can be shown as follows:

$$\exists T' = \sum_{t'=1}^{t'=T'} (\Delta I_{t'} - \Delta K_{t'}) \alpha t' \tag{4}$$

where $\Delta I_{t'}$ = the change of operation costs for users of the new technology in year $-t$ of the period of service excluding amortisation costs.

If one attempts perspective accounting, one must take into consideration that the normative rate of discount itself must be expected to undergo change depending on the changing conditions – including the relation between the value of capital investment resources and demands. This normative rate of discount, therefore, should be considered as function of time, that is as $E_H = f(t)$.

Since the complete effect is being considered for the whole period it makes it necessary to consider current costs without amortisation for purposes of replacement. This is done to avoid double counting of current costs and capital costs. Now, if we do not estimate the complete effect of new technology for the whole planned service period but take only a single year of the period – for instance the first year – then the change in the capital investments (ΔK) for a year is taken to be normative rate of discount (E_n) and the consumer effect is thus equal to the change of the whole sum of current costs with amortisation included. In this case there will be no double-counting of amortisation and capital costs. And then the formula will be:

$$\exists = \Delta s - E_n \Delta K \ldots, \tag{5}$$

where ΔS = change of current costs for one year with the total of yearly amortisation included.

There is a direct and logical relation between the calculation of the effect of the new technology in terms of time, in terms of its sphere of production and in terms of its sphere of consumption.

IV. THE INTERDEPENDENCE OF THE EFFECTS OF TIME AND SPHERE OF APPLICATION

The following interdependence can be found between these aspects of estimating the effect:

(a) the more fundamental is the process of technological innovation, the longer should be the period considered for calculation of its effects;

(b) the longer is this period, the broader is the relevant range both in forms of the new technology and, consequently, in terms of industries of its application; and the broader is this range, the larger should be the value indicators used for the purposes of technology planning and forecasting technology.

When change of the yearly outputs due to a new technology is under discussion within the limits of one and the same model, differences in technology are relatively inconsiderable and the whole period of production of a new model varies normally within a few years (on average up to ten years). Planning of such a model is limited within the comparatively narrow range of its varieties. At the same time this type of planning is performed in detail since particular sets of technological totals have to be taken into account in a single model.

When we consider the process of change from one model to another within a given level of technology, the complexity of technological estimation increases considerably with the length of the time period, since each level of technology lasts much longer than any particular model. Production planning and the utilisation of models belonging to a given technological level now cover a considerably broader range of technology vertically (representing parallel and simultaneous existence and development of various models belonging to a given level of technology) and horizontally (implying change of different models over time). Planning preference is inevitably given to proposals showing higher indicators (those for models and not for varieties of models).

When change of levels of technology is being planned or forecast within a given technological trend that is already considerable, so that major changes in technology are in sight, the length of the planned and forecast periods needs to be increased still further. The number of models greatly increases both vertically and horizontally: *vertically* – there is found parallel and simultaneous existence (at various stages of development) of levels of technology of production related to the given trend (e.g. in electro-energetics); it must be borne in mind that each level of technology is embodied in a multitude of simultaneously existing models; *horizontally* – in time within a given long period there is taking place a change of the levels of technology, i.e. there are a multitude of models based on different technological principles.

Planning and forecasting must, however, inevitably be carried out in less detail (individual models are handled in groups – by enlarged group indicators).

It is only when scientific and technological progress is considered as a whole complex, taking into account all of its trends, and involving considerable amalgamation of the indicators, that it becomes possible to define certain trends in this complex of the great multitude of technological applications belonging to different levels and spheres of utilisation, and that it becomes possible to determine the optimal structural changes both internally in particular industries, and of an inter-industrial character.

The range of the technology produced as well as the spheres of its application depend on the goals of planning at all levels including the national economy, the group of industries, the individual industry, the region, a technological enterprise or a group of enterprises.

Modern computers enable us to construct models of the most sophisticated interrelations, analysing each of the three factors in technology development: production, consumption and time.

The analysis of the total effect of the new technology can be shown as:

$$\exists = \sum_{j=1}^{i=J} \sum_{l=1}^{e=l} \sum_{t'=1}^{t'=T'} \exists jlt', \qquad (6)$$

where $\exists$ = the effect measured by all three aspects.

The effect measured by industries in which this technology or model is used is expressed as:

$$\exists^{J} = \sum_{j=1}^{j=J} \exists j,$$

where j = industry in which the new technology is used and $j = 1, 2, 3, \ldots J$.

The effect measured by types of technology developed is shown as

$$\exists^L = \sum_{L=1}^{l=L} \exists_l$$

where l = a type of new technology, $l = 1, 2, 3, \ldots L$.

And let the total measured in terms of time during which the new technology has been used be:

$$\exists^{T'} = \sum_{t'=1}^{t'=T'} \exists_{t'}$$

where t' = the years of use of the technology, $t' = 1, 2, 3, \ldots T'$.

The extent to which the analysis of the effect of the new technology is complete depends on the quantitative values of J, L and T'.

Thus, the analysis by industries in which new technology is in use can be made either with a broad or with a narrow definition of those industries. The extent of the coverage also depends on the extent to which examples of the new technology are included (L) as well as the length of the period over which the new technology T' is in use.

V. SOME VARIANTS

May I examine in greater detail some variants in the calculation of the effect in terms of the three aspects mentioned above.

Let us take as our starting point the formula of the effect measured by the annual output of technology in the whole sphere of its utilisation (by the total of the relevant enterprises and industries).

$$\exists^a = \sum_{j=1}^{j=J} \exists_j a_j, \ldots, \tag{7}$$

where $\exists_j$ = the effect of one unit in industry j

$\qquad a_j$ = the annual output of units produced for use in industry j.

Here the output of the technology is limited to the output of a single model, and the period of output to one year. Calculation in this case is made only in terms of industries in which this technology is in use. But since only one unit is considered during one year of output and utilisation, the sphere of application of the new technology as measured by such a formula is rather limited. All the values J, L and T' in this case are characteristic of very limited quantitative significance.

This formula neglects the total calculation through the lifetime of the model: the period of output and the period of service of all units relating to the given model; the output and service period of all models based on the given technology; the output and service period of all models belonging to basically different levels within the given technological trend; the output and service period of all models belonging to all levels and trends of technology, employed within the national economy.

Let us consider all these variables and the consequential broadening of the calculation of the effect, which progressively enables this process to achieve a full coverage of the national economy.

The formula for the effect of a given model ($\exists a$) measured not for a single year of its utilisation but for the whole service period (in all the relevant industries of its application) will appear as follows:[1]

$$\exists^{T'} = \sum_{t'=1}^{t'=T'} \exists t', \ldots \tag{8}$$

where T' = the period of use of the model
$\quad t' = 1, 2, 3, \ldots T.$

In this case the calculation of the effect will already include two aspects and not merely one: the time aspect (the whole service period of a machine is taken into account) and the sphere of application of the technology (since a long period of time is taken into consideration, which extends the number of industries and enterprises using this technology).

The next broad calculating formula considers the effect not by individual machines (the annual output) but by all annual outputs of the given model. The calculation of the effect in this case, therefore, is increased in all three dimensions: in time, because the period of output of the technology under consideration is increased and the period of application is successively estimated for all machines produced of the given model; the effect in terms of forms of the technology includes a number of models (horizontally and vertically) which form varieties of a model and not a single individual model; by spheres of the application of the technology (also horizontally and vertically) since there is an increase in the range of the forms of technology and in the period of their application.

The formula appears as:

$$\exists(T) = \sum_{t=1}^{T} \exists_{t'} {}^{T'} \ldots \tag{9}$$

where T = period of output of the technology (the total of annual outputs of the given model) $t' = 1, 2, 3, \ldots T.$

Let us extend further the calculation of the effect in all three aspects. It can be expressed in the following formula, which measures

[1] To cover the case in which the production of a given unit of equipment takes more than a single year, the symbol a denotes this period of time.

the effect for all models built on one and the same technological principle, and within a given trend of technological development:

$$\exists^y = \sum_{y=1}^{y=Y} \exists_y^{T'} (T), \ldots \tag{10}$$

where $y =$ the given technological level or principle, $y = 1, 2, 3 \ldots Y$.

Here, the number of embodiments of the technology produced which have to be taken into account greatly increases both vertically (the parallel operation 'in perspective' of a number of models belonging to the given technological principle) and horizontally in time (a change in the number of models based on a single technological principle). The length of the accounting period will also greatly increase because consideration will now have to be given not to the period of output and utilisation of one particular model but to the total of all such periods based on a given technological principle. And this results in a quite considerable extension of the sphere to be taken into account in which the technology is in use (the number of industries and enterprises).

VI. SUCCESSIVE TECHNOLOGICAL LEVELS

A still further extension of the estimated effect of a new technology is shown by the formula which measures this effect not in terms of one particular technological principle but in terms of a number of successive technological levels included in a single trend of technological development (e.g. electro-energetics):

$$\exists^m = \sum_{m=1}^{m=M} \exists_m \quad , \ldots, \tag{11}$$

where $M =$ technological levels involving differences of principle change of which is considered in the given fixed long-term plan or forecast for the time-period concerned. $m = 1, 2, 3 \ldots, M$.

It is clear that this formula involves considerable enlargement of the estimated scale of the effect of a new technology: in respect of time because consideration is now given to covering a number of periods including models belonging not to a single particular technological principle but to several of them; and in respect of types of technology produced both horizontally (a change of technological levels) and vertically (the simultaneous parallel operation of models belonging to different technological levels); and in consequence we have a huge extension of the estimated sphere in which a new technology is in use. The only limitation, as compared with covering the whole national economy, is that we are taking into account one trend of new technological development and not all trends.

VII. THE WHOLE TREND OF TECHNOLOGY

This illustrates the successive transition from the measurement of the effect in terms of a single model to its measurement in terms of a single technological principle and finally its measurement in terms of the whole trend of technological development. Analysis of each of these stages is of great importance, because each of them is characterised by specific economic principles which will be examined below.

Let us now consider the formula which gives the most complete calculation of the effect of the new technology in terms of the whole national economy.

This formula estimates the effect not only in terms of some particular trend of technological development but in terms of all such trends:

$$\exists^H = \sum_{h=1}^{h=H} \exists_H^M, \ldots, \tag{12}$$

where H = the trend of technological development
$h = 1, 2, 3, \ldots, H$.

In this case the estimate of the effect of the new technology is greatly expanded both in terms of types of technology and in terms of spheres of application. At the same time we obtain the total estimate of the aggregate national economic effect.

In this final formula measuring the effect, the estimate in terms of all aspects of all forms of a technology under production (L) in all spheres of application (J) in all relevant time periods (T'), leads to a highest estimate of the effect.

VIII. OPTIMISATION OVER A PERIOD OF TIME

The causal factor which led to the progressive increase of the estimate resulting from these formulae was principally the estimates in terms of time-periods and in terms of forms of application of the technology and the further transition to widening the sphere of application of the technology.

As the initial starting point one can take the consequences of the extension of the sphere in which the new technology is in use and then move on to estimating the effect in terms of time-periods and in forms of the technology. But whatever the initial starting point we need a complex approach which will reflect the interrelations between all these three aspects, thus enabling one to combine the analysis of the effect with dynamic balance calculations, including the perspective inter-industry balances.

To analyse the logical trends of scientific-technological progress, an optimisation of the possible effect through time is essential (including any change of technological principles). But it is comparatively seldom that this long-term, horizontal aspect of the analysis (in time) is needed. For many purposes, and particularly for purposes of medium-term and short-term planning, the calculation in terms of time is limited, for example, to a five-year period and the principal attention is transferred to a more realistic estimation of the vertical effect, in terms of forms of technology and in terms of industries of application, taking account also of the problems of regions and responsible ministries.

It goes without saying that such a national economic approach to the analysis of the effect of the new technology is far from being confined merely to such calculation of the aggregate economic results. By definition it requires that in any estimation in terms of forms of a technology, all industries in which this technology is employed have to be taken into consideration, and in addition it is necessary to estimate in terms of time-periods which enable one to measure not only the current but also the perspective (including long-term) effect which is of fundamental importance for the national economy.

The national economic approach can also be calculated by means of inter-industry analyses of the productivity of models, technological levels, and trends of technological development, involving comparative economic estimates and inter-comparisons in respect of all these aspects, with progressive allowance of the possibilities of mutual substitution of products.

Only a set of economic estimates of the comparative effectiveness of alternative lines of technological progress and their aggregate indicators for the whole national economy, calculated in terms of the long-term perspective, is capable of providing such a systematic national economic approach as will make it possible to co-ordinate technological policy with the basic structural changes in social production as a whole.

Economic estimates of the effects derived from some line of technology, from existing levels of technology, and even more so from particular models, even though directly relevant to any with analysis of intra- and inter-industry proportions, are of but limited importance.

The optimal solution possible of regional and individual industry problems of scientific-technological progress in given conditions can only be discovered by considering the national economic optimum.

IX. THE PLANNING OF TECHNOLOGICAL PROGRESS

The starting point for determining the complete national economic effect of a new technology must be the degree of structural optimisation of the national economy (the national economic ratios being used as the basis for the internal industry ratios) which is fully consistent with optimal rates of technological economic development of social production and the national industries.

On the basis of such optimisation of the ratios and rates of national economic development, as revealed in the relevant balance calculations, the optimal policies (resulting in the maximisation of the effect) for scientific and technological development, designed to achieve these optimal ratios and rates (with given possibilities for optimisation) are defined on the basis of the national economic effectiveness criterion for new technology by equation (6)

$$\boldsymbol{\ni} = \sum_{j=1}^{j=J} \sum_{l=1}^{l=L} \sum_{t'=1}^{t'=T'} \boldsymbol{\ni} jlt'$$

The leading role of scientific and technological progress as a factor determining the requirements and possibilities of economic development implies that the estimate of the national economic effect of new technology becomes one of the primary elements in the optimisation of the whole of social production. There is an inevitable interaction, derived by iteration, between the planning of the optimal national economic development and the optimisation in the sphere of scientific-technological progress.

Thus, the planning of economically applicable scientific and technological progress is both based on economic planning and at the same time appears an important factor in such economic planning. The basis for estimating the national economic effect of a new technology will be changing in accordance with the changes in the criteria for the national economic optimum. And at the same time we have to estimate this effect with the help of the existing initial data for national economic optimisation. At present a limited optimisation of production is adopted within an industry. For instance, we take planned indicators of rates and structures in an industry as our starting point and then solve the problem of how to achieve these indicators in the most economical way. The national economic optimum under consideration is still rather imperfectly attained, which explains the fact that the national economic criterion, applied to the given industry, is limited by the assumed conditions, which in turn are derived from the achieved level of partial optimisation by the industry. But in any case any industry-wide optimisation of the plan for scientific and technological progress must be based on the national economic criterion of effectiveness of this given technology.

X. THE RELATION OF SCIENCE POLICY TO ECONOMIC POLICY

Thus, the planning of an economically applicable technology must be based on the whole national economic plan which (taking into account planned conditions and results of the scientific and technological progress) uses balance accounts for the total dynamic co-ordination of demands and resources. It is the intention that the criteria by which the national economic effect of new technology is assessed (when applied to scientific and technological progress) shall discover the most economical means of achieving the objectives of socialist production, the growth of national welfare through greater saving of costs, which in its turn must influence the final calculation of rates and proportions in the national, industrial and regional plans for economic development.

The complete effect of the particular type of new technology in the particular industry is found through application of the national economic criterion for the effectiveness of this technology when applied to appropriate industry production plans which have been modified to take account of this criterion.

As time goes on, the national economic criterion of the effectiveness of a new technology will be taken increasingly into account in the national economic planning in the determination of economic proportions and the fixing of prices. Allocation of resources to one or another industry must, naturally, be made in terms of a forecast regarding the attainment of a given result and compared with a number of similar long-term forecasts. To put it differently, it is necessary to consider the likelihood that investments in the given industry will ultimately yield society, through inter-industry relations, the expected growth of national and net income.

Any far-reaching approach to scientific-technological progress implies a complete co-ordination of scientific-technological and economic decisions designed to achieve that development of new technology which will lead to a considerable improvement in economy of labour (covering both the reduction of capital and of current costs). This co-ordination includes the following basic elements. In the early stages there must be: economic estimation of the trends of scientific and technological progress; economic estimates of technological indicators, based on the calculated potential national economic effect (which is expected on the basis of existing actual forms of technology); selection of its production into the general production plans.

At the next stage it is necessary to solve the many difficult problems of creating the economic conditions in which potential effects can be converted into actual effects, that is to achieve in practice the national economic effect through utilisation of the new technology. This covers the creation of the necessary production capacities, the allocation of capital investment, and of the financial resources required, the development of the required production in the interconnected industries and enterprises, the punctual preparation of material and technological facilities for inputs and outputs, the training of the necessary personnel, and so on.

And, finally, it is necessary to ensure that the new technology as well as being economically effective from the point of view of the national economy shall also, in accordance with the principles of the recent economic reforms, prove economically effective for its producers and consumers: that is that the national economic effect shall be reflected in the profits of enterprises in accordance with the principles that 'everything that serves the interests of the national economy should be profitable for efficiently run enterprises'.

This requires a complicated implementation of the whole system of economic incentives and instruments (planned indicators, norms, prices, finance, credits, incentive funds, and the rest) on the basis of reconciling the interests of the national economy with those of enterprises, with priority attached to the former.

From this follows the system of distributing the total of additional profit from the new technology between producers, consumers (at the levels of the enterprise, the group of co-operating enterprises, and the industry) and the state budget.

The problem here is to enable producers and consumers of a new technology beneficial to the national economy (provided it works properly) to derive greater profits, as compared with those they would receive if they used the old technology, during the whole period of output and utilisation of the new technology down to its obsolescence, including the receipt of higher profits during the period of introducing the new technology into production. This is achieved, in particular, by the system of compensation required during this initial period of considerably added costs on condition that the additional profits received after the introduction period shall become a source for refunding the initial compensation.

This all implies that the national economic interests of the society as a whole are reconciled with the interests of enterprises in such a way that the scientific-technological progress can be accelerated.

Discussion of the Paper by Professor Gatovskiy

Professor Rasmussen, in opening the discussion, started out by drawing attention to a number of linguistic problems raised in the mind of an ordinary reader, many words being used in different senses from the ordinary.[1] The starting point was that improvements in technology meant saving of costs. This of course was true in a number of cases but might not be true in all cases, certainly not in those cases in which an innovation meant the introduction of a commodity or service which previously had been completely non-existent. But that was a minor, partly terminological question, of the construction of index numbers. More important, the essence of the paper was a discussion on how to measure for the nation as a whole the total benefits of potential innovations. Three problems arose. Firstly a given invention might be beneficial to a number of different industries. In evaluating the consequences, this became important in a disaggregated world *à la* Leontief or in the Soviet world of balance sheets. Secondly, a given industry would benefit from a number of different innovations. These two considerations of course immediately pointed towards a kind of input-output table and perhaps a related model. Professor Rasmussen suspected it would be quite easy to construct such a type of model by adding to the flow matrix of the ordinary model another flow matrix – 'below the line' – showing the flows of R & D from different R & D activities to the receiving industries, much in the same way as one could specify the labour input into different categories. He wondered whether these lines of thought had been considered by Professor Gatovskiy? The main difficulty, he suspected, would be that the basically weak point in the ordinary Leontief model, the assumption about fixed technical co-efficients now became almost ridiculous, since the essence of the R & D flows should be to change these coefficients. That, however, one could visualise being taken care of in some kind of systematic way. A number of other questions arose: just to mention one, the problem of the lags. The third problem arose because an innovation progressed over a number of years; this was the problem of time.

In evaluating potential technologies there were thus three levels of aggregation: aggregating each type of technology over industries; aggregating the different technologies; aggregating over time. Professor Rasmussen saw these issues as the main lines of the paper. He said that the time problem was slightly puzzling. It was indicated that this involved a 'time-factor' (α), when capitalizing the benefits and costs when,

$$\alpha_t = \frac{1}{(1 + E_H)^t}$$

[1] In fairness to Professor Rasmussen and other discussants a reader should be told that the version of Professor Gatovskiy's paper printed in this volume is considerably revised and rewritten from the English version available at the conference, which at a large number of points was obscure and difficult to follow. [Ed.]

and when Professor Gatovskiy informed us that

$$E_H = \frac{1}{T},$$

T being the 'payoff period'. (For some of the formulae the time factor seemed to have disappeared.)

It was inevitable that this raised the very old discussion between Soviet and western economists about the role of the rate of interest in planning. It might be unfair to reopen the discussion, but why hide the rate of interest in the determination of T (the payoff period) as many business firms did also in the Western World? *Inter alia*, it would also be interesting to know more about the time function of E_H. Was the payoff period an increasing or decreasing function of time? One might ask Professor Gatovskiy why he put such effort into this whole aggregation. Why not let each R & D project be judged on its own merit? The answer, Professor Rasmussen said, was that you would still have to perform the first kind of aggregation he had referred to, that over industries. Furthermore, on the overall planning level, you would have to decide on the share of G.N.P. going into R & D. And for that reason he could well imagine that some kind of operation such as that visualised by the author might be helpful, though more refined methods might come to one's mind. In this connection it might be added that no hint was given in the paper of the empirical difficulties. How were all these benefits going to be measured? We might usefully compare the series of difficulties presented by Professor Griliches. Moreover, the relations to the pricing system are not explained. Did we visualise all prices reflecting changes in technology – that they would be the shadow prices of a perfect market system? Towards the end of his paper, Professor Gatovskiy had mentioned the problem of how to induce individual plants to use the innovations found optimal at the overall planning level. This advice seemed to be akin to that well known in the non-socialist world: let profits be the temptation. It was interesting to meet the sentence 'Everything which serves the interest of the national economy should be profitable for efficiently run enterprises.' (This was in a way the inverted version of the saying that 'what is good for General Motors is good for America'.) Was that Libermanism? If so, Professor Rasmussen assumed that it would have to be carried into effect by way of the pricing system.

Professor Delivanis said there were two points he would like to stress. Firstly, Professor Gatovskiy had indicated that the longer and the more any new technology was applied, the greater was the result achieved. There was no doubt about this relationship, but one had to bear in mind that new technology was continually appearing and that we had to find when it paid from the general point of view to retain the old technology and when it paid to apply the new one. A great number of points had to be considered and examined before reaching the right decisions. Secondly, Professor Gatovskiy believed that the greatest share of the benefit derived from new technology had accrued to the treasury and not to the consumers or the producers. Professor Delvanis could not agree with this view, as the importance of technical progress was to be judged on the basis of the improvement of the living conditions of the poorest people and of all those whose lot was not considered to be satisfactory. The view of Professor Gatovskiy would apply in times of scarcity.

Professor Robinson said he would like to ask Professor Gatovskiy a few questions about the text, which he was not quite sure that he had fully understood. He said there were two separate problems – firstly the choice of techniques with existing and known technology. Here the technical performance of that technology and the probable capital cost of that technology could be established. However this was not easy if the technology had not been used in the U.S.S.R. before, and he asked what information they used to evaluate the techniques in terms of performance and capital costs. Professor Robinson said that working on this problem in developing countries, he had been able to use the material provided by the U.N. on the cost and experience of installing and operating capital equipment in different places, ensuring, of course, that this information was adapted to the country concerned and took into account the differences in capital construction in different countries. The problem was more difficult and interesting when one moved outside the field of existing and well-known technology, and where one was attempting to forecast potential savings resulting from an, at present, non-existent technology. In this situation one did not know the performance or capital costs of creating the equipment. Professor Robinson asked whether the U.S.S.R. was not in this position? He went on to ask how much did it pay to spend on R & D and how did one decide where to put money into developing technology in these more unknown areas? He suggested that this was an extremely practical problem of great importance. How did the U.S.S.R. make decisions about how much to spend on different branches of industry and how did they get these forecasts? Did engineers handle the assessment problems accurately enough for the calculations to be worth making?

Professor Gatovskiy said that on the question of whether it was possible to evaluate the effects of new technology on the whole of the national economy he hoped that most or even all of the participants would give a positive answer, as it was an essential question for economics. He suggested that it should be done by direct calculation, and by this he meant that we should evaluate the cost of each type of technique and the impact of each use, taking account of the consequences of the impact on all branches using the new technique and all categories of product to which it was applied. If one compared the contribution of the national economic effect of research to the growth of the national income with the total change in productivity, then the part played by technical progress in increasing economic production could be evaluated approximately at 55 or 60 per cent. This was the proportion they adopted in the U.S.S.R. They might be mistaken, but it was very different from the 0.16 that Professor Griliches seemed to suggest. Taxes had not been taken into account. If the state received these funds, this was a consequence of the way in which the state budget was distributed.

Professor Gatovskiy said he thought Professor Griliches was right in saying that there were conditions under which they should take the net effect. But their view was different. For national planning their view was more realistic. He did not say that the state should receive the greatest part. He said the effect was the profit derived by branches using and producing the technique. They might make use of the profit, but a great part of it remained with the firm. As regards the state, it used some of its share for basic sciences. On the question of prices, Professor Rasmussen was right. They could not evaluate the effects without assuming the price policy. But this was not the occasion to review the price theory in the U.S.S.R. Empirically it could be said that prices should guarantee sufficient normal profit for these firms which worked normally. The productivity of the firms should be improved and should correspond to the needs of the population. The firms should lower their own costs of production. They used the lever of prices to force firms to reduce their costs, and the period of assimilation of new techniques could be shortened and diffusion facilitated. On Professor Rasmussen's comment that what was profitable for firms should be profitable for the national economy, Professor Gatovskiy said he did not know whether Professor Rasmussen agreed or disagreed with his own quotation. But the slogan Professor Rasmussen quoted was the antithesis of what they said in the U.S.S.R. It all depended on the type of economy and the view one had of it. In the planned economy the criterion was based on the public interest. What he had said about prices and the needs of society was quite a different question. The criterion for choosing a suitable technique must be the public interest. They had to plan the direction to be taken by new techniques and what research was profitable for society as a whole. The state could not plan in detail. It was in the various firms that the details were worked out, and if the firms were not stimulated to do what was desirable, then the state could do nothing. What were the incentives? First the profits, or more accurately the net profits. On the relations between the public interest and the interests of the various firms, Professor Gatovskiy said that the state gave general guidelines. The criterion used was the national interest and, as he had said, the effect was the saving of costs in relation to the capital required.

Professor James said that, if he might speak as a non-specialist, it seemed to him that the paper by Professor Gatovskiy analysed the advantages of technical discovery wholly in terms of lowering the cost. If one analysed the effects of technical progress in this way, did one not leave out all the other advantages apart from saving cost, for example the improving of quality of products? Professor Gatovskiy seemed to minimise this point. At the same time technical discoveries led to new products. He thought Professor Gatovskiy's method underestimated the effect of technological progress because of this.

Professor Neumark said he would like to know what were the essential differences between the two economic systems which were at issue here. According to Professor Gatovskiy the major effect of technical progress consisted of cutting costs. In the socialist system was it the total of such economies that was passed on to the consumer? Or was it less, so that the benefit was used to help other enterprises or made available to the state to increase R & D expenditure? The second effect of technical progress was an increase in the quality of goods which was presumably reflected in prices. How did the state give effect to the change of quality in the price structure? Was it the price based on costs or was it a political decision?

Professor Mossé said he wished simply to ask one question which related to a point made in the paper about the planned norm of capital investment effectiveness. What was meant by this? Did it correspond to what western economists called productivity? Was it gross or net? In calculating it, to what extent was account taken of profits and amortisation?

Professor Tsuru asked whether Professor Gatovskiy had considered the contribution of technical progress separately from embodiment in capital. From the equations it appeared to be embodied. Secondly, the concept of the coefficient of relative effectiveness appeared to be equal to the reciprocal of the payoff period. This was a narrower concept than the rate of interest. He thought it was used for the allocation between alternative investments in particular industries and not for allocation between industries, because the coefficient was different for different industries. He asked whether this was consistent with other parts of the paper, particularly for national economic welfare. If differences in coefficients were allowed, on what ground were these differences approved?

Professor Munthe wished to raise a point concerning reduction of production costs as well as the reduction of utilisation costs. If one had a machine producing and a machine using industry, then should one take into account the reduction in costs of producing the machine and also of its use by a third industry? And what if an industry produced machines used by consumers? Washing machines saved two hours a week. What price did one put on the housewife's time?

Professor Khachaturov said in answer to Professor Tsuru on the coefficient of efficiency that he was right. This was a marginal coefficient, when one chooses between alternatives, but not only when the alternatives were inside a single branch of industry, because there were situations in which a particular problem could arise in different branches of industry. Metal, for example might be possibly replaced by plastics. Here two branches of industry were involved and the alternatives were inter-branch alternatives. They were of the opinion that in such cases they should use in general a unified coefficient of efficiency. But there could be exceptions. For example when they were concerned with longer-term technology which was not as yet fully efficient – atomic energy or thermo-nuclear energy. If one used the general coefficient used for contemporary alternatives, one could make a mistake. Professor Khachaturov said that these ideas were to be found in a paper presented by Professor G. Seaborg, a United States atomic energy scientist, who had applied them to an argument on variable discount rates. He accepted that when it was necessary to determine the effectiveness of long term projects one could not use the same coefficients as for contemporary alternatives.

Professor Gatovskiy said that many questions had been asked and that he could not hope to reply to all of them, but he intended to concentrate on two or three. Firstly, on the planning of investment. What were the criteria that the state used to determine the proportions between capital formation and consumption, and between the different sectors of the economy? That was the basic question. How did the U.S.S.R. divide up the national product or the national income? Was it only on the principle of maximum possible profit? The answer was 'no'. He did not believe it was 'yes' in non-socialist countries either. There were other considerations of which account had to be taken. They were aiming at a level of division on the basis of needs, and profit did not achieve this – needs for the comfort of the masses, needs for defence, needs for basic scientific research for the longer term. Did they have exact criteria for allocating between them? He did not think so. It was difficult to determine the exact criteria in mathematical and economic terms. The state produced a plan, and on the basis of this they dealt with the question of the allocation as economically as possible – allocation to achieve maximum effectiveness. This was the meaning of what he had termed here the 'rate of interest'. The same was true when they tried to measure the effect of technical progress between investment and consumption. It depended on the plan, on how one envisaged the future, on what one wanted to do – to increase profits or hold back funds for major investments in the future. These posed problems for economists.

Secondly, Professor Gatovskiy said there were problems of inter-sector relationships. Agriculture in the U.S.S.R. was backward in relation to other sectors. What ought they to do? They had given large sums in assistance to agriculture, but this kind of subsidy did not produce immediate returns. There had been a limitation of investment in industry in order to increase the investment in agriculture, but it was not possible to base this on any precise criteria. There were two levels here: allocation of income in a general way on the basis of the plan, and secondly choice of method to make use of the allocation. It is here that cost effectiveness became the major consideration.

Then there was the time factor. They were trying to plan for technical innovation in the long term, and there were two phases. Firstly the phase in which the effect was small – the stage of assimilation of the new technology. This was too lengthy in the U.S.S.R. One of their efforts must be to reduce the cost and length of time assimilating new technology. In the United States were there no such problems? After assimilation profits rose and costs were cut. At this stage supplementary profits arose which compensated for the loss in the first phase. Then there was the second phase in which an older technology had to be replaced by the newer technology. The effectiveness of the existing technology became less, and prices had to reflect this amortisation. There was also the problem of compensating the firm for its high expenditure during the first period of assimilation, which was difficult. There were funds to compensate for losses, which were based on credit, since firms had to earn their own way. The biggest profits were in the newest technologies, and history bore this out. When they had new technology fully in action the situation was ideal and produced the greatest profit flow. He was sure Professor Mansfield would agree on this.

To Professor Robinson Professor Gatovskiy said that his question was crucial, but unfortunately it was a very difficult one to handle. They were only at the beginning of their attempts to solve the methodological problems. Their biggest problem was forecasting. The government recognised this and had charged them to forecast to 1990. Forecasting the effects of technology was extremely difficult. There was great risk and uncertainty involved, and they had no real criteria. Formulae existed but the methods were not foolproof. They hoped they might have better means at their disposal to do this in two to three years, and to be able then to provide a more informative report. They felt that their forecasts were reasonably accurate for five to seven years ahead, but forecasting for fifteen to twenty years ahead was hazardous.

Professor Khachaturov said that this was over-pessimistic. *Professor Gatovskiy* replied that forecasting the achievements of fundamental science was extremely difficult to do. But they could forecast further ahead for technical achievements. Their principle was that lowering of prices was the criterion of technical progress. So far they had been reasonably successful in this field. With regard to his foreign trade prices, there were other laws that applied here than those that governed internal prices. Other factors come into their planning. It would take too long to list them. Competition was one factor, but not the only one.

6 Development of Science and Technology in the U.S.S.R.

T. S. Khachaturov

ACADEMY OF SCIENCES OF THE U.S.S.R., MOSCOW

I. SCIENCE PLANNING AND DEVELOPMENT

The great importance of science and technology as a factor determining the rate of growth of the national economy of the U.S.S.R. has always been recognised. Planning, therefore, takes account of the latest scientific achievements and their practical applications. With the improvement of the planning methods, scientific and technological development became a separate part of the plan. The preparation of this part of the plan, as well as the planning of scientific research, is based on a broad participation of research institutes and relevant state bodies responsible for the institutes.

The active participation of research institutes in solving the problems of economic development by advancing the applications of scientific and technological knowledge began shortly after the great October Revolution. V. I. Lenin, the founder of the Soviet State, early in 1918 prepared *A Draft Plan of Scientific Technological Tasks for the Republic*, in which he stressed a number of industrial and agricultural developments that needed to be tackled on the latest scientific-technological basis. Lenin in his speeches often underlined the necessity of using scientific-technological knowledge to achieve the most rapid recovery of the national economy and its further growth.

The first national perspective plan – the GOELRO plan[1] – drafted under the guidance of Lenin and based on his ideas, envisaged its task as that of introducing the most advanced technology into the national economy in the form of the electrification of the whole country.

Scientific-technological progress has been a major feature in the preparation and realisation of the five-year plans of the U.S.S.R. The pre-war five-year plans were in fact plans for a technological revolution, for transition from the contemporary dominance of manual labour to mechanisation. Under these plans the industrialisation of the national economy, and the establishment of national machine-building, metallurgical and chemical industries was

[1] The word GOELRO, is a composition of three abbreviated words, which means fully the name of the State Commission for Electrification of Russia.

accomplished. New plants were built, as had been planned, incorporating the most modern technology. In the process of industrialisation a technological revolution covering all branches of the national economy was achieved.

Concentration on technological progress required stimulation of scientific research throughout the country and shortly after the revolution research institutes were successively set up to tackle the theoretical problems of physics, chemistry, mathematics (for example, the Institute of Physics, the Institute of Chemistry, the Institute of Mathematics) and to tackle applied problems of energy, aviation, transport (for instance, the Institutes of Power Engineering, of Hydrodynamics, Aerodynamics and of Railway Transport). The best of the institutes became, as time went on, the principal research centres; progressively a number of more specialised institutes and laboratories were hived off from them.

Great attention has been given to the development of the Academy of Sciences, under which a number of research institutes for each of the sciences have been set up.

During the post-war years the development of science and technology has progressed still faster. Soviet achievements in the exploration of space and the development of astronomy, in the utilisation of nuclear energy for peaceful purposes, in rapid improvements in aeronautics, in the building of quantum generators and many other innovations are the best witnesses of the advance of Soviet science. These achievements also reflect the high level reached in a number of other branches of science and technology in the country. Space flights, for example, have only become possible through numerous achievements of scientists in creating new materials capable of meeting great tensions, new fuels, new radio-electronic devices, new computers and methods of making calculations.

Science and technology are today of great importance to the economic development of the U.S.S.R. and thus their planning has also become proportionately more important.

The time is now long past when the possibility and expediency of planning the development of science and technology was doubted; when one could often hear it said that scientific activity was purely intellectual and was too personal an activity to be capable of being planned; that planning would only restrict the initiative of the scientist; that planning would be an obstacle to the scientist's research and eventually would have a negative impact on the development of science. These now obsolete views belong to the time when scientific progress was made by individuals; when its growth was slow; and when there was a long time lag between scientific discoveries and their practical applications. The situation today is quite different. Those secrets of nature which could be investigated and discovered by individual scientists depending on their own resources have long ago been discovered and carefully analysed. The next stages of scientific investigation become more and more complex and difficult for individual scientists to undertake and in most cases require the work of teams of scientists or institutes and the use of sophisticated and expensive equipment. The large expenses involved are worth while only if one knows the justification for them, and knows the expected results of the research. And as has also been shown in practice, considerable expense has often yielded the expected valuable results. The economic effect of such projects can be measured not only by the direct gains but also by the indirect creation of opportunities for further industrial expansion and for further gains in the future.

In these days, even in capitalist countries the development of science is proceeding in certain deliberately chosen directions in pursuit of some expected goal and with the aid of financial support from the state. If this is how capitalist countries do it, it is natural that a planned socialist economy should to an even greater extent plan and direct the development of science and technology in accordance with the needs of the national economy and with the logic of scientific progress itself. And it is the socialist state itself which undertakes such planning and financing of scientific research.

The scope of the planning is the scientific activity and work of the various institutes and laboratories attached to the Academy of Sciences of the U.S.S.R., to the Academies of Sciences of the Soviet Republics, to the Ministries and functional organisations of the U.S.S.R. and the Soviet Republics and also to the Universities. The training of specialists for these institutions is also planned by the state and undertaken by universities and technical colleges. The total number of scientific institutions can be seen in the following table:

TABLE 6.1

NUMBERS OF SCIENTIFIC INSTITUTIONS 1940–69

	1940	1950	1960	1969
Number of scientific institutions[a]	2359	3447	4196	4953
Number of scientific research institutes[b]	786	1157	1728	2388

[a] Including higher educational establishments.
[b] Including affiliated organisations and branches.

These figures show the very considerable growth of scientific establishments – especially research institutes – the number of which has tripled between 1940 and 1969, and has grown 1·5 times between 1960 and 1969. Similar growth is shown by the figures of scientific research workers.

1950	163,000
1960	354,000
1969	883,000

In each ten years the number of research workers has more than doubled and the rate of growth is steadily increasing. The majority of the research workers are employed by higher educational establishments.

Of the 883,000 research workers in 1969, 189,000 are specialists in physics, mathematics, chemistry, biology and geology; 391,000 in engineering technology; 35,000 in agriculture; 48,000 in medicine. The rest are working in the social sciences; 53,000 of these are economists.

II. RESPONSIBILITIES FOR SCIENCE PLANNING

The bodies responsible for the planning of science and technology in the U.S.S.R. are: the U.S.S.R. State Planning Committee; the State Committee for Science and Technology; the U.S.S.R. Academy of Science; the Ministries concerned. The functions of planning and management of scientific and technological development are divided between them as follows:

The State Planning Committee, when preparing the five-year national economic plan takes into consideration the tasks of solving the principal scientific-technological problems involved, the introduction of new types of products and new technologies (including methods of mechanisation and automation) and also the training of scientific and science-teaching staff. The annual plans include the introduction of new types of production of new technology. The justification for including these objectives in plans is derived from proposals made by the State Committee for Science and Technology and the Academy of Science. It is the intention that the data regarding the introduction of new technology shall serve as the basis for calculating the indices of the national plan itself, since the productivity of labour, production costs, volumes of production are affected by the application of new technology. But in practice this relation between the introduction of new technology and the production indices of the plan has not yet been put fully into effect.

The State Committee for Science and Technology, jointly with the Academy of Science, defines the main trends of scientific and technological development in the U.S.S.R. Other important functions of this Committee are: the organising the investigation of inter-disciplinary scientific-technological problems (at the time of the Committee's creation about 250 such problems had arisen); respon-sibility for measures to raise the general effectiveness of scientific research and to ensure the speedy introduction of research results into the national economy; exchange of scientific-technological in-formation and co-operation with other countries.

The U.S.S.R. Academy of Science, as the highest scientific research body in this country, is responsible for the investigation of all problems in the natural and social sciences carried out in its insti-tutes, and co-ordinates the scientific activities of the Academies of the Republics, the Universities and other research bodies.

Ministries and other State Departments are responsible for indus-trial research institutes and design offices under their control. They put into effect the general technological policy in the relevant branches of industry; they introduce into practice the scientific and technological results, including scientific methods of labour organisa-tion and management.

The basic division of functions between the academic and the industrial research institutions is that the former are expected to investigate the fundamental problems of science-mathematics, physics, chemistry, biology, geology and other natural sciences as well as economics, philosophy, history and other social sciences, and to develop these ideas to the point where they can be put into practical application. Industrial research institutions, on the other hand, on the basis of their own research and the research results of the fundamental sciences investigate scientific-technological problems which are important for the technical progress of the particular industry, for the design of modern equipment, apparatus, materials and technical processes. For example, fundamental in-vestigations in such fields as the physics of solid substances, the theory of crystals, the physics of durability and plasticity serve as the basis of work on the production of highly-durable materials in industrial institutes and laboratories; discoveries in the field of high pressure physics find practical application in industrial production of artificial diamonds; fundamental studies on problems of low temperatures and super-conductivity permit the creation of super-conductive electro-lines; research results in quantum radio-physics create opportunities to develop new methods of communication, and so on. The list of such repercussions could be considerably extended in other branches of science – in chemistry, biology, geophysics, geo-chemistry, for example. Work on these applied problems, based on fundamental research, is often conducted not only in industrial institutes but also in academic institutes, which pass on their results directly to industry. It is natural that in some cases applied research carried out in an industrial institute proves to be capable of being theoretically generalised and formulated and thus becomes a scientific contribution to a fundamental research. This all illustrates the close relation between fundamental and applied research. But in general, measured by the number of institutions, their staff, the cost of their equipment and similar criteria, industrial research accounts for a considerably greater share in the total than fundamental research.

III. THE METHODS OF SCIENCE PLANNING

Today nobody in the U.S.S.R. is in any doubt about the necessity and expediency of planning the scientific research activities. This planning, in terms of subjects, durations of projects, and their financing – is based on several initial considerations. The first is the need of the national economy to solve this or that economico-technical problem affecting the growth of the volume and quality of production and consumption – for example, increasing the durability of metal, speeding up the delivery of goods, cutting down time spent in shopping, and so on. In proposing these objectives the plans take account of the existing levels of technology and production. In addition the possibilities of technical progress, the available resources of savings and capital investment are of great importance as prerequisites for the actual realisation of scientific and technical progress. One must also take into consideration the logical development of the scientific inquiries themselves as new ideas and new opportunities appear in the process of research; science has no limits and its possibilities are unbounded. The personal interests of scientists, their ambitions, aspirations and attitudes are also very important and cannot be ignored. And finally, the research plans must reflect also world progress in scientific knowledge and the discoveries made by foreign scientists.

Taking all these underlying factors into account, the plans for scientific research activities are worked out. But it must be remembered that these plans are not elaborated in a vacuum. The composition and contents of these plans are largely predetermined by the completed and current research that is proceeding and this limits the degree of freedom of those planning the new research. At the same time it is necessary to bear in mind that a proportion of investigations of new problems will fail to produce results, though negative results are often of great practical importance. In the process of planning, moreover, one sometimes has to overcome inertia and to stop particular research projects which have been overtaken by events and have lost any prospect of success.

The planning of scientific and technical development includes several stages and covers different periods of time: fifteen to thirty year forecasts represent the pre-planning stage; perspective plans for science are compiled for five years to correspond with the five-year national economic plan; at present there is a possibility of preparing a general plan for fifteen years. Annual plans are also made, representing detailed applications of the five-year plan as it relates to the particular year.

IV. FORECASTING TECHNICAL PROGRESS

Forecasting scientific and technical development is of great importance both for planning the directions of scientific-technological development and for forecasting national economic development. A peculiar feature of forecasts is their probability character. They can be made with this or that degree of exactness only for those lines of scientific and technological development which have already become well established. The task of these forecasts is to define possible time-periods for solving given problems (for example, the use of thermonuclear synthesis to obtain energy, or the discovery of the origin and treatment methods of malignant tumours) with known possible methods of investigation. It is obvious that one cannot plan completely new lines of scientific research, not to mention fundamentally new discoveries or inventions; we are concerned here not with science fiction but with reasonably firm scientific forecasting. The methods of such forecasting are expert estimation, questionnaires and in some cases analogy and extrapolations.

It has become common practice in the U.S.S.R. in recent years to entrust forecasts concerning the principal scientific problems under investigation to special commissions formed of the most prominent scientists in each given field. These commissions prepare scientific-technological problems in which they assess the existing level of science and technology and possible directions and steps in its further development. Of great importance here is the economic estimate of the problem concerned – that is the estimate of the possible national economic effects of successful solution of the problem. They may consider such problems as raising labour productivity with the help of the new technology (for example, lasers), the saving of materials that would result from the introduction of new synthetic substances, and so on. This possible estimate of the effects – direct and, if possible, indirect – can be compared with the costs required for the research, experimentation and industrial introduction of the product. Economic estimates and calculations, even if only very approximate, are of great importance because only they can make it possible to measure with a certain degree of probability the extent of the increase in labour productivity that may result from the application of the new technology. And on the basis of the increase of labour productivity one can try to forecast the general growth of production that is likely.

The long-term forecasts of scientific and technological development, fifteen to twenty years and more, need to be defined more precisely in relation to the following five-year period. The State five-year plan for scientific research and the incorporation of scientific and technical results into the national economy are made on the basis of this forecast; they are based on estimates of the national economic needs and on proposals made by scientific research institutes, approved in the first place by the Academy of Sciences. This plan is the basic one in the system of plans for scientific and technical progress and forms an integral part of the five-year plan of national economic development.

A plan for scientific research is wholly different from a forecast of scientific progress. The plan sets out concrete lines of research and their expected results; it also covers the time periods involved, the staff required and discusses the volumes and sources of finance required, as well as the eventual application of the expected results in actual practice. To prepare such a plan requires certain calculations and justifications for including this or that project. And only the most promising projects which are well worth all the labour, finance and materials invested in them, get included in such a plan. To select the most promising projects economic appraisals are carried out for each of them. A second difference between planning and forecasting is that at the planning stage more exact estimates of the expected results and of the expenditures are provided in place of the preliminary estimates used for the purpose of forecasting. Of great importance for these calculations is the estimate of economic potential – that is to say the sum total of the economic effects with maximum possible volume of production designed to meet fully demand for a new or improved type of product. The economic effect is measured by the difference between the full costs of the unit of production (including interest) required by the scheme under consideration and the former existing situation.

For this purpose it is necessary to take account not only of the expenditures on scientific research but also of any necessary expenditures on prototypes, design, experimentation, testing and organisation of production. It is obvious that the economic effect cannot be attributed exclusively to any one particular part of the operation – to research, or to design, or to production. It should on the other hand be distributed between the various elements in the operation in proportion to the comparative contribution of each element, and this must be at least approximately evaluated.

The comparison of the economic effect with the expenditures makes it possible to measure the effectiveness of the proposed capital investment. Economic calculations show that the effectiveness of capital investments in applied scientific research projects as such is very high and ranges from 0·8 to 1·4 roubles per 1·0 roubles of investment (excluding expenditure on training scientific staff).

For each problem set by the State five-year plan, a controlling body is appointed, which is responsible for the preparation of the co-ordinating plan, composed of all the executive participants both in the research project itself and in the designing, experimental and other activities necessary for the successful introduction of the scientific results into actual production. It is also responsible for determining the timetable and means of financing. Necessary finance is drawn from Ministries, from other State bodies of the U.S.S.R., and from Councils of Ministers of the Soviet Republics, depending on who is principally responsible for the given problem as a whole.

A series of separate stages, tasks and projects are also established in this co-ordinating plan, with most of them on a contract base (the finance being drawn from the total allocated fund) and the rest financed by the executive participants themselves. Large and important problems attract a number of separate organisations and industrial enterprises which sometimes can be counted in dozens. The whole project is headed by a well-known scientist or designer.

For complex problems, which require a lengthy period of research, the Academy of Sciences and the State Committee for Science and Technology organise Scientific Committees composed of prominent scientists and persons working in practical fields, who represent various scientific and economic organisations or are selected on a personal basis. These Scientific Committees co-ordinate the projects of the given problems, criticise its plans, discuss reports on work completed, organise scientific sessions and publish regular reports about their activities.

In the course of the five-year period, the annual State plans for scientific research and for the introduction of scientific-technical progress into the national economy are worked out. These plans appear as integral parts of the annual national economic plans. They define the volume of work to be completed within the given year, include various new projects and eliminate certain projects which have reached finality.

Industries and republics make their own five-year and annual plans for scientific research and the application of new technology. These cover organisations and enterprises attached to the various Ministries of the U.S.S.R. and to the Councils of Ministers of the Soviet Republics. These plans define the scope of work to be done in relation to the State five-year and the annual plans for scientific research and the co-ordinating plans. They also are responsible for problems which are important to particular industries or regions and which have been proposed by the corresponding organisations.

The U.S.S.R. Academy of Science makes five-year and annual plans of scientific research in natural and social sciences. The actual preparation of these involves not only the academic research bodies of the U.S.S.R. and Soviet Republics but also the industrial institutes and state organisations.

Finally it should be added that universities and other higher educational establishments, as well as individual research bodies and design offices make their own five-year and annual plans which are related to subjects entrusted to them by the State and co-ordinating plans or undertaken on their own initiative.

Thus the planning of scientific and technical progress embraces all the necessary stages: fundamental research; the introduction of new technology into the national economy; and all stages of supervision and control from the U.S.S.R. State Planning Committee and the State Committee on Science and Technology down to the individual research institutes, laboratories and industrial enterprises. Of great importance in this are the careful organisation of all facilities for carrying out the plans, a systematic control, the examination and evaluation of the work as it is completed, and timely action in all cases in which certain stages and projects are for some reason expected to go wrong.

V. THE RESOURCES AVAILABLE FOR RESEARCH

The successful fulfilment of the planned scientific research depends greatly on the extent to which the material resources for scientific research and for the training of necessary scientific staff are expanded and strengthened. Scientific research is financed partly through the State budget and partly through the budgets of various State, co-operative, trade union and other bodies and organisations. The State budget finances the fundamental research, which is mainly carried out by the Academies of Science. It also allocates finance for a considerable proportion of the applied scientific research. This money is allocated to ministries and State bodies and distributed by them between the various scientific organisations in the form of payment for a piece of completed work or a part of one. It should be made clear that such payment is made only after the customer-agency has received and approved the completed work.

There is similar contract-type financing when a research organisation makes a contract with some ministry or an industrial enterprise to carry out certain research. In such cases an industry (an enterprise or a group of them) finances this work through its own funds, derived from profits, and the research organisation receives the payment after the work has been completed and approved by the other contract-party. This contract-type financing serves as an incentive to better quality and shorter periods of research work and at the same time it emphasises the purely applied character of the research and its relation to the current needs of production.

The growth of expenditures on science year by year can be seen in Table 6.2.

TABLE 6.2

EXPENDITURE ON SCIENCE (1940–70)
(billions of roubles)

	1940	1950	1960	1965	1968	1969	1970
Total expenditure	0·3	1·0	3·9	6·9	9·0	10·0	11·0
Expenditure borne by the State Budget	0·1	0·5	2·3	4·3	5·5	5·9	6·5

The average rate of growth of expenditure on science during the past ten years has amounted to 11 per cent. That is considerably more than the average growth rate of the national income. As a proportion of national income the expenditure on science amounts at the moment to about 4 per cent. About one-eighth of the total goes to fundamental research; the rest to applied research and development.

The greater part of all scientific expenditure goes to the salaries of research staff. Material costs – equipment, apparatus, materials – form currently a smaller part of the whole but in the future this part is certain to grow. Every successive step in scientific advance, in conquering the secrets of nature, becomes more and more difficult, requiring more complicated investigations and more sophisticated equipment. And this sophistication of research in practice increases the relative share of expenditures on equipment and materials.

It is estimated that about 7·5 billion roubles have been spent on material costs during 1960–7. Technical equipment for scientific work not only increases its productivity but in some cases serves as the basis for further scientific development. But though costs of equipment steadily grow, their volume is still insufficient and does not increase proportionately to salaries.

With the progress of scientific research, scientific bodies require more and more sophisticated apparatus and instruments. It is often difficult to satisfy these demands because such apparatus, though needed by a limited number of projects, has to be designed and made completely from the beginning. The solution of this important problem is closely related to the development of the manufacture of scientific equipment (which will probably become an independent sub-industry) and with the establishment of special design offices, at least for such apparatus and instruments as are needed in a sufficient scale to set up series production of them.

The creation of new research bodies and the expansion of existing ones requires capital investment. Such types of scientific equipment as particle accelerators or radio-telescopes are very expensive and depend on industrial resources. The total volume of annual capital investment in establishing research bodies and their experimental laboratories amounts to hundreds of millions of roubles and will obviously reach the figure of one million roubles a year during the present five-year period, representing a little less than 2 per cent of the present annual total of centralised state capital investment in production. About eighteen per cent of this goes to the Academies of Science; and the rest is distributed between the industrial branch research institutes.

VI. THE TRAINING OF RESEARCH STAFF

Great importance is attached to the timing of the training of scientific staff in view of the fact that this training takes considerable time. In accordance with the forecasts of scientific development and their plans for scientific research the various institutions estimate their measurements for scientific staff with new special qualifications – for instance, for mathematical economic methods in planning and the management of the economy, for theoretical physics, for chemistry, biology and so on. In close relation to these new requirements, the universities and colleges prepare and amend curricula and syllabuses and start teaching new subjects and disciplines. The data on graduates during the past ten years show a particularly noticeable growth of graduates with new special qualifications which are essential to technical progress. For example, included in the 65 per cent general increase of specialists from 1960 to 1969 (representing 565,000 graduates) the output of specialists in electronics, the construction of electrical equipment, and automation in 1969 as compared with 1960 was 4·5 times higher in 1969 than in 1960 (37,000 graduates); in radiotechnology and communication it was almost three times higher (18,000 graduates); in chemical technology it was 2·5 times higher (14,000 graduates). It should be added here that a quarter of all post-graduates are trained in physics and mathematics, chemistry and biology; and about 10 per cent in economics; a large proportion, that is, are in fields which are of particular importance for modern economic development.

VII. THE CONTRIBUTION OF RESEARCH TO ECONOMIC PROGRESS

The effectiveness of scientific and technological progress depends not only on the success and achievements of the scientific research itself but also on the application of these achievements in production and their wide dissimination through the national economy. As has been mentioned above, the effectiveness of capital investment in science, according to the available estimates, is very high; the investment is recovered in about a year. Other comparisons also indicate the high effectiveness of science. For example, the growth of production in the U.S.S.R. is attributable to the extent of 85–90 per cent to growth of labour productivity, and this in turn is due to the extent of 70–75 per cent to technical progress.

Thus at least 60–67 per cent of the growth of industrial productivity is attributable to new technology. A number of industries, including machine-building, oil-processing and the chemical industry among others, have been undergoing constant changes and have immensely improved their technology during the past five to ten years. But it should also be stressed that in some cases the introduction of new technology does not proceed fast enough. The problem of the industrial application of new technology arises not only in the U.S.S.R. but in other countries also. To judge from the available data, only 30 to 50 per cent of all completed research projects have currently become embodied in actual production. This means that a very considerable number of projects are still awaiting application. A proposal has been advanced to make some kind of 'inventory' of all completed but as yet unapplied research projects with the purpose of identifying those of them which are potentially valuable and capable with modification of being usefully applied in the national economy.

A very significant contribution comes from the process of development and experimentation. Improvement of the experimental activity which connects research and design projects with industrial production will, in my view, speed up the process of the introduction of completed research projects into practical application. This development work tests the design and technology, helps to correct inevitable miscalculations and improves the project itself. Thus the development work influences greatly the process of the introduction of new technology.

The ever-increasing sophistication of new technology requires better quality and this in turn requires better development and experimental centres. The time-periods needed for experimental tests become more and more lengthy and expensive, but their effect also becomes more and more important. Apart from traditional development and experimental centres, for instance in the aircraft and automobile industries, a number of other industries have begun to develop experimental facilities, which already are beginning to yield valuable results. For example, development and improvement of experimental and research facilities in the field of the manufacture of building and contruction plant, road-building and community services has reduced the lag in producing new models of equipment from five to six years to two to three years. About 800 new models of machines for construction, road-building and communal services have been produced during the years 1966–9 – that is almost three times the number during the preceding four years. This serves to emphasise the obvious economic value of development and experimental centres.

VIII. INCENTIVES TO INTRODUCE IMPROVED TECHNOLOGIES

To make the process of introducing new technology into production more effective it is important to improve the incentives to both producers and consumers. When new technology is being introduced, there are usually increases of costs, practical difficulties and problems of training the necessary personnel. This has the effect of lowering economic indicators of the enterprise concerned, reduces its profits and consequently its incentive funds. To overcome these adverse factors, which often result in adverse attitudes of industrial enterprises towards the introduction of new technology, compensation funds and corresponding incentives have been created for those enterprises which propose to introduce some new technology. To achieve this aim a so-called 'new technology fund' is available which partly covers the additional costs. The enterprise receives some credit, to be repaid later out of future profits. At the same time so called 'stage prices' are used. Such prices allow for increased costs during the initial period and a subsequent reduction. As a result, by the time that what was formerly the new technology becomes obsolescent, the enterprise begins to earn lower profits and is thus economically encouraged to move to a newer technology.

The problems of introducing new products and new technology are quite different when research, design and development organisations are all combined under industrial enterprises. These combined units are jointly interested in designing and using new technology. At present such combined systems are going through an experimental stage. The Ministry for the electro-technological industry has been selected as the subject of this experimentation. In this Ministry since 1969 the former method of financing its research, design and experimental activities and the introduction of new technology through State budget, with a certain percentage of internally borne costs, has been replaced by financing entirely through a fund for scientific and technological development. This fund is derived from a certain percentage of planned profits, and this helps to establish the connection between profitability and the effectiveness of technological progress. All research institutes of this Ministry as well as some research institutes belonging to other Ministries have been changed over to the new system of economic incentives which makes it directly dependent on the actual economic effectiveness of the application of scientific results to production. Such a form of economic incentives is, in my opinion, appropriate in the case of industrial research institutes engaged mainly in applied research.

One of the principal ways in which science and production have been closely associated is through the so-called scientific-production corporations. These are based on scientific research institutes, but include industrial enterprises which provide general facilities for testing and experimentation. These do not, of course, affect the necessity to fulfil their production plans and profitability quotas.

In these scientific-production corporations there are opportunities for more rapid completion of the whole operation – science-technology-production. This can be achieved by early correction of possible mistakes and quick application of research results. Equally important is the establishment of necessary incentives for designers, experimenters and industrial staff. At present there already exist such units in the U.S.S.R. and this trend will be developed further.

There remain a great many problems to be solved in the U.S.S.R. in the planning of science and technology. These problems are regarded as of very great importance, because their solution determines the rate of introducing scientific and technological progress into practice and thus the rate of growth of the whole national economy.

Discussion of the Paper by Professor Khachaturov

Mr. King, in introducing the paper by Professor Khachaturov, said that it was in many ways complementary to that given by Professor Gatovskiy and that many of the questions raised in the previous paper, particularly those on technological development, were raised again in this one. It was a very interesting descriptive report of Soviet scientific policy, and of the way in which the Soviet planning process was constructed, starting from the belief in the use of science and technology, which had been basic from Lenin himself. He agreed that in the present day, when it was held that the day of the individual scientist and of the individualist was over, stress was laid on organised science and teamwork and multi-disciplinary groups. It was suggested that the nature of post-war problems required this approach because of their size and cost. The information given in this paper on the expansion of Soviet research institutes, the numbers of workers employed, and the rest was very interesting. It was also interesting to know also how these institutes were included in the central planning process through such bodies as the state committees on economics of science and the Academy of Science, although the latter body was a very different thing from the academies in the West. It was an *élite* body, much like the Royal Society, but with strong governmental functions. Mr King suggested that no academise in the Western world would accept willingly such official responsibilities. He said that no mention had been made of the universities in relation to basic research and he would be interested to know whether Soviet scientists thought this had any repercussions on the teaching of science at the university level, and especially the postgraduate level.

The State Committee for Science and Technology acted in combination with the Academy, but it was not easy to see the distinction in function of the two, and the relation of the State Committee to the Academy was obscure. It appeared that departments and ministries had dominant responsibilities for applied research while the Academy was concerned with fundamental research, but this was essentially orientated towards national objectives. As regards the criteria for selection of projects, these were much as in the West, based on scientific and economic promise, but also taking into account the views and interests of individual scientists. He was interested that the criteria also included taking account of world events in science and technology. On the question of planning, there appeared to be a great deal of interest in forecasting. His own impression was that forecasting was less of a separate function in the U.S.S.R. than a prerequisite of the planning function, in contrast to the United States where it appeared that many attempts at technological forecasting were rather in the air, without leading from forecasting into planning. It appeared to him that the forecasting function, both technological and economic, was more systematically and deliberately undertaken than in most Western countries through the use by committees of procedures such as Delphi. There were various long term forecasts of the order of fifteen or twenty years and even some longer than this, and the plans derived from these forecasts were given to special commissions to follow up in detail. The forecasts of technological promise led to the selecting of projects and priority problems which was dealt with at various levels, in an attempt to see how the five-year plan could take account of the stage reached in particular research fields. After disaggregation into annual budget terms institutes were given responsibility for projects. There were also corresponding five-year plans in individual industries, republics and universities. With regard to the fulfilment of plans he thought that the research depended greatly upon provision of the material basis and training of necessary personnel. Apart from general planning there was a good deal of contract-type financing with the research organisations being able to give contracts, and this gave a certain flexibility. On the question of the cost of research he noted that the main expenditures were on the salaries of personnel, but that the capital cost were gradually increasing with the use of more sophisticated instrumentation. He would like to know how much of this equipment was of Russian origin and how much of it was from other countries.

He felt that it would be very valuable to have more information on the effectiveness of capital investment in science, on the question of payback in about a year, and on the statement that 60 per cent to 67 per cent of industrial productivity growth was caused by new technology. The latter would not appear to be so different from Denison's figures for the United States. Mr. King said he would like to know how they have been arrived at. On the figure of 30–50 per cent of the projects introduced into public use, he thought this was a rather high proportion, and was very much higher than that in Western countries. The proportion of scientifically interesting research reaching a productive level was rather small in the West, and he wondered what 'completed' really meant. One of the final points which was most interesting was the stimulation of innovation in a planned economy. This appeared to be taken very seriously, the question of providing incentives being considered in detail, and he would very much like to know more about this. The introduction of improvements would tend to reduce output temporarily and when enterprises were trying to achieve quotas people who had to meet targets would tend to be afraid of too much innovation. Some of the methods mentioned in the paper for overcoming this were most interesting, in particular the idea of a new technology fund which could be used to cover additional costs, the credit given having to be repaid out of future profits. Also the method of fixing prices was interesting, where the prices included the R & D costs in the first stages, but gradually become lower and made less provision for this as time went on. On the question of material incentives for designers and management staffs, Mr. King said he was interested to know how these were selected. On the question of links between research institutes and individual enterprises which used the results, this was like the British research associations, and he would like more information on this. He was particularly interested in the levy on projected profits from new technology which could be used as a development fund: Mr. King said that many of the points raised in this paper were very interesting and important for other countries, especially where the proportion of government research was high and he would like to hear more about this and to know how successful the methods had been in practice.

Professor Khachaturov thanked Mr King for his most interesting introduction to the paper and for raising so many interesting questions. He would try to answer them very briefly. On the question of universities and research institutions he said this was the result of historical development. In old Russia they did not have research institutes. Science was concentrated in the universities. The Academy had been founded 250 years ago, and the development of special research institutes occurred only after 1917, arising out of the needs of the state to resolve important scientific and technological questions. He felt that the problems of connection between university science and teaching were important. In universities and technological institutes they had special laboratories for scientific research but not to such a great extent within the Academy and the Ministries. On the question of scientific apparatus, the development of science was very much dependent now on very expensive equipment, and they could not meet all their own requirements, importing a significant proportion of equipment from aboard. Referring to Mr. King's point about the effectiveness of scientific work and the figures in this paper, the 85–90 per cent quoted was derived from the growth of the level of productivity and was fairly simple to calculate. But the second calculation on the rate of technological progress was very much more difficult and he could not in the time available go into the methodology of the calculations. The work had been done in the Institute of Labour and many factories had been investigated. On the question of the application of completed research Professor Khachaturov admitted that there were people in the factories who wanted to live quietly and who would not be bothered with new technology because its introduction tended to be expensive. At the same time a considerable proportion of research became obsolete before it could be introduced, and this was a particularly important factor in those cases in which completed research failed to be applied.

Professor Oelssner said that long-term planning of science and technology was an issue of vital importance for his country – East Germany. Without the help of this they could not achieve the aim of economic development. During the coming five years – 1971–75 – industrial production was expected to increase by 37–38 per cent and national income by 26–28 per cent. Since there was no reserve of manpower (rather the opposite, they had a serious shortage) the extensive type of expansion was impossible. They had to achieve their objectives wholly by increasing productivity and during the five-year plan this was expected to lie between 35–38 per cent. This was impossible without the development and diffusion of science and technology; and the government provided substantial resources for the development of science and technology. The amount had increased from the DM. 1.2 billion in 1960 to DM. 3.5 billion in 1970, corresponding to 3.2 per cent of national income. With this they were on the level of the Federal Republic of Germany and of Great Britain. The problem however was not only financial. The efficient use of the many was more important for them. They had recently introduced new measures to ensure the more efficient use of their financial means, their material resources and their scientific manpower. First of all there was a new and very far-reaching reform in higher education. Its aim was to bring teaching and research closer to production needs. In 1975 the number of university graduates would be 225,000. In the final years of their studies students were participating in the research activities of research institutes and of industrial organisations and the industrial concerns and enterprises were obliged to sign contracts with the students to provide the necessary working facilities for their further scientific progress. The second measure was the reform of the Academy of Sciences. The German Academy of Science in Berlin was transformed into a large centre of scientific research and it now employed more than 13,000 people. The basic objective of the reform was to limit the number of research projects. A relatively small country could not undertake the solution of all scientific problems. They had to concentrate their scientific resources on the most important task from the point of view of their national economy. A further objective of the Academy was to establish direct relations between research institutes and industrial practice. To do this the institutes were concluding contracts either with firms or with national concerns and enterprises, and they were financed by them. This reform also aimed to strengthen the co-operation between basic research and industrial practice and special working groups had been formed for this purpose with participation of scientists, designers and engineers. The third important measure was close co-operation with other socialist countries, and above all with the great research potential of the Soviet Union. This co-operation was becoming closer and was yielding them better results from their expenditures on research.

Professor Matthews said that the concept in economic theory underlying the discussion in most of the previous sessions had been the production function. Professor Khachaturov had carried us into a different area, the theory of industrial organisation. A central question here was the optimum degree of vertical integration between R & D on the one hand and production and marketing on the other. In the United Kingdom for historical reasons there existed numbers of large government research establishments and it had been, and was currently being argued that this was an inefficient way to deploy R & D resources, because the researchers were bound to be out of touch with problems of marketing and of large-scale production. As against this it was argued by some that the spectator saw most of the game and that in-house fully integrated research might not be radical enough. The issue could be expressed in terms of economic theory–the cost of communication, the costs of transactions, and so on. But the balance of considerations was ultimately an empirical one. It might be expected that because Soviet production was less sensitively geared to consumer preferences, the need for vertical integration from research to marketing would be less than in the West. But it was apparent from passages in Professor Khachaturov's paper that in the U.S.S.R. too the need had been felt to decentralise research and integrate it more closely with production. It would be interesting if he could say more about Soviet experience with regard to this problem.

Academician Stefanov said he would like to stress the very simplistic conceptions of the development of science found in many countries. On the one hand scientific research was considered only worth while if immediate practical use could be made of it, and people with this simplistic conception would not recognise the real importance of fundamental research. The other conception was that only such fundamental work was significant and not the researches related to its practical application. It had been very well explained by Professor Khachaturov that the stimulation of both was important. In reality we had the development of science taking place at an increasing tempo and with increasing complexities. During the next fifteen to twenty years we would have as much innovation as in the whole period until now in the history of almost every country. In all the developed countries and still more in the developing countries that state took upon itself the function of organising scientific work. This was not only for defence or research for moon travel, but also the main direction of development of science, and for the creation of new branches of industry. We could see in all countries the state creating more research institutes and institutions for the special purposes of scientific investigations. The problems were becoming more and more complex and therefore there was a need for co-ordination between an ever greater number of institutes in order to solve these problems satisfactorily. In socialist countries even the Academy of Science was not able to co-ordinate the work and therefore the state had built up special state councils of science and technology. However the Academy remained the chief scientific body in every state. Academician Stefanov said he would like to stress that the co-ordination of scientific work was an important task of our time. It began with the co-ordination of plans of research and we now had the problem of the management of science. In the last decade there had been a very intensive development in the United States, for instance, where business schools had been created in almost all the universities. Academician Stefanov wished finally to point out that research that never reached application might nonetheless be extremely useful because it demonstrated what was a blind alley and not worth further investigation.

Professor Griliches said he had two brief questions to put to Professor Khachaturov. Firstly was it fair to say that a certain fraction of increased labour productivity was attributable to technological change? The calculations were basically similar to Denison's. One subtracted from the observed growth various directly indentifiable causes, such as the deepening of capital. To this *Professor Khachaturov* replied "No', and *Professor Griliches* went on to ask whether the rates of return to scientific investment were for selected investments or for the economy as a whole. Was he interpreting him correctly that the rate of return was 80–140 per cent or that the benefit to cost ratios were 1·8–2·4?

Professor Tsuru said that Professor Griliches had asked his question for him. He expressed surprise that the U.S.S.R. has 53,000 economists when Japan, with more than half the population, had only 4000 economists and managerial economists. He went on to say that if the calculation was similar to Robert Solow's pioneering work then he had basic disagreements with this approach for the Soviet-type economy. He would like more detailed explanation of the 70–75 per cent contribution to productivity by technology.

Professor Freeman said that he was very interested in the points on linking research institutes and productive enterprises, as this interface always seemed the critical problem in industrial innovation. He was also interested that $12\frac{1}{2}$ per cent of that total research expenditures were for basic research, and asked whether we could have further breakdown of scientific expenditure, for example on the allocation between various objectives of public policy, the allocation between space research (that is prestige research), and the growth-orientated civil R & D? He asked how much of the scientific expenditure was growth orientated and what proportion was medical and environmental.

Professor Triantis said he would like to repeat Professor Robinson's question from the previous session when he had asked Professor Gatovskiy to explain how research funds and personnel were allocated between the needs of the U.S.S.R. state and industry, and wondered whether Professor Khachaturov could tell us something about this? Picking up Professor Freeman's point, Professor Triantis said the problem was not just in allocating funds between science and industry but was equally difficult in deciding how to allocate them between industrial sectors – for example how did one decide how much to allocate to the automobile industry as compared, say, with the electrical industry? These were very important and difficult questions.

Professor Khachaturov replying to Professor Matthews' question on vertical integration said that they felt the connection between applied research and industry was not strong enough even now. They had 50,000 enterprises in industry in the country as a whole and all of these were separate. He felt that such a large number of enterprises was not good for many aspects of management of the economy, especially for the absorption of scientific research. They were coming to think that it would be good for management of industry to introduce corporations which would give them certain advantages, for example more flexibility for allocating the labour force and for the absorption of research, and they believed it might be of advantage in some cases to create scientific production corporations in which the manager of the scientific institute was the manager of the corporation. Research was not finished when the work of the research institute was finished. Many modifications were required when the products came to be made, and when the director of the research institute had the responsibilitites of managing the project right through it made the research better. At present they had 200 such corporations and expected to have more.

Replying to Professor Griliches Professor Khachaturov said that the 70–75 per cent figure was the result of factor analysis and did not make use of the Denison technique. He went on to distinguish between two types of new technology: firstly what was essentially new technology, not yet

168 Summary Record of Discussion

applied in the U.S.S.R. or abroad; this was not the greatest problem; secondly technology which was known but had not been used before. They still had a very great deal of manual work; in fact 45 per cent of all workers were manual workers. If they changed the situation and used machinery it was not new technology, but it was included in the 70–75 per cent figure. On the question of the effectiveness of research Professor Khachaturov said that he could not claim that these figures were precisely right, and referred to an article by Komsin in the UNESCO journal which had pointed out that they had not solved this problem. On the one side they had the whole sum of expenses for the whole chain from the beginning of research to the final product, and on the other side they took into account the effect – what they would receive if they realised the whole quantity of product which could be sold. With this comparison they could calculate the real effect on the whole system. Referring to Professor Freeman's point, he made it clear that the calculation was not for the whole economy but only for certain branches which had the necessary information. On Professor Triantis' point about the allocation of funds between different branches of industry, Professor Khachaturov said that there could be great changes from year to year. The allocation was made on the basis of developments over a period of several years, but where they could see that some institutes or lines of research were becoming obsolete they did in fact make arrangements to shut them, and in other cases they had made arrangements to provide additional resources where the sector looked as though it justified them.

7 Financing the Generation of New Science and Technology

A. Nussbaumer
UNIVERSITY OF VIENNA

I. PROBLEMS OF DEFINITION AND MEASUREMENT

Science and technology represent one of the most important sources of economic growth, together with the accumulation of capital and the formation of human capital. Their generation may even be considered a prerequisite for the application of other factors of growth, since science and technology affect not only the technological processes involved but also the possible objectives of production to be realised by factor input.

On the other hand we know that today the pace of technological change depends on a variety of factors – on the irresistible growth of scientific knowledge; on a momentum within technology itself which leads to more and more invention; on the competition for markets between industrial corporations; and on growing consumer demand, be it induced by individual incomes, commercial competition, or a spread of knowledge and education. Furthermore, the growth and distribution of world population both require and determine technological change and where and how it is implemented. Finally, the continued expansion of population and production depends on whether new science and technology permit us to make more economic use of resources and limitational factors and help to prevent environmental spoliation as a consequence of the expansion of industry and of waste connected with consumption.

Expenditure on science and technology is defined to include salaries of scientists and technologists, and the cost of equipment and auxiliary personnel. Moreover, the impact of the general infrastructure of an economy must be kept in mind, at least to the extent that its contributions to the development of science and technology are noticeable and measurable. We are thus in line with definition no. 3 given by R. C. O. Matthews in his paper [1].[1]

Furthermore, agreement must be reached on what is to be understood by science and technology. Following the definitions used by the National Science Foundation [2], we shall define *research* as the 'systematic, intensive study directed toward fuller scientific knowledge of the subject studied', the subcategory *basic research* receiving

[1] References are to the bibliography on p. 189.

the additional qualification that its aim be knowledge *per se* rather than its practical application. *Development* is defined as 'the systematic use of scientific knowledge directed and toward the production of useful materials, devices, systems, of methods, including design and development of prototypes and processes'.

If R & D are seen as an historical process, 'development' unquestionably represents more closely the practical application of science than does 'basic research'. Sometimes, therefore, a distinction is made between basic research and engineering (or development). This is also relevant to the problem of financing research, since discoveries in basic science may be said to have an unpredictable and accidental character (some hold that all we can say is that some researchers seem better able to make them than others [3]). If that were true, success in basic research would depend more heavily upon the selection of those people who have a reputation of being 'discovery prone' rather than on the organisation and financial endowment of research. In that case only engineering would show any close connection between financial resources used and the output of investment in science and development. This explains why expenditures on science and on development are generally separated into sub-groups for purposes of analysis. It indicates, also, that basic research can be carried on only in long term projects, since much time may elapse between discovery or invention and innovation resulting from the engineering process. Present contributions of science and new technology to economic growth may be the result, therefore, of expenditure in a relatively distant past. This may help to explain why there is hardly any empirical evidence of a close correlation between present expenditure on development of science and technology and economic growth. Finally, there are problems of statistical classification in measuring and comparing expenditure on R & D. Most European countries today follow the standard classification as developed by O.E.C.D. [4]. Even so, it is sometimes difficult to decide the exact division between expenditure on research and on production or on the treatment of expenditure on projects which contribute to scientific development as a by-product. Problems also arise in finding objective criteria for distinguishing between expenditure on basic research, research, and development, or between expenditure on industrial research programmes and non-profit government programmes.

Technological progress has been recognised as one of the most important factors contributing to economic growth, even if modern theories of growth still find it difficult to embody it into a formal analytical system. At the same time, the financing of the generation of new science and technology generally is regarded as an important form of investment, even if it is extremely difficult to measure exactly the benefits resulting from such investment – not only *ex ante*, but equally *ex post*. In both cases uncertainties about the speed of technological progress and its influence on production and on consumer wants make exact theoretical solutions difficult.

Technological progress is a very complex process demanding considerable investment, both public and private. It is difficult to isolate parts of this investment for detailed analysis. Cost-benefit calculations for specific research projects are confronted, for instance, with the problem that technical knowledge is treated frequently as a free commodity available to everyone and that, as a consequence, the investor in technological progress will never be credited with the total benefits resulting from his invention, nor will it be clear what contribution has been made by earlier investment in other research projects.

Some help in clarifying this issue may be obtained by distinguishing between *basic research*, *applied research*, and *development*. Current expenditure by the Austrian government on R & D in 1967, for example, was divided: 18 per cent for basic research, 56 per cent for applied research, and 26 per cent for development, with a total of 102·2 million Austrian Schillings [5]. Because of the great amount of time which some research projects take, current expenditures cannot be compared directly with the effects of research on production and growth; judging which past expenditures are relevant introduces, however, another *a priori* decision into the cost-benefit estimates.

There is a further problem of time-cost trade-offs: how much development time can be reduced by allocating additional resources to development [6]. If there is an alternative between time-inputs and resources (i.e. financial inputs), decision makers have a definite choice problem: they must decide how far to carry the sacrifice of scarce resources in order to reduce development time, or *vice versa*. In many cases fast development even at the price of high financial input is the more economic alternative since it reduces interest on funds already invested and earns a dyamic rent for the investor by giving him time advantage in competition.

On the other hand, introducing new products or methods of production involves considerable risk which only strong firms are able to carry. If a research process has to be cut short, most of the capital invested in it is lost, or only a small part of the benefits originally envisaged accrue to the investor. At least part of this risk will have to be carried by the government if small and medium-sized enterprises are to be encouraged to take part in scientific research. Another alternative would be co-operative ventures uniting groups of firms to do research together. There is, of course, always the possibility that a government itself may set up or at least finance institutions engaged in expensive long-term research projects. The latter method is justified especially if the results of investment in research and development are to be made available without charge to a large number of users, as may often be the case in basic research, or when the chances of limiting the use of the innovation to certain authorised firms are slight. Furthermore, a beneficial effect on growth and competition may result from permitting technical knowledge to be disseminated freely; unless the government finances and organises the research itself, compensation for the costs and a reasonable profit will have to be allowed to the firms engaged in research. In any case, problems of external economies and diseconomies will arise, and the process of research and innovation will contribute to concentration in industry.

In the same way, the government will have to finance or at least provide incentives for research and development designed to reduce the adverse secondary effects of our industrial civilisation on human environment and natural resources. Some of these adverse effects are already with us, partly a legacy of the past. Others are seen on the horizon as an irrevocable consequence of current trends. A considerable financial effort will have to be made not only to develop techniques of production which result in a minimum environmental waste and pollution, but also in designing remedies for damage which already exists or which cannot be avoided by the individual producer.

II. SECTORS OF PERFORMANCE AND SOURCES OF FUNDS FOR RESEARCH AND DEVELOPMENT

Activities in research and development are usually classified under the following sectors of performance: public sector, enterprises, higher education, and non-profit institutions. The sectors supplying the funds are the same. They include, however, the rest of the world or international contributions as an additional source of finance. Financial contributions to other countries or to international organisations doing research are generally included in the funds provided by the public sector. Sectors of performance and sources of funds have to be distinguished since transfer payments such as financial contributions of one sector to the cost of research and development in another one are frequent. Table 7.1 gives the relevant figures for Austrian expenditure on R & D for the fiscal year 1966–67. It shows clearly that the chief contributors toward the expenses of R & D projects are enterprises (58·0 per cent) and the public sector (38·2 per cent). But only 9·0 per cent of the total funds is spent on research within the public sector itself; 63·4 per cent is spent on R & D in the industrial sector, and 27·5 per cent on R & D in the sector of higher education. Non-profit institutions are relatively unimportant in Austria both as a source of funds and as a performing sector; they are, however, net suppliers of financial means to other sectors.

TABLE 7.1

SECTORS OF PERFORMANCE AND SOURCES
OF FUNDS FOR RESEARCH AND DEVELOPMENT
IN AUSTRIA 1966–7

Sectors of Application Sources of Funds	Public Sector	Enter-prises[1]	Higher Educa-tion	Non-profit Institu-tions	Total
(a) *Schillings* (000's)					
Public sector	127,372	83,411	406,100	1,495	618,378
Enterprises[1]	5,430	931,366		261	936,997
Higher education	5,963	—		—	5,963
Non-profit institutions[2]	57	1,487	39,000	58	40,602
Abroad	5,994	8,218		137	14,249
Total performances	144,816	1,024,422	445,100	1,951	1,616,289
(b) *Percentages*					
Public sector	88·1	8·1	91·2	76·7	38·2
Enterprises[1]	3·7	91·0	—	23·3	58·0
Higher education	4·1	—	—	—	0·4
Non-profit institutions[2]	—	0·1	8·8	3·0	2·5
Abroad	4·1	0·8	—	7·0	0·9
Total	100·0	100·0	100·0	100·0	100·0
% of total performance	9·0	63·4	27·5	0·1	100·0

[1] 1966.
[2] including all institutions which could not be included in other sectors.
Source: Central Statistical Office.

The sectors already mentioned may be subdivided. The public sector generally includes: federal government, state government, local government, and specialised governmental agencies; corporations organised under public law by government in some countries, providing for rail transportation, postal, telephone and telegraph services (sometimes, also, for city transportation, water supply and electric energy); finally, trade organisations in those countries where the working population is organised by law in self-governing associations (for instance, in Austria, the Chambers of Commerce and Industry, of Labour, and of Agriculture). The sector of enterprises consists of private as well as of publicly-owned enterprises, which have in common that they are organised under private law and are profit-oriented. Higher education draws its funds mainly from the public sector. As a performing sector, it may be divided into universities, academies of science and specialised affiliated research institutions. Non-profit institutions are either bodies set up and financed by the government outside the public sector and education, or research institutions created by enterprises for common ventures, or charitable bodies created to do research in the common interest.

Whenever a variety of institutions and organisations are concerned with financing and implementing scientific research and development, co-ordination is necessary. This applies especially to countries in which the promotion of scientific research is considered a part of the activities of the government, whereas the practical application of its results is left to industry; we may speak of a dualist concept of R & D. Financial means for such projects are not only allocated directly by various authorities, but a large part of R & D is also financed out of business profits. This implies that public policy directed towards financing R & D is not restricted to direct contributions, but extends also to the distribution effects of taxation. Investment benefits in the form of fiscal measures designed to diminish the burden of taxation where income is spent on science and research may be more important in many countries than direct contributions. This makes co-ordination of financial measures even more difficult.

There are thus various ways in which a government may promote scientific research and development. First, a government may directly provide funds for science and research. Data collected for the United States by the National Science Foundation show that in 1963 in the six N.S.F. industry classifications responsible for approximately $300 million or more worth of company-financed R & D, the government-supported R & D represented over 25 per cent of total expenditure, with the exception of chemicals and allied products and of petroleum refining. In these two industries it amounted to 21 and 6 per cent respectively. On the whole, the government tends to rush in wherever businessmen fear to tread in all cases in which a sector is considered important enough.

The reason for government support is frequently the necessity to support expensive programmes because of economies of scale in research and in production, or because of external economies, or because of special conditions of consumption, especially where monetary charges cannot be collected easily [7]. The government is also by far the most important single contributor to the finance of research carried on at universities even in those countries in which the universities themselves are privately operated. The government can finance special research projects as well as general research in some selected fields in addition to providing the basic equipment and even the personnel of research institutions. Finally, research may be carried out directly by government institutions incorporated into the general system of public administration. This is frequently the case with defence, medicine, social welfare, agriculture and education. There is one problem, however, which at least the last two forms of direct government expenditure have in common: benefits resulting from these expenditures cannot be measured easily. No market system exists which might permit benefit imputation or estimation in those cases where the product of research and development does not enter the market. This poses problems for a rational allocation of resources for research projects which are increasing with the increasing share of direct government contributions to the total of available resources.

Sometimes the government itself does not directly finance expenditure on research and development, but leaves this to parafiscal institutions. These may be organisations in the form of professional groups and endowed with a public status (for example, in Austria the Chambers of Commerce, of Labour and of Agriculture) mainly concerned with such activities. Similarly, institutions responsible for social security may finance R & D, especially in medicine but also in the social sciences. Special funds are constantly being set up to administer money dedicated to research in particular sectors of the economy or by particular groups of firms.

In addition, the government may provide incentives to employ private funds for science and research. Tax benefits, government risk-sharing in research and innovation, and an economic policy favouring high profits, are methods frequently used. This may imply the stabilisation of business incomes at a high level, even at the expense of the consumer, by being lenient towards oligopolistic price increases or by permitting seller's markets to result from overfull employment and creeping inflation. Concentration of firms and a policy friendly to big business may lead to higher investment in R & D, since it is believed widely that larger firms are more frequently responsible for new inventions and their practical application than smaller ones. Data collected by the National Science Foundation have shown that this hypothesis originally presented by Schumpeter is supported by empirical evidence at least to the extent that firms below certain threshold sizes do very little R & D. If the results obtained by the National Science Foundation

[8] are generally valid – that firms with 1000 employees or less accounted for 60 per cent of total industrial employment, but performed only 5 per cent of company financed R & D – this would in fact imply that small countries must make a choice between enforcing market competition and encouraging scientific research, unless they are prepared to participate in some regional economic integration. In Austria, of 676 million schillings spent on R & D by firms in 1966, only 2·5 per cent was spent by firms with less than 100 employees, 10·7 per cent by firms with 101–500 employees, and 86·8 per cent by firms with more than 500 employees [9]. Fortunately, however, threshold sizes vary significantly between different branches of industry. In Austria small and medium-sized firms with less than 500 employees often concentrate their efforts on developing and refining the results of basic research already done by others, but are hardly active in the initial field themselves. Medium-sized firms are more exposed to risks consequent on R & D than either small or big business; for they lack frequently the flexibility of small firms and they do not possess the financial reserves of big ones.

If smaller firms are to participate in R & D, certain special provisions are necessary: (1) free access to the results of basic research conducted by the government and of research financed by government; (2) co-operation between big and small firms, so that results can be made available to the smaller partners for further development; and (3) provision for joint ventures in which small firms may co-operate in common projects.

Finally, there are certain other general conditions which have to be created in order to establish a climate favourable to R & D. They include a high technical background to the economy, an adequate stock of trained experts, sufficient funds for R & D to make longer term projects possible, a high social standing and appropriate financial remuneration for persons engaged in R & D, and legal protection of innovations so as to allow firms that engage in research to recover their costs and make a reasonable profit. In addition, a sufficient climate of competition must prevail to induce firms to use some of their profits for research in order to stay ahead of or to keep up with their competitors in technical knowledge and its application.

III. FINANCE FOR RESEARCH AND DEVELOPMENT

International comparisons reveal remarkable differences in total expenditures on R & D as well as in sources of finance. The United States is not only by far the largest investor in R & D in absolute terms in the Western World (see Table 7.2), but by devoting 3 per

TABLE 7.2

EXPENDITURES ON RESEARCH AND DEVELOPMENT AND PERSONNEL EMPLOYED 1967

Country	Expenditure on R & D ($m.)	as % of G.N.P.	Expenditure per head (£)	Total R & D Personnel	Qualified Scientists and Engineers	R & D Personnel per 10,000 of population	Qualified Scientists and Engineers per 10,000 of population	Sources of Funds Government (%)	Sources of Funds Other (%)
United States 1966	23643·0	3·0	114·0	(1,375,000)a	524,776	69	26	62·9	37·4
France	2506·8	2·3	50·3	193,457	49,224	39	10	53·5	46·5
United Kingdom	2480·1	2·3	44·9	(228,000)a	(63,000)a	(42)	(11)	49·6	50·4
Germany (Fed. Rep)	2084·3	1·7	16·3	205,866	61,559	34	10	41·3	58·7
Japan	1684·1	1·5	16·9	356,275	157,612	36	16	30·2	69·8
Canada	828·3	1·5	40·6	51,790	19,350	25	9	53·4	46·6
Netherlands	513·8	2·3	40·8	50,200	15,700	40	12	39·0	41·0
Italy	447·1	0·7	8·5	49,939	19,670	10	4	35·2	64·8
Sweden	336·1	1·4	42·7	25,172	6,605	32	8	42·1	58·9
Switzerland	304·0	1·9	50·7	(21,000)a	10,954b	(35)	18	21·1	78·9
Belgium	176·0	0·9	18·4	19,750	7,945	21	3	18·8	81·2
Denmark	90·4	0·7	17·5	8,378	3,919	17	8	55·5	44·5
Norway	80·7	1·0	21·3	7,357	2,958	19	8	57·9	42·1
Austria	62·2	0·6	8·5	6,620	2,401	9	3	38·4	61·6
Finland	50·6	0·6	18·4	5,154	2,026	11	4	42·7	57·3
Ireland	17·2	0·5	5·8	2,800	1,121	10	4	50·6	49·4
Greece	11·3	0·2	1·3	2,730	1,217	3	1	61·0	39·0

a Estimate by the O.E.C.D. Secretariat.
b Including technical university graduates.

Source: O.E.C.D.: *International Survey of the Resources Devoted to R & D in 1967*. Paris.

cent of G.N.P. to it, her total investment in it also exceeds substantially that of all other heavily industrialised member countries of O.E.C.D. combined, which invest between 2·3 per cent and 1·7 per cent of their G.N.P. in R & D. Small countries and those where agriculture still contributes largely to G.N.P. spend no more than 1 per cent (Italy and Austria, for instance, 0·7 per cent and 0·6 per cent respectively).

Comparing dollars spent per head yields similar results. Again, the United States devotes to R & D more than twice as much money per head ($114 in 1967) as the next biggest spender, while Switzerland and France come next with investments just of over $50, followed by the United Kingdom with $45, Sweden with $43, and the Netherlands and Canada with $41. At lower levels, German and Japanese investment per head is relatively small ($16, and $17), while Italy and Austria lie still further behind with an investment per head of only $8·5. These figures alone show that in many European countries financial contributions towards expenditure on R & D must be increased substantially if they are to keep up with technological and scientific progress in the most advanced countries.

The number of qualified scientists and engineers engaged in R & D varies as widely between countries as the amount of money spent. As can be expected, countries with high expenditure on R & D also employ more qualified personnel in this sector, but proportions are subject to wide variation. Differences seem to depend on the sectors of the economy in which research is being done, on the amount of capital and the number of scientists available, and on wage rates.

Sources of funds also differ widely. Extreme positions are held by the United States, where 63 per cent of all funds made available come from the federal government, and by Belgium and Switzerland, where 81 and 79 per cent, respectively, come from private sources. It is obviously not the degree of industrialisation which makes the difference. Certainly the total volume of expenditure on R & D is larger when the government gives more financial assistance, but in Switzerland research expenditure is high, despite a relatively small government contribution. The factors determining the relative share of government funds seem to depend on the kind of research projects receiving the most funds. In the United States, 62 per cent of all funds spent on R & D is devoted to atomic energy, space projects and defence – explaining why total R & D expenditure by the United States is so high. France and the United Kingdom devote only 45 and 40 per cent of their R & D expenditure to these sectors. Western Germany, Austria and Belgium spend only 17, 12 and 4 per cent on the projects mentioned and at the same time a relatively small contribution by the government towards the total cost of R &

TABLE 7.3

EMPLOYMENT OF FUNDS BY PURPOSES AND SECTORS 1963

Country	Employment of Funds (% of Total)			Industrial Research by Sectors of Industry (% of Total)				
	Atomic Energy, Space and Defence	Purely Commercial Motives	Other	Iron and Metal Manufacturing[d]	Chemicals	Metallurgy	Other[e]	Total
United States	62	28	10	79.1	10.9	1.4	8.6	100.0
France	45	41	14	56.0	14.1	4.7	25.2	100.0
United Kingdom	40	51	9	78.6[a]	9.7[a]	2.6[a]	9.1[a]	100.0
Germany (Fed. Rep.)	17[c]	62[c]	21[c]	53.1	32.4	6.6	7.9	100.0
Belgium	4	82	14	35.7[b]	44.6[b]	8.2[b]	11.5[b]	100.0
Austria	12	62	26	31.9	22.2	17.1	28.8	100.0

[a] 1960–61.
[b] 1961.
[c] 1964.
[d] Including aviation, electronics, electrical appliances, vehicles, machines, etc.
[e] Including food, textiles, wood, paper, crude oil, etc.

Source: O.E.C.D.: *The Overall Level and Structure of R & D Efforts in O.E.C.D. Member Countries*, Paris, 1967.

D (see Tables 7.2 and 7.3). A close relationship between research for purely commercial motives and a non-governmental origin of funds is also shown in the case of Belgium, where 82 per cent of the funds originate in the non-governmental sector and 82 per cent are correspondingly employed for purely commercial motives.

In the United States, while 70 per cent of the research took place in industrial R & D establishments, less than half of this industrial R & D was funded by industry itself. The major part of the money came from a few big federal agencies: the Department of Defense (DOD), the National Aeronautics and Space Administration (NASA), the Atomic Energy Commission, the Department of Health, Education and Welfare, and the National Science Foundation. This government-sponsored 'big science' has not only contributed to the objectives of these agencies, but has also had a broader economic impact, which is widely believed to be the main cause for the technological gap between the United States and other industrialised nations.

Finally, the structure of research by sectors of industry is closely connected with the size of firms and with the relative importance of these sectors in total production and employment. R & D tends to concentrate in those sectors in which big firms prevail, as is shown in Table 7.3. This is very obvious in the case of Austria where, by international standards, an unusually high percentage of all industrial research is done in the field of metallurgy, which is dominated by four corporations owned by the Austrian government; this is, however, equally true for larger countries such as the United States, France, or the United Kingdom. There also it is relative and not absolute size which matters. Thus, a relatively high proportion of funds for R & D is spent in France and in Austria in such sectors as food production, textiles, paper and wood, which are important for these countries, but in which even large firms have relatively small labour forces as compared with industrial giants in such sectors as iron and steel, electronics and aviation.

Even so, differences in R & D between industrial sectors are to be explained only in part by factors that are of universal validity. As the statistical evidence on R & D expenditure in Austria shows (Table 7.4), there are very big differences between industrial sectors not only in total expenditure, but in the proportion of money spent on R & D as well. In the case of Austria, it is only in the iron mining and steel producing industries that almost as much as a half of all R & D expenditure is devoted to actual research (49 per cent); these industries are dominated by large government-owned companies. More than a quarter of all R & D funds are devoted to research in the foundry and metal industries (30 per cent), the crude

TABLE 7.4

INDUSTRIAL RESEARCH AND DEVELOPMENT IN AUSTRIA, 1969

Industries	Total of Research and Development schillings m.	Distribution between		Distribution between	
		Research %	Development %	Intramural %	Extramural %
Mining and production of iron and steel	216·898	49·33	50·66	97·37	2·62
Crude oil	49·091	29·62	70·38	91·75	8·25
Non-metallic minerals and ceramics	29·091	24·00	75·99	91·75	8·24
Glass	570	—	100·00	100·00	—
Chemicals	290·252	28·81	71·18	95·65	4·34
Paper	21·460	25·32	74·66	93·82	6·17
Timber products	15·944	7·87	92·19	90·97	9·02
Food products	40·499	21·00	78·99	86·47	13·52
Leather	8·892	14·03	85·96	93·87	6·12
Foundry and metallurgy	22·461	30·25	69·74	91·61	8·38
Metal construction and machinery	126·292	17·00	82·99	87·20	12·79
Vehicles	80·162	17·00	82·99	99·28	0·71
Manufacturing of iron and steel	68·145	13·99	86·00	93·50	6·49
Electrical goods	196·367	14·75	85·24	89·49	11·50
Textiles	14·686	8·00	91·99	93·99	6·00
Clothing	1·441	1·59	98·40	93·96	6·03
	1,182·251	26·11	73·89	93·38	6·61

Source: Austrian Statistical Office and Federal Chamber of Commerce.

oil industry (30 per cent), the chemical industry (29 per cent), and the paper industry (25 per cent) where technical change is fast and firms are relatively large by Austrian standards. In sectors of industry where medium-sized or small firms are prevalent and where technology either is relatively static or where the development of new techniques is dependent upon size, the vast majority of funds is spent on development rather than on research. This is the case with the glass industry (100 per cent on development), the clothing industry (98 per cent), timber products (92 per cent), the textile industry (92 per cent), the leather industry (86 per cent) and the metal goods manufacturing industries (86 per cent), the latter consisting in Austria of relatively small firms.

Despite the predominance of small firms in Austria, intramural R & D is preferred to extramural expenditure. Only in the case of three industry groups does extramural expenditure exceed 10 per cent of total expenditure on R & D. This indicates that competing firms are generally hesitant to transfer their funds for development to other companies or even to common research institutions since they are afraid this may lead to disadvantage in competition. Considerations of this kind may prove very unfavourable to R & D, especially in countries in which small firms dominate.

Finally, there are important differences between different sectors of R & D regarding sources of funds made available, as is shown by Austrian expenditures in 1966/67. Total expenditure on R & D in that year amounted to 1·8 billion Austrian schillings or U.S.$73·5 millions, of which 89·1 per cent was used in the natural sciences and technology, 5·5 per cent in the social sciences, and 5·4 per cent on R & D in philosophy, humanities and the arts. While only 39·4 per cent of the funds for R & D in the natural sciences and in technology came from governmental sources, 90·4 per cent of R & D expenditure in the social sciences and 95·3 per cent in philosophy, humanities and the arts were financed by the government.

There is great concern on the part of the Austrian federal government about the insufficiency of funds devoted to R & D. Only 0·6 per cent of the country's G.N.P. is spent on R & D today, targets for the near future being 1 per cent and later 2 per cent. The unfavourable trends in the Austrian balance of payments in respect of patent rights, licences and know-how, as well as the rapid increase of Austrian patents held by foreigners in comparison with those held by Austrians, demonstrates that more R & D needs to be carried on, especially in natural sciences and technology, and that industry alone cannot finance it. Measures have been envisaged, therefore, to stimulate expenditure on R & D with tax benefits for people making inventions and companies financing innovations, by increasing the funds for scientific research made available to universities, by increasing the money appropriated to special funds or agencies financing research, and by giving preference to new products when the government makes purchases. Better co-ordination of projects and of means available for R & D is to be achieved, and priorities are to be fixed.

Past experience has shown that government programmes and finance, the technical standards that exist and the size of firms exert great influence on the volume of R & D, quite apart from the special conditions of any particular industry. There are, moreover, a number of other hypotheses about factors influencing the size of funds made available and of expenditure on R & D.

One widely held view is that a firm's decisions on R & D are strongly influenced by the behaviour of competitors and that a great deal of imitation exists among firms with respect to R & D allocations [10]. Most R & D is performed by firms operating in oligopolistic market structures; this may indicate that both higher profits than are possible under conditions of perfect competition and the pressures exercised by competition must be simultaneously combined to stimulate a big effort for R & D. Mutual imitation in policy directed towards R & D will, furthermore, tend to minimise risk in a situation characterised by competition amongst the few.

The size of a country may also be important for expenditure on R & D, at least in such sectors as atomic energy, space research, aviation and defence, where the government is the main financial contributor, since research has to be started and carried on on a sufficient scale often referred to as 'big push'. As a consequence of this investment, a technical reaction may develop which will create extra possibilities of growth in other sectors of production and in exports if firms are interacting so as to capture and internalise one another's external economies. In small countries this may be an argument for concentrating R & D wholly on some specific sectors despite the disadvantage that selecting only a few sectors of the economy for stimulated development implies the additional risk that other sectors may later prove to be more ripe for growth. In large countries government expenditure on R & D is also stimulated by political and military ambitions. Since the benefits of the programme involved tend to spread rather widely, R & D generally will benefit.

One further comment is still necessary on the effects of the size of firms: empirical evidence suggests that there is no clear-cut relationship between size measured by output and research intensity; a study by Mansfield found a significantly positive relationship for the chemical industry, whereas the drug and petroleum industries exhibited no clear relationship [11]. Better results are obtained if the size of a firm is measured not simply by output, but if product diversification and the availability of funds generated internally and measured as a percentage of total sales are also taken into account. A suitable formula for doing so is suggested by H. G. Grabowsky [12].

$$\frac{R_{i,t}}{S_{i,t}} = b_0 + b_1 P_i + b_2 D_i + b_3 \frac{I_{i,t-1}}{S_{i,t}}$$

... where $R_{i,t}$ is the level of R & D expenditures of the ith firm in the tth period, $S_{i,t}$ is the level of sales of the ith firm in the tth period, $I_{i,t-1}$ is the sum of after-tax profits plus depreciation and depletion expenses of the ith firm in the $t-1$ period, P_i is the number of patents received per scientist and engineer employed by the ith firm in a prior four-year period (1955–9) and D_i is the index of diversification of the ith firm (the number of the separate five-digit S.I.C. product classification in which it produces).

As to the generation of funds, governments may be expected to contribute more toward expenditures on R & D, firstly because sectors such as space and defence in which finance comes almost exclusively from the government are increasingly important and governments tend to get more and more involved in those sectors of heavy industry in which R & D is especially important; inventions and innovations in these sectors will influence other sectors also, technical knowledge showing a strong tendency to spread. Secondly, the government will have to set aside a larger part of its resources in order to preserve or improve the human environment under conditions of an industrial civilisation that exploits natural resources and environment on an increasing scale; a larger part of the funds for R & D will need to be used, therefore, in developing knowledge and techniques in this sector.

As far as R & D financed by industry is concerned, much will depend upon the trend of profits. There is considerable evidence that retained earnings and other internally generated funds have an especially significant effect on investment expenditures. This may be attributed to the general reluctance of firms to raise funds for R & D externally because of the higher risks entailed in this type of financing, as compared with those involved in internal finance. Financial risks must be kept small if the risks involved in the projects themselves are considerable. There is hardly any other kind of project in which there is greater uncertainty about results, the resources required and the time needed than in R & D.

IV. SOME UNSOLVED PROBLEMS OF FINANCING RESEARCH AND DEVELOPMENT

What, then, is the relationship between expenditure on R & D and economic growth? Modern growth theory does not generally take account of changes in technology or productivity, but rather operates under the assumption of a constant input/output ratio and attributes all economic growth to capital accumulation; for example, the Harrod-Domar models of growth. Statistical evidence, moreover, does not show any simple relationship between growth and expenditure on R & D, even in the case of economically motivated R & D, as Professor Matthews has shown in his paper. This may not only be due to the fact that R & D has been a relatively unimportant source of growth or that its influence was overshadowed by other factors, as Matthews has stated; it may also be because the contribution of R & D to economic growth frequently consists in the creation of new products or of products of better quality (generally referred to as hardware) and of 'software' – new insights into techniques of organisation or new scientific discoveries. The results of R & D are frequently, therefore, not accounted for correctly, even if they consist in hardware; in the cases in which they consist in software it may take a relatively long time before they lead to an increase of G.N.P. For this reason in particular direct statistical evidence of the effect of R & D expenditure cannot be produced.

Furthermore, growth may be due to the joint action of a variety of factors. It may depend upon investment in R & D as well as on accumulation of capital and on changes in the quantity or quality of labour inputs. To separate the influence of these factors by techniques of imputation is in any case near to impossible. At the same time, however, price effects have been shown not to be neutral to growth in real terms. Finally, the scale of economic operations and external economies arising out of mutual interactions of firms concentrated in poles of growth may be important. On the other hand, R & D itself often leads to external economies for firms which have not contributed to the financial effort involved because there is frequently diffusion of useful results of research or development without financial compensation. All of this makes it extremely difficult to measure the effects on growth of expenditure on R & D.

There is also the problem of whether it is more profitable for a government to spend money on its own projects or on those of industry, and whether criteria exist to judge and select projects on the basis of their contribution to economic welfare. It is sometimes held that government programmes absorb resources which otherwise might be employed in privately sponsored research to increase industrial productivity and create new products. But, on the other hand, it is recognised generally that government expenditure on research on space projects or defence has stimulated industrial production and technical progress even in sectors of industry not directly related to these government programmes and that this has contributed to the technological gap between large and small nations. All the more important is it, therefore, to develop a system which will make it possible to distinguish between more and less efficient allocations of resources on government projects. The Planned Program Budgeting System (P.P.B.S.) frequently is under discussion in this regard [12]; but the basic obstacles to exact measurement (the problems of how to evaluate targets scientifically and of how to assess all the various benefits of government R & D in all sectors of the economy) have not yet been overcome. Finally, results of R & D may be beneficial to growth and social welfare as long as they are applied on a restricted scale only. But once their application has become universal they may create new problems and require new expenditures to offset their adverse effects. Something which at first may lead to external economies may at some later point turn into a factor generating external diseconomies; the effects of the widespread use of motor cars or of vast urban concentrations are two striking examples of this.

Are there sufficient criteria for choosing between alternative R & D projects and for deciding upon their scope? Obviously, this question concerns micro-economic theory, since the effects of a chosen set of R & D projects have to be maximised subject to the limitations of budgetary constraints. Before the usual techniques of cost-benefit accounting can be applied, however, three decisions have to be made: what projects are to be taken into consideration; how large are the resources that can be allotted to R & D; to what extent are external effects due to the wider diffusion of technical knowledge to be included in the calculations. It is with regard to these questions that cost-benefit accounting for R & D projects at the national level relies much more heavily on *a priori* decisions than it does at the level of the firm, even if the basic problem of how to value the objectives of the projects is disregarded. If it is true that the debate about benefit-cost analysis centres on the question whether the social value of benefits can be estimated reliably enough to justify the trouble and effort involved in a benefit-cost computation [13], this applies especially in the case of R & D.

So far as private expenditure on R & D is concerned, it is much easier to solve these problems. Objectives may be derived from an identifiable set of consumer preferences. The volume of finance largely depends upon the firm's capacity to accumulate capital by way of profits, since credit financing generally involves risks considered to be too high. External effects need not be taken into account since to the firm it is only revenue and expenditure which matter. Information about alternative goals and means generally is sufficient, and time limits are set by financial plans. All of these simplifications of the problem do not apply to government expenditure. It is understandable that this has led to special emphasis on the development of better criteria and methods for project selection, such as cost-benefit analysis, the cost-effectiveness approach, or P.P.B.S.

What is the relationship between general economic policy and the financing of expenditure on R & D? Certainly, the economic system chosen influences the sources of funds available for R & D, since research carried on by firms motivated by expected profits in market economies generally accounts for about 60 per cent of total funds, this percentage only being lower in cases of considerable government expenditure in the fields of atomic energy, space and defence. In centrally planned economies, however, where profit incentives are missing, the vast majority of funds for R & D is contributed by government. Governments using central planning also have definite advantages in collecting funds, since government income relies to an extent considerably smaller than in market economies on direct personal taxation, and dis-incentives due to the burden of taxation are less relevant.

Funds for R & D must meet two requirements. First, their volume must be sufficient to finance investment projects of the necessary minimum size, this being sometimes an important obstacle to research carried on by small firms; investment in R & D should be financed out of the security capital of firms and not by credit, so as to avoid extra risks. Secondly, capital must be available on a long-term basis, the plans for investment in R & D must stretch over a number of years, and there must be the possibility of extending the time available for research, if results do not come as quickly as originally expected.

The first set of conditions frequently poses problems for industrial R & D in market economies. The second also makes for difficulties, especially for small firms trying to finance rather ambitious R & D programmes; it also poses serious problems for research financed by government, since the budgetary period is only one year and there is usually no guarantee that finance will be continued at the scale envisaged, even if programmes for R & D have been adopted officially. Funds for investment of all kinds are more prone to cuts than current government expenditure, since there are generally no legal obligations to make them available and political consequences are less severe than if allocations for the cost of administration in the widest sense were cut. This has provoked attempts to make funds for R & D more independent from current budgetary situations and to provide a more stable financial basis for long-term R & D. For this purpose, independent research institutions or public funds with a legal standing in their own right have been created to act as financial intermediaries, with government pledges to make funds available to them for more than one financial year. Thus, legal obligations for financing R & D are created also on a long-term basis, and continuing finance is assured.

This, however, is contrary to some opinions about government investment held mainly by protagonists of a counter-cyclical Keynesian investment policy: that government investment should vary with the fluctuations of the business cycle. Since it is probably valid in most cases that the gains from research increase as time needed for research decreases and since equipment and personnel employed in research cannot be laid off easily and then hired again, counter-cyclical government expenditure on R & D in general will not influence total expenditure on R & D much. If government sources for R & D are cut as part of a restrictive counter-cyclical policy firms will be forced to replace them by bank credit or by a reallocation of their own funds. This change in the composition of funds may, however, have a restrictive effect, since government funds will have to be replaced by those from other sources and total money and credit available will diminish. This, then, will lead to a general decline in liquidity, it may also increase the risks involved in the process of finance itself as well as in the projects to be realised. Limited anti-cyclical changes of government expenditure on R & D may be regarded therefore as a suitable means of counter-cyclical policy and not very harmful to R & D itself, as long as the general volume of expenditure of R & D can be maintained and only private sources of finance are being substituted for government ones.

Finally, there can be no doubt that the objectives of R & D will be influenced by those who provide the necessary financial basis. This is obvious in those cases where a government devotes huge resources for R & D in space projects or defence. Research projects financed by industry will be profit-oriented in general. Institutions carrying on research on a private basis (be they non-profit or profit-oriented) will generally depend on credit from large banks or on funds from government. This influences at least the process of selection between projects originally envisaged: in some cases research may even be carried out upon orders of the financing government or group. While the influence of government on R & D is rising with an increase in the relative contribution of government to total finances available, it should not be forgotten that industry exerts considerable influence on government-sponsored programmes. To the extent that R & D concentrates on big firms, and that the growth of G.N.P. and of tax revenue depends on the production and sales of big industry, and that government shows increasing interest in guiding company policies, including those of the big banks, a political-industrial nexus is developing which determines the major characteristics of R & D and the ways in which its financial needs are to be met.

REFERENCES

[1] R. C. O. Matthews, 'Contribution of Science and Technology to Economic Development', in *The Role of Science and Technology in Economic Development*, UNESCO Sc. 70, XIII. 18A (UNESCO 1970), p. 29. (Printed as Chapter 1 of this book.)

[2] National Science Foundation, *Federal Funds for Science*, XI, Fiscal years 1961, 62, 63 (1964), p. 93.

[3] S. Peter Burley and Oskar Morgenstern, 'Insiders and Outsiders in Industrial Research', in *Zeitschrift für Staatswissenschaft*, Vol. 125 (1969), p. 199.

[4] O.E.C.D., *Proposed Standards Practised for Research and Development*.

[5] Figures for 1964 have been 25·3 per cent, 39·7 per cent and 35·0 per cent respectively. For details see W. Klappacher, *Lage der Forschung und Entwicklung in Österreich 1963–1964* (Wien).

[6] See F. M. Scherer, 'Government Research and Development Programs', in *Measuring Benefits of Government Investment*, ed. R. Dorfman (Washington, D.C., 1965), pp. 34–7.

[7] R. Dorfman, *Measuring Benefits of Government Investment* (Washington, D.C., 1965), pp. 4–6.

[8] National Science Foundation, *Federal Funds*, p. 6.

[9] *Betriebliche Forschung in Österreich*, Bundeskammer der gewerblichen Wirtschaft.

10] S. P. Burley and O. Morgenstern, *Zeitschrift*, pp. 193, 196.

[11] E. Mansfield, 'Industrial Research Expenditures: Determinants, Prospects and Relation to Firm Size and Inventive Output', *Journal of Political Economy*, Vol. LXXII (1965), pp. 333–7.

[12] See for instance D. Novick, 'Das Programmbudget: Grundlage einer langfristigen Planung', in *Nutzen-Kosten-Analyse*, ed. H. C. Rechtenwald (Tübingen, 1970), p. 155.

[13] R. Dorfman, *Measuring Benefits*, p. 8.

Discussion of the Paper by Professor Nussbaumer

Professor Perkins, in opening the discussion, said that Professor Nussbaumer had brought together many aspects of a wide range of topics. He could summarise and comment on only a few of these, and he intended to concentrate on those aspects that related to methods of financing R & D expenditure by private industry in market economies. But one could not separate discussion of the means of financing from that of the considerations that determine the desirable levels of the various forms of R & D expenditure, and Professor Nussbaumer has rightly considered these two issues together. Since we had not in our discussions hitherto considered in any detail the important issues relating to the capital market and its role in influencing R & D expenditure, he would try to suggest a few lines of discussion of the topic that seemed to him to follow from those parts of the paper that touched especially on these aspects.

Any imperfections in the capital market made it less likely that those firms that could make best use of the capital would obtain it. As Professor Nussbaumer had pointed out, the risks attached to R & D expenditure made it especially difficult for firms to borrow externally for financing it. This made it likely that any imperfection in the capital market would make it especially hard for many of those firms that could usefully undertake that expenditure, and especially the new, small and medium-sized firms, to obtain the capital to finance it. The essence of the problem was that the individual investor, especially the private investor but also the institutional investor, found it difficult or impossible to assess the prospects of success of a firm that was borrowing for purposes of this kind. When a firm was borrowing for other forms of investment, it was usually somewhat less difficult for the lender to assess the risks involved in the project and its prospects of success. The main issue, then, was to overcome this crucial imperfection in the capital market that made it difficult to finance R & D expenditure from sources external to the firm.

One way mentioned by Professor Nussbaumer was that the government should itself finance such expenditure. But Professor Perkins suggested that one difficulty with this method was that government departments were not usually well-equipped or temperamentally suited to assess risky projects and to decide between them. For this reason governments had sometimes set up intermediary institutions financed partly or wholly from government sources with considerable freedom to choose where to invest their funds in private industry. It was important that governments should not themselves generally take the decisions about which firms should be financed from government sources, for often such arrangements became subverted to purely political ends and were used for helping 'lame-dog' industries over stiles. If, on the other hand, such arrangements left the maximum of discretion to the individuals working for the financial intermediary, who were specialists in assessing the risks and prospects of potential borrowers, there was a much greater chance of such an institutional arrangement being used to channel the funds to those firms with the best prospects of success. The role of such an intermediary was, then, to provide for the individual lender a body of expertise in assessing particular risks which individual lenders were not themselves well-equipped to assess and compare when deciding whether to lend to individual firms. In this respect its role in relation to the individual investor was somewhat analogous to the role of a consumer association in comparing the qualities of various products which required some expertise to assess them. There were other means of financing such intermediaries. They included the provision of finance by groups of banks and non-bank institutions, and a combination of bank finance, government finance and privately-raised finance. It was also perfectly possible for such intermediaries to be set up by private capital alone, and where this could be done it might well be the most satisfactory solution to the problem. In some places there had been privately financed intermediaries set up, co-operating with scientists and economists and others (sometimes from universities), who might be equipped to assess the complicated combinations of risks that had to be considered if the finance was to be channelled in those directions which seemed to have the best prospects of success. As equity financing was, in any case, the appropriate form, it was important that any intermediaries set up for this purpose should be able to borrow on an equity basis, for otherwise they would feel inhibited from lending on such terms. Professor Nussbaumer had also mentioned the role that might be played by joint ventures (whether of several small firms or of large and small ones together) in overcoming the imperfections in the capital market that impeded borrowing by small and medium-sized firms.

Professor Perkins said that more generally anything that reduced the imperfections in the capital market should facilitate the flow of funds towards those firms that could best apply them, whether for R & D expenditure or anything else. It was therefore especially important that governments should refrain from interfering with the flow of funds through the capital market by extensive controls over particular interest rates and over the portfolios of different financial institutions. It was also important that there should be adequate institutional arrangements and appropriate regulations to ensure, as far as possible, that investors could operate in the capital market with confidence and with as much knowledge as possible about the affairs of the firms in which they were investing. It was also important that due attention should be given to the free flow of capital between different countries as capital markets were becoming more integrated internationally, and policies that regulated in detail particular international capital flows were likely to reduce the perfection of the international capital market, and to this extent make it harder for those borrowers with the best prospects to borrow internationally. This might be especially important for small countries whose capital markets were not large enough to reap the economies of large scale, for example those arising from the offsetting of risks involved in some projects against those of others, and the employment of skilled experts to assess risks.

Professor Perkins said that imperfections of the capital market appeared to be a considerable obstacle to the appropriate channelling of funds for R & D expenditure even in such a large and well-developed capital market as the United States. In Britain and Australia – the two countries with which he was most familiar – such imperfections had also been found to exist, even though their capital markets were fairly well developed by world standards. But for the vast majority of the countries of the world such imperfections must be even more considerable, for the vast majority of the world's countries had capital markets less well developed than the three he had mentioned. The case for both direct government intervention in financing R & D expenditures and also for improving the capital market generally, might therefore be strongest in countries with less well-developed capital markets. On the other hand, one argument for government financing and encouragement of R & D expenditures was probably weaker in smaller countries than it was in the United States or even in Britain. Where there were economies external to the firm arising from a given expenditure on R & D these economies were obviously more likely to be gained within the country where the expenditures were being undertaken when that country was a large one. In the extreme case of a government responsible for economic policy for the whole world, all such economies would of course be internal to the economy for which the government was responsible. On the other hand, in a very small country the probability was that the flow of new knowledge that went outside the firm undertaking the R & D would go predominantly to other countries, once it became generally available. Professor Perkins suggested that it might also be true that once knowledge became public property it was more likely that large and well-developed countries would be able to obtain it and apply it than that small countries would be equally well able to appropriate for themselves scientific knowledge that became generally available in the United States or Britain, for example. There might therefore be a good case for governments in smaller countries concentrating the flow of their finance towards the application and adaptation of the knowledge that was becoming available in the outside world. To some extent the flow of knowledge towards smaller countries was facilitated by the operations of multinational firms, and in countries where these firms were operating on a large scale one would expect the knowledge of technological developments in the major countries to flow more readily than to those small countries where such direct investment by American and European firms was not widespread. In industries in which such direct investment was negligible or non-existent there might be a stronger case for smaller countries to take steps to facilitate the financing of R & D expenditure.

Finally Professor Perkins touched upon the issue of macro-economic policy raised by Professor Nussbaumer, and his reference to the difficulties and disadvantages of varying government R & D expenditure anticyclically. Professor Perkins suggested that a preferable device might be the anti-cyclical variation of tax-incentives, that is to say of subsidies for private R & D expenditure. These could take the form of concessions that would be available for only a limited period in a recession, the aim being to prevent reductions in private R & D spending that were otherwise likely to occur, especially as the financing of R & D from internally generated funds made them especially likely to be cut in a recession. The increasing of government R & D spending in a recession was, however, definitely a second-best to trying to stabilise private R & D spending, if only because it would usually employ rather different resources from those released by the fall in private R & D spending.

Professor Nussbaumer, commenting, said that the background of Austria had had a considerable influence on his paper. Austria had a very restricted capital market and there were difficulties because there were very many small firms, with R & D normally confined to the big firms. In addition the government intervened to a considerable extent by tax discrimination incentives for R & D. Thus the emphasis in other countries might be different. As far as the government was concerned Professor Nussbaumer thought there were three possibilities. First that the government might help the financing not by giving money directly to firms but by carrying the risk, while leaving the financing to banks and other financial institutions, and simply guaranteeing that the government would pay if the company undertaking the project ceased to carry on in business. The second possible way was more direct, and he agreed with Professor Perkins that this should not be done by government agency because they were not equipped to judge the problems of R & D. They might make funds available to a specific group of banks, leaving them to decide the projects which they would want to back, or alternatively they might set up a specific body to distribute the money, though in this case political as well as commerical reasons were more likely to come in. The third method which might be adopted was to use the banks as intermediaries, and Professor Nussbaumer said that this was his preference. Tax benefits could be given on a broad scale with profits spent on R & D exempted from tax. On the question of changes in government tax policy as a means of stabilisation, Professor Nussbaumer was sceptical because it was difficult. It was not too difficult for a government to grant and revoke benefits on the expenditure side but it was much harder to make tax changes at short notice, because the necessary legislation took a long time. If one allowed the government the right to make tax changes without going to the legislature, it could lead to insecurity, and in some countries it would be unconstitutional, since control of taxation was the first right of the democratic assemblies.

Dr. Sedov said that he felt that Professor Nussbaumer had overemphasised the role of the market as the mechanism controlling the development of science and technology, and that in suggesting that in centrally planned economies profit incentives were lacking he had to a certain extent misrepresented the relationship between planning and economic incentives in a planned economy. It was well known that profit was one of the main instruments of the system of economic incentives which existed in a socialist planning economy. The plan was supplemented by a system of economic stimulation for its fulfilment. At the same time planning was a process of the choice of the level of profits which would provide the economic stimulus to concentrate resources in the desirable

194   *Summary Record of Discussion*

directions for the development of science and technology. Many, if not the majority of those present, believed that the stimulation of technology and scientific progress could be based on a combination of planning and economic incentives, and this was the direction now being taken in socialist countries.

Professor Delivanis would confine himself to mentioning only his points of disagreement with Professor Nussbaumer's very interesting paper. Firstly, he felt that the distinctions between fundamental research, applied research and development were rather nebulous and did not support the conclusions which had been based on them. Secondly, on Professor Nussbaumer's argument in favour of financing research through taxation and not through subsidies, he did not think this was the best policy because it was then not only the firm that decided on the usefulness of research but also the government. Thirdly, considerable importance had been attached to the actual expenses of research and this seemed an easy way to make comparisons. But in fact what really mattered were the results. Fourthly, Professor Nussbaumer had been worried about the Austrian balance of payments and the high payments for foreign technology. This was only natural. But he should not forget that the development of many countries was based principally on the application of the results of other countries' research. Finally, Professor Nussbaumer appeared to have concluded that private financing was easier because it was not submitted to annual budgetary restrictions. He was certainly right about this according to the letter of the law; but United States experience was that a great deal of government funded research was carried on over a long period of years despite this apparent limitation.

Professor Sargent said that on the question of the role of government in financing high risk projects he agreed with the reason put forward in the paper concerning externalities. But he was mainly interested in the case where risk existed and where it affected the capital market. There were two reasons why the government might interfere in risky projects. Firstly, the government might be better able than the capital market to assess correctly what the real risks were. The experience of the United Kingdom might, however, make one sceptical whether governments could improve on assessments by capital markets. The latter might be bad, but the government could do it worse. Secondly, governments were very often turned to in largely private enterprise economies because they had the money. There were situations in which the project had the best assessment possible and it was argued that the government should step in because of the sums involved. This was an argument one should guard against unless it was within the power of the government by taking on the risks to reduce the risks. For example, to bring together a number of firms to make a single firm which might share, say, a basic core of research for the projects concerned and also to provide information from research which might be useful in other projects. The government had the power to enforce mergers between firms to reduce the real risk of risky projects. This sort of question should always be asked by governments taking on risky projects which the capital market declined to take. If government aid was sought we must ask if intervention enabled some reduction of the real risk – if the answer was 'no' we should be very sceptical of the view that the government should step in merely because it had the means.

Professor Matthews suggested that it could be argued that a case could be made in the opposite sense on the grounds that the principle of diminishing marginal utility meant heavy utility losses for an individual whose risky enterprise failed, whereas for the economy as a whole the curvature of the marginal utility of income schedule might be negligible over the relevant range. One way of dealing with this was by having larger companies that could support risks. If on the other hand larger companies were objected to for other reasons there was a conflict and we had to be content with some compromise balance of optima.

Professor Sargent said that he was aware of this point, but what he had in mind was something subtler, namely that the entrepreneurs or companies might be richer, or less risk averse, or both than the generality of the population.

Mr. Economides said he agreed with Professor Matthews that the government could afford to accept risk, even when the probabilities for good results were small. He quoted the instance of Cyprus, where the probability of finding petroleum was small, and no individual enterprise could afford the necessary outlay, but in this case the government had taken the initiative.

Professor Tsuru said he would like to quote some experience in Japan which was relevant to the present discussion. Several years ago a group of distinguished applied scientists had formed a guild to monopolise the rent income of the results of research and they had been rather successful at this. The cost of research had been far less than the revenue and the difference was rent income to the quality of the people concerned. This had created some controversy, since many regarded scientific knowledge as an externality. Professor Tsuru said that private business firms tended to employ scientists and engineers and to monopolise their results and did not pay enough to the creators of the results. This was precisely why this guild had been formed. There had been some criticism of it in Japan. He would like to hear some opinions on this sort of activity.

Professor Dupriez said he would like to draw attention to another form of co-operation and stressed the fact that research should not necessarily be organised in one firm. In Belgium they had a number of research units on an industrial scale receiving regular subsidies from industries through a special institute for subsidising applied research. This assessed projects and submitted them to industry. The advantage of this arrangement was that it was profitable to everyone. Firms were not obliged to participate but if they did not share in the subsidies they might not use the results.

Professor Perkins wished to clear up a misunderstanding that appeared to have arisen about a point he had made relating to anti-cyclical policy. He had commented on the suggestion made in the paper about counter-cyclical government spending, comparing this with the encouragement of private R & D in recessions. He had not been suggesting that tax incentives were better than subsidies as means of encouraging R & D expenditure. Nor was he denying that variations in government spending might be worth while in order to offset fluctuations in private R & D expenditure. He was merely pointing out that measures (whether by subsidy or by tax concessions) to try to stabilise private spending might be preferable.

Professor Eastman said that he did not quite understand Professor Nussbaumer's point about government guaranteed loans. He thought this absolved the government from making evaluations. This was most certainly not so. The government must have criteria for assessing loans. On the question of tax incentives against subsidies, there was good experience in Canada, not necessarily in R & D but on giving accelerated depreciation to desirable investments, especially on a counter-cyclical basis. This was a very powerful tool which was now being used by the United States. But it could be a dangerous one because it was difficult to remove when once given on a discriminative basis to different industries. This tended to lead to different tax rates in different industries and could distort the long-run allocation of resources. *Professor Neumark* asked whether Professor Eastman was speaking of tax credits or deductions from tax base. *Professor Eastman* replied that it was the latter, and added that one could have tax 'holidays', which had in fact been given to mining. But the general method was by accelerating the depreciation.

Professor Triantis asked Professor Nussbaumer whether there was any government policy whereby three or four ways of supporting R & D (e.g. guarantees of loans, subsidies and tax reductions) could be co-ordinated. He also asked whether it was possible to distinguish the way in which an R & D loan was being used. It might be taken from R & D and used for something else. How would one know when the company failed? He also felt that a tax deduction on profits spent on research would be awkward, and asked whether there was any experience of how these were assessed.

Professor Nussbaumer, replying, proposed to take Professor Perkins' points first. He had suggested that equity capital might be one way of limiting risk, and Professor Nussbaumer said that he thought this was a good idea, though he had not mentioned it explicitly. It would be better than loan capital, if it were possible. But this was not very practical where the capital market was small. On the question whether the government should interfere or not in the capital market, Professor Perkins had said a general 'no'. However there might be a second point. If we talked about the international capital market we needed to distinguish between the short and long term. If one raised money through a national bank the international aspect was only between the bank and the capital market, and no concern of the company itself. There was no problem if the country was in surplus. There were no problems of re-finance in Austria. As to the problem of whether the government should interfere or not in the process of financing R & D he believed that even medium-sized companies, for example companies employing about 500 people, in small countries must have access to sufficient financial resources, otherwise they might be forced to sell out to big international corporations. This might be undesirable if medium-sized companies with potential for development had to be bought up by international companies simply because they could not get backing for projects. International companies tended to buy at the point where small companies had just completed research but could not carry the risk. In this situation the provision of finance from government sources would be of assistance. On the question of competition, Professor Nussbaumer said that he had made the point for competition amongst the few. We needed competition, but at the same time bigness. Concentration in industry was the result of this. On the question of centrally-planned economies he had not said that central finance was all-important, but that it was easier for a government to transfer funds to whichever project they liked, and they had less need to be worried about the burdens of taxation. In answer to Professor Delivanis, Professor Nussbaumer said he wanted to re-emphasise that there were various methods by which a government might assist in financing R & D – by direct contributions, risk subsidising and tax benefits on investment. However, whenever a government had to allocate money to R & D it was important that finance should be ensured over a longer period of time, and not just for the budgetary period of one year. Various legal devices had been developed for that purpose. As regards Professor Sargent's comments, Professor Nussbaumer said that he did not believe that a government was better equipped to assess the risks involved in R & D. However it was likely to be more aware of market externalities and therefore more willing to finance projects which were not likely to derive profits from this wider diffusion. This was especially true in the case of basic knowledge. On his second point, the government should indeed carry risk if it was in a position to reduce it by bringing firms together. We should not forget that in some cases socialisation of risk was considered desirable. But this had to be justified on political not economic grounds, for example, in regional development problems. On Professor Tsuru's point, Professor Nussbaumer said he was generally sceptical about extending this concept of ownership externality too far. If R & D had to be financed privately the profits must also accrue to the inventor and the investing firm. But if the dynamic rent was so big, then more should go to the inventors rather than assume that knowledge was an externality in the market system. This could also be regulated with different legal protection for different types of inventions through the patent system. But if knowledge was very basic, it was impossible to prevent it from spreading in any way. Regarding the questions that Professors Eastman and Triantis had raised, he shared their opinion that government was not well qualified to decide which projects to finance. This should be left to the bodies generally concerned with such problems.

In reply to Professor Dupriez, Professor Nussbaumer said that in small countries the great difficulty was to devise forms of finance for those firms which were too small. It was often very difficult to get such firms to co-operate because they did not see how they could divide the benefits. Others might secure the benefits anyway, and in these circumstances they might simply be prepared to sell out to larger firms and make immediate profits. In reply to Professor Eastman, Professor Nussbaumer said that as regards government guaranteed loans, he had not implied that the government guaranteed the effects of the research, but only the bank credit to finance it. It gave the bank the means to provide money for specific projects. As far as the other alternative was concerned, that of accelerated depreciation, this was only one of various methods, and he agreed with the scepticism of Professor Neumark, that it had cumulative effects, and it might not always be good that those companies which had higher profits should get accelerated depreciation. Professor Nussbaumer said he was not in favour of policy which affected the tax base but preferred one which gave tax credits. One then got a fixed relationship between the investment and the government subsidy. But in the end, how did one co-ordinate the various instruments of policy effectively? If aid was provided both by a government and by trade associations, co-ordination became much more crucial.

Professor Tsuru said he had not proposed that science should be a free good. The result of science and technology were often regarded as ownership externality and the type of guild he mentioned was devised to market this externality. Once initial funds were available such a venture could become self-financing. An alternative would be to make it free. The question was which was economically more rational.

Professor Nussbaumer said this raised most interesting problems. How much monopoly control of the results of R & D was necessary? Should we have anti-cartel legislation to control the activities of such enterprises?

Professor Griliches said that this type of problem had been dealt with in a chapter the book by Nordhaus. It was necessary to have the promise of monopoly in invention to provide an incentive to invent, But this could result in social loss. A major question was what should be the optimal length of a patent on order to maximise social benefit. He pointed out that there had also been some interesting papers recently by Peter Swan on the relationship between durability and market structure.

8 Determinants of the Speed of Application of New Technology[1]

Edwin Mansfield
UNIVERSITY OF PENNSYLVANIA

I. INTRODUCTION

Technological change is a fundamental factor in the development of the poorer countries, in the growth of the industrialised countries, and in the improvement of the well-being of all mankind. It is a tremendously powerful force that, used with wisdom, can promote great strides in man's welfare. What are the basic factors that influence how rapidly new technology is applied? This is a very important question, and one that has been in the forefront of discussion and controversy in recent years. There has been a widespread feeling in many countries that new technology is being applied at an inappropriate rate – too slowly in the eyes of most, too rapidly in the eyes of some. Consequently, there has been considerable interest among economists, scientists, technologists and policy makers in the determinants of the speed of application of new technology. My purpose in this paper is to discuss the basic factors influencing the speed of application of new technology, the emphasis being on industrialised countries and much of the material being drawn from experience in the United States.

II. THE RATE OF APPLICATION OF NEW TECHNOLOGY

To begin with, I will try to summarise briefly what we know – or think we know – about how rapidly new technology is applied in the United States. This is not a simple matter, since it generally is difficult to define exactly what one means by a particular piece of technology, to describe exactly when it was first available, and to describe when it was first used. None the less, enough information is available to support the following propositions: First, the time interval between the invention of a major new process or product and the first commercial introduction of this process or product seems to average about a decade for the cases where data exist. Second, this time interval seems to vary from industry to industry: for example, it averages about eleven years for important petroleum

---

[1] The research on which this paper is based was supported by grants from the National Science Foundation and the Ford Foundation.

refining processes but only about five years for important new pharmaceutical products. Third, within an industry, this time interval varies considerably: for example, its standard deviation was about five years both in the case of the petroleum refining processes and the pharmaceutical products. Fourth, the time interval is shorter, on the average, when the inventor himself attempts to innovate than when he does not attempt to do so. Fifth, this time interval seems much shorter, on the average, than it was fifty or seventy-five years ago.[1]

These findings are based on the traditional distinction between invention and innovation. Although this distinction is useful, economists have too often tried to use this simple two-step model – invention and innovation – to characterise the entire process that, if successful, leads to the first commercial introduction of a new product or process. The obvious truth is that this process contains a number of stages, and that for many purposes, it is important that these stages be sorted out and recognised. As a first step, let us recognise that the time interval between the discovery or establishment of technical feasibility of a process or product and its first commercial application can be broken down into two parts: (1) the time interval between the establishment of technical feasibility and the beginning of commercial development, and (2) the time interval between the beginning of commercial development and the first commercial application of the new process or product. According to the available evidence, the first part of the time interval has been about as long as the second part in the post-World War II period, and the decrease in the total interval over the past seventy-five years has been due much more to a decrease in the first part of the interval than to a decrease in the second part.[2]

Next, let us recognise that the second part of this interval – the time elapsing from the beginning of commercial development to the first commercial introduction of the new product[3] – consists of a number of distinct stages – applied research, preparation of product specification, prototype or pilot plant construction, tooling and construction of manufacturing facilities, and manufacturing and marketing start-up. Judging from the available evidence, the stage

---

[1] See F. Lynn, 'An Investigation of the Rate of Development and Diffusion of Technology in our Modern Industrial Society', *Report of the National Commission on Technology, Automation, and Economic Progress* (Washington, D.C., 1966); E. Mansfield, J. Rapoport, J. Schnee, S. Wagner and M. Hamburger, *Research and Innovation in the Modern Corporation* (W. W. Norton, 1971); and J. Enos, 'Invention and Innovation in the Petroleum Refining Industry', *The Rate and Direction of Inventive Activity* (Princeton, 1962).

[2] Lynn, 'An Investigation'.

[3] The stages would be different, of course, for a new process.

that generally goes on for the longest period of time is the prototype or pilot plant stage. Typically, this stage may go on for about half of this part of the interval. Applied research and tooling and construction of manufacturing facilities also go on for long proportions of this part of the interval, each of these stages continuing for about 30 per cent of this part of the interval, on the average. Of course, these various stages often overlap.[1]

When a new technique or product is introduced commercially for the first time, the battle is by no means over. Indeed, in many cases, the battle has just begun. According to the available data concerning the imitation process, it takes about eight years, on the average, before one-half of the major firms in an industry begin using an important innovation. And in many cases it takes longer. The rate of imitation varies widely. Although it sometimes takes decades for firms to install a new technique, in other cases they follow the innovator very quickly. For example, it took about fifteen years for half of the major pig-iron producers to use the by-product coke oven, but only about three years for half of the major coal producers to use the continuous mining machine. Even when a firm begins using a new technique, this is not the end of the story. It generally takes a number of years before a firm completes the substitution of the new technique for the old. For example, once it began to dieselise, it took about nine years, on the average, for an American railroad to increase its stock of diesel locomotives from 10 per cent to 90 per cent of the total. And there was wide variation among firms in the intrafirm rate of diffusion. For example, some firms took only a few years to go from 10 to 90 per cent, whereas others took fourteen years or more.[2]

III. SCIENCE, EDUCATION, AND THE APPLICATION OF NEW TECHNOLOGY

Judging by the previous section, it often takes a long time for new technology to go through all the stages leading from the first recognition or discovery of technical feasibility to widespread use. Moreover, the length of this process is highly variable. What determines how rapidly the various stages of this process are carried out? Clearly, the nature and extent of a nation's scientific capability, and the size and quality of its educational system are of fundamental importance. With regard to the influence of a nation's scientific

---

[1] See Mansfield, Rapoport, Schnee, Wagner and Hamburger, *Research and Innovation*.

[2] E. Mansfield, *Industrial Research and Technological Innovation* (W. W. Norton, 1968).

capability on its rate of application of new technology, the first thing that must be said is that science and technology are two quite different things, and that they have drawn together only recently. Until the twentieth century, it simply was not true that technology was built on science. Even today, many technological advances rely on little in the way of science. However, in more and more areas of the economy (such as aircraft, electronics, and chemicals), the rapid application of new technology has come to depend on a strong scientific base. Merely to be able to imitate or adapt what others have developed, it is necessary for a firm to have access to high-calibre scientists.

However, this does not mean that leadership in basic science is a prerequisite for leadership in technology. Unless a country is able to exploit its leadership in basic science, this leadership may not have a great effect on its technology. For example, according to recent studies, the United States has been more adept than Western Europe at the application and translation of the findings of basic research into economically significant innovations. Moreover, it is important to add that, according to many observers, industrial innovations are based to a considerable extent on old science. They often involve science that is twenty-five or thirty years old when they are developed, due largely to the fact that the technologists depend heavily on the science they learned in school. Thus, it is not at all accurate to assume that innovations, when they are science based, must rely on or exploit relatively new science.[1]

A nation's educational system also has a fundamental influence on the rate of application of new technology. First, and perhaps most obviously, it determines how many scientists and engineers graduate, and the level of competence of these scientists and engineers. Clearly, the rate of application of new technology depends upon the quantity and quality of scientific and engineering talent available in the society. Second, it influences the inventiveness and adaptability of the nation's work force. Whether or not modern technology can be used effectively depends on the skills and educational level of the work force. Equally important, the extent to which inventions and improvements are forthcoming from the labour force depends on their skill and educational level. The fact that technology is becoming more closely linked to science should not lead one to

[1] See National Science Foundation, *Proceedings of a Conference on Technology Transfer and Innovation*, Washington, 1967; H. Bode, 'Reflections on the Relation between Science and Technology', *Basic Research and National Goals* (National Academy of Sciences, 1965), IIT Research Institute, *Technology in Retrospect and Critical Events in Science*, prepared for the National Science Foundation, (1968); and J. Ben David, *Fundamental Research and the Universities* (Organisation for Economic Co-operation and Development, 1968).

believe that workers and independent inventors are no longer important sources of invention. On the contrary, they remain very important in many areas.[1]

A third way that a nation's educational system influences the rate of application of new technology is through its impact on the nation's managers. There is considerable evidence that better educated managers tend to be quicker to adopt new technology than poorer educated managers.[2] Unfortunately, little beyond this seems to be known. For example, we know practically nothing about the sort of education that is best suited to develop, and meet the needs of, managers who are able to perform well as leaders of the innovation process and as adapters to technological change. One of the most important ways that a nation's universities might contribute to rapid application of new technology is through the development of curricula and methods to help nurture managers of this type.

IV. INDUSTRIAL ORGANISATION AND THE NATURE OF MARKETS

Besides a nation's scientific and educational capabilities, the speed of application of new technology depends on the state of the economy. For example, a high rate of investment in plant and equipment will encourage the rapid use of new techniques, since some techniques are capital-embodied. However, this does not necessarily mean that innovations are most likely to occur at the crest of the business cycle. In the steel, oil and coal industries, there is some evidence that the maximum rate of introduction of new processes has occurred when they were operating at about 75 per cent of capacity. Contrary to the opinion of many economists, the available data indicate no tendency for innovations to cluster at the peak or the trough of the business cycle. Apparently, process innovation at the trough is discouraged by the meagreness of profits and the bleakness of future prospects; at the peak, it is discouraged by the lack of un-utilised capacity that can easily and cheaply be used for 'experimental' purposes.[3]

The organisation of industry and the nature of markets are also of great importance. There is a considerable literature at this point on the effects of firm size and industrial concentration on the rate

---

[1] See E. Mansfield, *The Economics of Technological Change* (W. W. Norton, 1968).

[2] For example, see J. Bohlen *et al.*, *Adopters of New Farm Ideas*, North Central Regional Extension Publication No. 13 (October 1961). For further discussion of innovators, see E. Rogers, *Diffusion of Innovations* (Free Press, 1962).

[3] E. Mansfield, *Industrial Research and Technological Innovation*.

of utilisation of new technology. Although there is by no means complete agreement, the evidence seems to indicate that, for firms to respond quickly to certain types of new technology, they must be beyond a certain threshold size. However, there is no evidence that this threshold size is so great that corporate giants – by United States standards – are required in all or even most industries to promote rapid technological change and the rapid utilisation of new techniques. Moreover, this threshold size varies from one aspect of an industry's technology to another, allowing complementaries and interdependencies to exist among large and small firms. There is often a division of labour, smaller firms focusing on areas requiring sophistication and flexibility and catering to specialised needs, bigger firms focusing on areas requiring larger production, marketing, or technological resources.[1]

Turning to the effects of industrial concentration, as distinct from firm size, there is no evidence that highly concentrated industries are quicker to introduce new technology than less concentrated industries. Although a certain amount of concentration probably promotes the rapid use of new technology, increases in concentration beyond a moderate level probably do more harm than good. Among economists who have carried out empirical work in this area, there seems to be considerable agreement on this score.[2] For example, a careful study of the diffusion of numerically controlled machine tools shows that, allowing for differences in the profitability of the innovation, there was no tendency for numerically controlled machine tools to be accepted and utilised more rapidly in highly concentrated industries than in less concentrated industries.[3] Despite the fact that a vocal band of economists, led by John Kenneth Galbraith, claim that an 'industry of a few large firms [is] a perfect instrument for inducing technical change,'[4] the facts do not seem to support them.

Another factor that can influence the rate of utilisation of new technology is the size and sophistication of available markets. The scale of the market influences the extent to which the firm can spread the fixed costs of developing and introducing an innovation. Also, the sophistication of market demand obviously will influence the kind of technology adopted by producers. For example, the large

[1] E. Mansfield, *The Economics of Technological Change*.

[2] For example, see F. M. Scherer, *Industrial Market Structure and Economic Performance* (Rand McNally, 1970).

[3] This work is being carried out by A. Romeo, a graduate student at the University of Pennsylvania, who is writing a doctoral dissertation on numerical control under my direction.

[4] John Kenneth Galbraith, *American Capitalism* (Boston: Houghton Mifflin, 1952), p. 91.

government demand for advanced electronic components, particular types of scientific instruments, and electronic computers certainly promoted the rapid utilisation of new technologies in these areas in the United States. Of course, it should not be assumed that the relevant market is only the national market. Tariffs and other barriers can be overcome, although it is often a difficult task to do so.[1]

Still another factor that must be mentioned is the availability of risk capital. The application of new technology is typically a risky business, and, as indicated above, it is often begun by small businesses run by technical people with little business experience and little knowledge of the market for venture capital. On the other hand, the banks, wealthy individuals and others who are in a position to put up the money typically have no appreciation of the technical matters (and their staffs are often of little use in this area, either). There are enormous problems in making a proper appraisal of such new ventures. Without question, an important determinant of the rate of application of new technology is the size of the pool of venture capital – and the efficiency and imagination with which it is lent out.[2]

V. MANAGEMENT AND ORGANISATION OF FIRMS

Having discussed some of the broad social and economic determinants of the rate of application of new technology, we turn at this point to look at the role of an extremely important agent in this process – management. To begin with, it is obvious that the rate at which new technology is applied is dependent on the way in which firms make decisions as to which R & D projects to carry out, which ideas to develop, and which new processes and products to commercialise. Since these decisions must be made under considerable uncertainty, particularly if the new technology is a major advance in the state of the art, the rate at which new technology is utilised is dependent on the extent to which firms are willing to accept risk. In practice, the available evidence seems to indicate that the commercial risks typically are greater than the technical risks. That is,

[1] According to a recent study, by the O.E.C.D., there is a relatively low correlation between various countries' rates of innovation and the size of their national markets. See Organisation for Economic Co-operation and Development, *The Conditions for Success in Technological Innovation* (unpublished, 1970), p. 33. However, it should be noted that this analysis rests on a relatively small number of data points and that other important factors are not taken into account.

[2] See U.S. Department of Commerce, *Technological Innovation: Its Environment and Management* (1967).

there is a much greater chance that a firm can achieve the technical goals of the R & D projects it picks than that it will have proved economically worth while to have achieved these goals. In part, this may be due to inadequate co-ordination between R & D, on the one hand, and marketing and production, on the other. In part, it may also be due to improper tradeoffs between technical and commercial risks.[1]

It is important to emphasise that the proper management of innovation – and the proper utilisation by the firm of new technology – is much more than the establishment and maintenance of an R & D laboratory that produces a great deal of good technical output. In many industries, the bulk of the significant innovations are not based in any significant degree on the firm's research and development. For example, a National Planning Association study found that only about 17 per cent of railroad innovations, 27 per cent of housing innovations, and 44 per cent of computer innovations that occurred in a recent period were based directly on R & D. It is also important to note that many ideas come from external sources. Even in industries like chemicals and drugs, where R & D plays an important role, a large proportion of a firm's innovations are based on inventions made outside the firm. Moreover, even when the basic idea comes from a firm's own R & D, the coupling of R & D, in the one hand, and marketing and production, on the other, is of enormous importance. For example, many good ideas are not applied properly because the marketing people do not really understand them, and many R & D projects are technically successful but commercially irrelevant because they were not designed with sufficient comprehension of market realities. Typically, technological innovations seem to be stimulated by perceived production and marketing needs and requirements, not by technological opportunities. In other words, in the bulk of the cases, it takes a market-related impetus to prompt work on a successful innovation.[2]

Unfortunately, there is little data concerning the relationship between a firm's organisation and procedures, on the one hand, and the rate at which it utilises new technology, on the other. However, many observers are impressed by the difficulty of carrying out radical innovations in the large, established firm, the feeling often being that such innovations tend to be spearheaded by new firms and 'invaders' from other industries. Also, there is some evidence that firms where problems and tasks are broken down into speciali-

---

[1] E. Mansfield, Rapoport, Schnee, Wagner, and Hamburger, *Research and Innovation.*

[2] S. Myers and D. Marquis, *Successful Industrial Innovations*, National Science Foundation (1969); National Science Foundation, *Proceedings of a Conference on Technology Transfer and Innovation* (Washington, 1967).

ties and where there is a strict vertical chain of command are less likely to be innovative than firms where there is no strictly defined hierarchy, where communication resembles consultation rather than command, and where individuals have to perform tasks in the light of their knowledge of the tasks of the whole firm.[1]

As noted above, one of the biggest problems of the firm is to effect a proper coupling between R & D and the rest of the firm. To help solve this problem, firms often promote frequent personal contacts between research workers and other parts of the firm. There is considerable evidence that person-to-person contacts are the most effective means of transfering technology.[2] Also, there is evidence that the existence of a champion, a missionary, a zealot who fights strongly – and sometimes not in strict accord with 'the rules' – for an innovation is an important contributor to the application of new technology. Thus, it is sometimes argued that organisations that can somehow tolerate, perhaps even nurture, such individuals may be most likely to utilise new technology. Finally, a number of large firms – Du Pont, I.C.I., 3M, and others – are trying to obtain some of the advantages of the small firm by creating a number of teams in their development department that operate somewhat like small firms. New venture management – as this is called – is an interesting development.[3] It will be some time before its success can be estimated at all accurately.

VI. A SIMPLE MODEL OF THE IMITATION PROCESS

In the previous three sections, we discussed some of the general determinants of how rapidly new technology is applied. Let us turn

[1] D. Schon, *Technology and Change* (Delacorte, 1967); and T. Burns and G. Staulker, *The Management of Innovation* (Tavistock, 1966).

[2] It should be noted that the movement of people seems to be one of the most effective ways to transfer technology, whether this transfer is 'horizontal' (from one firm or type of application to another) or 'vertical' (from the general to the particular, as when science leads to technology). People in basic and applied research move, with new technologies, into application and operations. People in one firm move to other established firms and form new firms. Publications seem to be a much less important means of technology transfer than direct person-to-person contacts. Judging by a number of studies of both military and civilian innovations, personal contact seems to have resulted in the transfer of the relevant knowledge in about four-fifths of the cases. See Office of the Director of Defense Research and Engineering, *Project Hindsight* (Washington, D.C., October 1969); W. Price and L. W. Bass, 'Scientific Research and the Innovation Process', *Science* (16 May 1969); M. Tannenbaum, *et al.*, *Report of the Ad Hoc Committee on Research–Engineering Interaction* (NRC Materials Advisory Board, 1966); Myers and Marquis, *Successful Industrial Innovations* and H. Brooks, *et al.*, *Applied Science and Technological Progress*, (GPO, 1967).

[3] D. Schon, *Technology and Change;* and R. Peterson, 'New Venture Management in a Large Company', *Harvard Business Review*, (May–June 1967).

now to the diffusion of a particular technological innovation. What determines how rapidly the use of a particular new process spreads from one firm to another? Given that one firm has begun using a new technique, what determines how rapidly other firms begin using it? Some years ago, I suggested that the following simple model might be of use:[1] Letting $\lambda_{ij}(t)$ be the proportion of firms not using the innovation at time t that introduce it by time $t + 1$, I proposed that

$$\lambda_{ij}(t) = f_i(P_{ij}(t), \Pi_{ij}, S_{ij}, \ldots), \tag{1}$$

where $P_{ij}(t)$ is the proportion of potential users of the innovation that have introduced it at time t, Π_{ij} is the profitability of installing this innovation relative to that of alternative investments, and S_{ij} is the investment required to install this innovation as a percentage of the average total assets of the firms. In other words, the model assumes that the probability that a non-user will use the innovation between time t and $t + 1$ is dependent on the proportion of firms already using the innovation, the profitability of using the innovation, and the investment required to install the innovation.

Assuming that $\lambda_{ij}(t)$ can be approximated adequately by a Taylor's expansion that drops third and higher-order terms and assuming that the coefficient of $P^2_{ij}(t)$ in this expansion is zero, it can be shown that the growth over time in the number of firms having introduced the innovation should conform to a logistic function. Specifically,

$$P_{ij}(t) = [1 + e^{-(L_{ij} + \phi_{ij}t)}]^{-1}. \tag{2}$$

It can also be shown that the rate of imitation depends only on ϕ_{ij}, and on the basis of our assumptions,

$$\phi_{ij} = b_i + a^1 \Pi_{ij} + a_2 S_{ij} + Z_{ij} \tag{3}$$

where the a's and b's are parameters and Z_{ij} is a random error term.

This model has been tested against data for over a dozen innovations in five industries, the results being quite favourable. In general, the growth in the number of users of an innovation can be approximated by a logistic curve. And there is a definite evidence that more profitable innovations and ones requiring smaller investments had higher rates of imitation, the relationship being similar to that predicted in equation (3).[2]

---

[1] E. Mansfield, 'Technical Change and the Rate of Imitation', *Econometrica*, (October 1961).

[2] Ibid. Also, see Zvi Griliches, 'Hybrid Corn: An Exploration in the Economics of Technological Change', *Econometrica* (October 1957), for an excellent study of the diffusion of an agricultural innovation.

VII. APPLICATION OF THE MODEL: FORECASTING THE RATE OF APPLICATION OF NUMERICALLY CONTROLLED MACHINE TOOLS

Although this model is over-simplified in many respects, it has proved useful in forecasting the rate at which the use of particular innovations has spread. For example, it was employed in a study I carried out for the Small Business Administration at the beginning of 1968 to forecast the percentage of firms in the tool and die industry that would be using numerically controlled machine tools in 1970. When the study was carried out, about 20 per cent of the firms in the National Tool, Die, and Precision Machining Association were using numerical control. Two sets of data were obtained: interview data and mail survey data. Based on the interview data, the model forecasted that about 33 per cent of the firms would be using numerical control at the beginning of 1970. Based on the mail survey data, the model forecasted that about 37 per cent of the firms would be using numerical control at the beginning of 1970.[1]

To see how these forecasts compared with those obtained on the basis of other methods, I constructed two alternative types of forecasts. First, I asked the firms – both in interviews with a carefully selected sample of firms and in the mail survey of the industry – whether they planned to begin using numerical control in 1968 or 1969. Since there was considerable lead time required in obtaining numerical control and since firms plan reasonably far ahead, it seemed reasonable to suppose that their replies would have some forecasting value. The results of the interviews indicated that about 16 per cent of non-users planned to use numerical control by 1970; the results of the mail survey indicated that this was the case for 28 per cent of the non-users. Thus, the forecast was 33 per cent, based on the interview data, or 43 per cent, based on the mail survey. Second, I obtained forecasts from the machine-tool builders, the firms that presumably are closest to and best informed about the market for numerically controlled machine tools. About 25 of the 150 members of the National Machine Tool Builders Association provided forecasts. The results showed a considerable amount of variation, but the median forecast was about 30 per cent.

How accurate were these forecasts? Which forecasting approach was most accurate? Table 8.1 shows that the model's forecast based

---

[1] The mail survey resulted in data for about 300 firms. The interview data were based on a carefully designed two-stage sample of all firms. In both cases, eq. (2) was used to forecast the 1970 figure on the basis of the time series up to 1967. E. Mansfield, *Numerical Control: Diffusion and Impact in the Tool and Die Industry*, report to the Small Business Administration (1968).

on the data from the mail survey was almost precisely correct and that the model's forecast based on the interview data was off by only 4 percentage points. Regardless of whether we look at results based on the interview data or the mail survey, the model forecasts better

TABLE 8.1

ALTERNATIVE FORECASTS (MADE IN 1968)
OF THE PERCENTAGE OF FIRMS IN THE
UNITED STATES TOOL AND DIE INDUSTRY
USING NUMERICAL CONTROL BY 1970,
AND THE ACTUAL PERCENTAGE

Type of Forecast	Based on Interview Data	Based on Mail Survey
	(per cent)	
Model	33	37
Plans of tool and die firms	33	43
Median forecast by machine tool builders	30	30
Actual percentage[a]	37	37

[a] See note 2.

than the other two techniques. Moreover, it forecasts better than simple extrapolation by the usual 'naive' models.[1] Certainly, this is encouraging. Based on these and other results, it appears that this simple model may be of use in forecasting the rate of imitation. Of course, this does not mean that it is anything more than a crude device – or that it can be applied in situations where its basic assumptions do not hold. But it does mean that, used with caution, it may perform at least as well as other commonly-used forecasting devices.[2]

VIII. EXTENSIONS OF THE MODEL

Although the basic model outlined in Section VI is sufficient for some purposes, it omits many important variables. This section indicates several directions in which this model can be – and is

[1] Both for the interview data and the mail survey data, the model forecasted better than naive models that assumed that the increase in the percentage of firms using numerical control would increase by the same amount, or the same percentage, during 1968–9 as it did during 1966–7.

[2] To estimate the actual figure for 1970, we assumed that the ratio of the national percentage of firms using numerical control to the percentage of firms in the Philadelphia region using numerical control was the same in 1970 as it was during 1965–7. This assumption seems reasonable. This is the best estimate one can make. The N.T.D.P.M.A. directory is unreliable in 1970, according to A. Romeo's results.

being – extended. First, it is important to recognise that the dispersion of the profitability of the innovation among firms is likely to have a significant effect on the rate of imitation. Holding the average profitability of the innovation constant, the rate of imitation is likely to decrease as the variance among firms in the profitability of the innovation increases. For example, suppose that, on the average, firms can get a 30 per cent return from a new process. If there is wide variation of firms about this average, the firms where the rate of return is much lower than 30 per cent will be much slower to adopt the new process than the firms where the rate of return is much greater than 30 per cent. On the other hand, if there is little variation of firms about the average, there is likely to be less variation in how rapidly firms begin using the new technique, with the result that the rate of imitation will be higher. Empirical studies confirm the significance of this factor.[1]

Second, another factor that can have an important influence on the rate of imitation is the length of time that the innovation has been in use in other industries. Many innovations come to be used in a variety of industries; for example, numerically controlled machine tools are now used in the aircraft, automobile, electrical equipment, farm machinery, instrument, printing press, tool and die, and many other industries. Since the information and experience that is gathered in one industry can help to resolve uncertainties regarding the innovation in other industries in which it is introduced subsequently, one would expect the rate of imitation in an industry to be directly related to the length of time that the innovation has been in use in other industries (before this industry begins using it). Empirical studies seem to bear out this hypothesis; for example, our studies confirm that this has in fact been true in the case of numerical control.

Third, science-based industries that spend a great deal on research and development may tend to have higher rates of imitation, holding other factors constant, than industries based on technologies more remote from science and that do little R & D. This hypothesis is often advanced, but there have been no systematic studies indicating whether or not it is the case. We are currently engaged in a detailed study of the diffusion of chemical process innovations which should help to throw new light on the question. Also, an industry's market structure may have an effect on the rate of imitation. As noted in Section IV, some economists believe that innovations spread more rapidly in more highly concentrated industries. To

[1] For example, unpublished work carried out under my direction by A. Romeo (who is writing a doctoral dissertation on numerical control) indicates that this is the case.

test this hypothesis, I have tried in various ways to introduce this factor into the model. The results do not seem to bear out the contention that a high level of industrial concentration leads to a more rapid rate imitation. The effect of this variable is generally not statistically significant, and it freqently seems to be in the opposite direction.[1]

Finally, it is important to look at other measures of the rate of diffusion of an innovation besides the rate of imitation. For many purposes, it is more important to know how rapidly the percentage of output produced by the new technique increases, or to know how rapidly the percentage of new equipment that embodies the new technique increases, than to know how rapidly the percentage of firms using the new technique increases. We have been looking at all of these measures and comparing the results based on them. The results to date indicate that the same sorts of factors which seem to be important determinants of the rate of imitation are also important determinants of these other measures of the rate of diffusion. In part, of course, this is because there is a positive correlation between the rate of imitation and these other measures of the rate of diffusion.[2]

IX. TECHNICAL LEADERS AND FOLLOWERS

What are the characteristics of the firms that are relatively quick to begin using new techniques? What are the characteristics of those that are relatively slow to begin using them? Based on studies of over a dozen major industrial innovations in a number of quite different manufacturing industries, it is reasonably clear that firms where the expected returns from the innovation are greatest tend to be quickest to introduce the innovation and that firms where the expected returns from the innovation are lowest tend to be slowest to introduce the innovation. This, of course, is what we would expect: a firm's introduction of a new technique is delayed if the return is not deemed adequate to offset the risk involved. Another factor that influences how quickly a firm begins to use an innovation is its size. Holding constant the profitability of the innovation, big firms tend to introduce an innovation before small firms. In some industries, this may be due to the fact that larger firms – although

[1] E. Mansfield, Rapoport, Schnee, Wagner and Hamburger, *Research and Innovation*.

[2] For example, in the case of 21 chemical processes, unpublished work by Frank Husic, a graduate student at the University of Pennsylvania, shows that the correlation coefficient between the rate of imitation and the rate of diffusion (based on the rate of increase of percent of output produced by the new process) is over 0·70. Of course, this correlation may be much lower in other industries.

not necessarily the largest ones – tend to be more progressive than smaller firms. But even if the larger firms are not more progressive and do not introduce more than their share of the innovations, one would expect them to be quicker, on the average, to begin using a new technique than smaller firms. Thus, there is no contradiction between this finding and the conclusion in Section IV that the largest firms do not seem to introduce more innovations, relative to their size, than somewhat smaller firms.[1] Also, there is no contradiction between this finding and the conclusion in Section VIII that the rate of diffusion of innovations does not tend to be higher in more highly concentrated industries.[2]

Other factors that often seem to be associated with how rapidly a firm begins using a new technique are the education and age of the relevant management personnel. Both in agriculture and in the tool and die industry, there is evidence that firms with better educated and younger managers tend to be quicker to begin using new techniques than firms with less educated and older managers. In industries with bigger firms, no such tendency is found, but this may be because the data used pertain to the wrong managers – the presidents of the firms, who are quite far removed from the decision to introduce many innovations.[3] Also, based on interview studies that I have carried out, my impression is that the firms that are quicker to begin using innovations are run by managers who are perceptibly brighter and less averse to risk than the managers of firms that are slower to begin using innovations. But this, of course, is merely speculation based on the most casual empiricism.

Before leaving this subject, it is important to note that the firms that are relatively slow to begin using an innovation tend to substitute it for older techniques more rapidly than those that are quick to begin using it. In other words, the late starters tend to 'catch up'. For example, railroads that began using diesel locomotives in the fifties made the transition from 10 per cent of full dieselisation to 90 per cent of full dieselisation much more quickly, on the average, than did railroads that began using diesel locomotives in the forties. The same sort of phenomenon is found in the case of numerical control in the aircraft, electrical equipment, farm machinery, printing press, computer, tool and die, coal-mining equipment and other industries. It is also worth noting that, holding constant the date of first use of the innovation, small firms tend to be quicker to

[1] See Mansfield, *Industrial Research and Technological Innovation*, pp. 171–2.
[2] Ibid.
[3] E. Rogers, *Diffusion of Innovations*; Mansfield, *Industrial Research and Technological Innovation*; and Mansfield, Rapoport, Schnee, Wagner and Hamburger, *Research and Innovation*.

substitute the new technique for the old than large firms. In part, this is due to the greater homogeneity of conditions and the smaller number of quasi-autonomous decision-making centres in a small firm.

X. PUBLIC POLICY AND THE RATE OF APPLICATION OF NEW TECHNOLOGY

Finally, a few words should be added concerning public policy. There are many ways that public policy influences the rate of application of new technology. For example, governments often promote the use of new technology by encouraging and financing research and development. This role of government has, of course, been very significant: for example, in the United States the federal government finances about 60 per cent of the nation's research and development. The composition of government spending on R & D – the extent to which it is geared to defence or civilian needs, the extent to which it is directed at one industrial or technological sector rather than another – is clearly an important determinant of which kinds of new technology tend to be introduced particularly rapidly.[1] Recognising that their scientific and technological resources are limited, many countries have decided to specialise in particular areas of science and technology. For example, in France, criteria have been established for the support of science and technology on the basis of market and technological opportunities, and it has been suggested that the nation be 'active' in certain fields, but only 'vigilant' in others.[2]

In addition, the government's procurement policies – both in R & D and elsewhere – can be important in other ways. For example, it has been suggested that performance criteria, which specify the desired end product without limiting the design to existing products, be substituted where possible for product specifications in government procurement. Proponents of performance-based procurement argue that it will free industry to innovate, limited only by the requirement that it perform certain specified functions; encourage cost reduction for the government; and encourage the government to serve as a pilot customer for technical innovations in areas where it represents a big enough market or a market sufficiently free from local restrictions, codes, etc., to make it worth industry's while to innovate.[3] Also, in the R & D area, the way in which the government

---

[1] Also, there are various programmes of loans by government to industry to promote the application of new technology, as well as tax incentives. Such programmes have been adopted by many countries.

[2] O.E.C.D., *The Conditions for Success.*

[3] E. Mansfield, *The Economics of Technological Change.*

procures R & D can have important effects. For example, many governments have tended to reconvert government laboratories and increase the amount of government-financed R & D done in industrial firms in order to bring R & D in closer contact with application and commercialisation. Thus, in the United Kingdom, the percentage of R & D carried out in government laboratories was reduced from 27 in 1961 to 22 in 1966.[1] In addition, in the United States, the government has made efforts of various kinds to stimulate 'spillover' of technology from defence and space projects to the civilian economy. However, the extent of their success to date is by no means clear.[2]

Another important factor determined by government policy is the educational system – the importance of which was discussed in Section III. It need hardly be pointed out that the rate at which new technology is applied is dependent on the number of scientists and engineers that a nation educates. But this is only part of the story. The general educational level of a nation's work force, not simply the number of scientists and engineers, will surely have an important effect. Moreover, the educational level – and type of education – of a country's managers will have an extremely important effect too. This latter point is emphasised in recent O.E.C.D. reports and elsewhere, but it cannot be over-emphasised. Too often in the past – and at present – the educational system's role in this area has been viewed almost entirely in terms of scientific and technical education, whereas in fact the proper education of managers may be at least as important.[3]

Still another way that governments influence the rate of application of new technology is through their policies regarding industrial competition and monopoly. In general, what evidence we have seems to indicate that competition and rivalry promote the application of new technology, whereas monopoly tends to slow down its application. Moreover, the government's attitude towards union policies aimed against the application of new technology are important. So, too, are its attitudes towards building codes and other devices that have been used to obstruct the use of new technology. In addition, since they influence the incentives for innovation, the tax laws are of some importance. The patent laws also have an effect, but it is doubtful that changes in the patent system of the sort proposed recently in the United States would be of primary importance.[4]

[1] O.E.C.D., *International Statistical Year for R & D.*
[2] For some discussion of the extent and type of such 'spillover', see Denver Research Institute, *The Commercial Application of Missile-Space Technology* (Denver, 1963).   [3] See O.E.C.D., *Gaps in Technology: General Report* (1968).
[4] See E. Mansfield, *The Economics of Technological Change*, ch. 7.

Finally, governments influence how rapidly new technology is applied by their policies regarding full employment, the distribution of the social costs of technological change, and the assessment of the social costs of new technology. Reasonably full employment, coupled with a high rate of investment, tends to promote the application of new technology and the rapid and relatively painless adaptation of the labour force to technological change. Programmes that spread more equitably the social costs of technological change – for example, retraining programmes that help displaced workers and programmes of continuing education that prevent professional obsolescence – reduce resistance to new techniques. The effects of the newly-proposed technology assessment programmes are themselves difficult to assess, although the desirability of an effective mechanism for technology assessment – if one could be devised – is clear enough. However, given our very limited capacity for technological forecasting and the tremendous problems of evaluating the various effects of a new technology, it is important that, while exploring the usefulness of various types of technology assessment programmes, we guard against the possibility that these programmes are turned into devices for the retardation of needed technological change.[1]

[1] See National Academy of Sciences, *Technology: Processes of Assessment and Choice*, U.S. House Committee on Science and Astronautics (July 1969); and National Academy of Engineering, *A Study of Technology Assessment*, U.S. House Committee on Science and Astronautics (July 1969).

Discussion of the Paper by Professor Mansfield

Professor Lundberg, in introducing the paper, said that it raised many interesting and fundamental questions. In particular Professor Mansfield asked at the very beginning of the paper whether new technology was in general being applied at an appropriate optimal rate or whether governments should try by various means to speed up or slow down this rate. Professor Mansfield came back to this issue at the end of the paper, when he discussed government policy and even expressed some anxiety that programmes for evaluating the various effects of new techniques might be turned 'into devices for the retardation of needed technical change'. This was certainly a fundamental question that could not be answered. But the issue should unquestionably be raised and discussed on the basis of whatever evidence and subjective valuations might be used. Professor Lundberg said that he would later return to this issue. First we needed, following Professor Mansfield, to understand what factors appeared to determine the speed of technological change. And Professor Mansfield had many extremely interesting observations to make on this problem – observations to a large extent based on a combination of theory and econometric research.

Professor Lundberg said that he could only dwell on some few of these points, and he had tried to pick out certain points on which one would like more information from Professor Mansfield or where his arguments seemed doubtful and needed to be discussed. Professor Mansfield stressed the importance of a nation's scientific capability and the size and quality of its educational system. Everything he said on these factors was very much to the point and he presented a great deal of most interesting evidence. What Professor Lundberg missed in his presentation of it was any consideration of possible negative effects of higher education. The rising awareness of bad environmental effects, the rising social costs (in many respects) of rapid industrial development, could be regarded as results of better general education. Selection of new techniques would to an increasing extent be affected by considerations of environmental effects (for example on air, or water pollution). The people who wrote about these things were highly educated and saw the negative impacts. There are also the problems of redundancies and retraining – many older workers were not prepared for retraining, and in comparing the speed of transmission of new technology the attitude of trades unions had to be considered. In this respect there were important comparisons to be made between Sweden and the United Kingdom. In Britain the trades unions had traditionally had a great impact on the introduction of new technology. The Swedish attitude has been quite different – much more positive to new techniques and new machines, but he thought that the present more negative attitude would have a slowing-up impact on the speed of application in the future.

On the question of government policy Professor Mansfield emphasised the importance of the state of the economy, and pointed out the inadequacy of many of the generalisations of business-cycle economists like Galbraith and Schumpeter. Empirical facts did not substantiate their statements; in fact innovations were not predominantly initiated at the top of the boom. However, Professor Mansfield did not mention how things go in a depression. Did a depresssion tend to eliminate the long tails of inefficient firms which were using old technology, implying a raising of the average standard? Were there any data on this old issue?

On the question of concentration, Professor Lundberg said he was delighted to see old theories like those of Galbraith being killed. Professor Mansfield said that there was no evidence that concentration had an impact on the speed of adoption of new technology. Further clarification of this would, however, be helpful, as in his necessarily short paper there seemed to be a little confusion. In one particular section there was a suggestion that big firms introduced innovation before the small, but Professor Mansfield maintained that this was no contradiction between this and his statement about concentration. He also said that competition and rivalry promoted the application of new technology and that monopoly slowed it down. Again, it would be valuable to have some amplification of this point.

Another issue which was raised was that of the availability of risk capital. In this respect it was necessary to ask about the measurement of profitability of a new invention. The questions of the supply of risk capital for innovation and profitability were concerned with two opposing forces. There might be an ample supply of risk capital from high profits but there might not be sufficient pressure to make innovations; therefore in boom periods one might not necessarily get high rates of innovation. For example, in Swedish industry in the 1960s there had been a heavy squeeze on profit margins from rapidly increasing wage rates which forced the firms at the lower levels of efficiency to rationalise and innovate if they were not to die. In the 1950s with inflationary profits and a relaxed atmosphere there had been a much slower introduction of new technology. All this was very important from the policy point of view. We had also to remember the relationship between the volume of production and the rise of productivity and to remember that if profit margins were squeezed inefficient firms tend to be squeezed out. Productivity might rise rapidly but the volume of employment might go down and labour had to be shifted and retrained. Professor Mansfield mentioned the question of marketing and the problems of co-ordination. He had also said that commercial risk was higher than technical risks. It would be very interesting to know how this difference was measured, and what more could be said about it. He also presented in his paper a model of the imitation process which he had used for forecasting and found that it worked well. Perhaps this could be discussed later in relation to the paper by Dr. Nabseth. One factor which was stressed was the profitability of using innovation and the dispersion of this profitability. He would like to know how such expectations of profitability were measured. Was it done by questionnaires, and did one tend to get optimistically biased estimates? Was there any information on this and had the results been tested *ex post*? There appeared to be a wide variation around his 30 per cent and Professor Mansfield had suggested that if there was wide dispersion the rate of adoption was lower. It would perhaps have been more interesting if he had reached the opposite result. One must also remember that profitability was not an easy criterion. If in a certain branch one firm did not do what most other firms had done with regard to innovations there were likely to be losses for this firm. If an innovation succeeded in the branch there might be lower profits or higher costs for the firms which did not make the innovation in question. It was therefore important to look at the whole structure of the branch of industry and to ask what costs had to be covered for the older plants to survive. One was likely to find that the best plants had recently introduced new technology but that other firms might well be using ten- to twenty-year-old technology because they could still cover the direct costs. This had a great deal to do with the question of the optimal speed for the introduction of new technology. This type of lag might thus be the result of good management, not bad, and risks might be reduced by waiting. It was therefore important to note that the structure of a branch of industry would have a bearing on the speed of new technology introduction that was to be expected.

Professor Mansfield had mentioned the importance of management and how managers were educated, and suggested a correlation between the introduction of new technology and 'brightness' of managers. Professor Lundberg said it would be very interesting to hear a little more about this, because he must admit that he himself had made great mistakes in judging the brightness of managers. Many that he thought to be good because they had rather academic minds had often turned out in practice to be bad entrepreneurs. It would be interesting to know what Professor Mansfield's criteria of brightness were and to ask if he had managed to avoid a circular definition.

On the question of government policy Professor Mansfield had discussed various ways in which it could make an impact. Some might appear as rather vague assertions but one would assume that these were based on some empirical evidence. In discussing the returns to social costs Professor Lundberg said he would like to ask what the error was likely to be from calculating private profitability of introducing innovations, when regarded from the point of view of the whole economy. As had been pointed out, education might mean reduction of resistance to new technology, but it might also imply more critical attitudes to new technology with reference to high social costs in a number of respects. At the present time there was an aversion among many young people in Sweden to taking employment in industry. They preferred service occupations because of the supposed monotony and dirtiness of industrial work, which in many cases had to be done by immigrant workers. Compensation could be and in many cases was given to people who suffered from the introduction of new technology but these costs should be included in a social profitability measurement. Was it not probable that consideration of social costs would have an impact on the speed of introduction of new technology?

Professor Mansfield, replying to Professor Lundberg's question on industrial organisation, said that at first glance there might appear to be a contradiction amongst the findings. But in reality there was no contradiction. To clarify this he suggested we should consider two alternative approaches. The first was to look at the rates of diffusion on the basis of the percentage of firms which were using the innovation. If one then used as explanatory variables profitability and if one included measures of concentration, the latter were not statistically significant. In fact, the sign often came out opposite to that expected on the Schumpeterian hypothesis. The second approach looked at individual firms in respect of which ones were quick and which ones were slow at introducing innovation. If one held profitability constant, the bigger firms tend to be quicker, on the average. This was quite compatible with the first result; in fact, the second proposition was almost trivial, but interesting. If one considered that a large firm might be taken to be a large number of small firms then one could assume that the distribution of its present equipment would be wide in terms of age, degree of obsolescence and need for replacement. Therefore a large firm was likely to have at least one section within which it could take advantage of an innovation at a fairly early stage. In the case of the large group of small firms perhaps only a small number of these would be in a similar situation. Thus when we calculated the proportion of large firms that innovated in the first stages it would inevitably be bigger than the proportion of small firms, simply because the expected value of the minimum of a group of items decreased as the number of items in the group increased. If one asked what effect industrial organisation had on the amount spent on R & D, on the productivity of R & D, on the nature of R & D projects carried out and on how quickly results were put into practice, one had only scraps of information. But these scraps seemed to present the following kind of picture: In the United States we had no evidence that if one allowed the mergers of large firms to make giants there would be any increase in R & D. Money would be no more efficiently spent in terms of output. If inventions required large amounts of money to implement, then big firms might have advantages here. There was no evidence that the biggest firms did more basic research and there were data from a study of a sample of a couple of dozen firms in the chemical and petroleum industry to back up this statement. On the question of the commercial risk being greater than the technical risk, Professor Mansfield agreed that this was a rather slippery dichotomy and many people might not think it was useful, particularly if R & D had not got specific objectives. However, their experience was that most R & D was aimed at specific targets. In this case the technical objectives could usually be clearly defined, and many firms did use this dichotomy in many branches of United States industry and made a clear distinction between the technical and commercial success. There were, however, some problems about definition here. For example, technical success with how much spent and over how long? We might define as within the orginal budgetary constraint. Or we might define as before technical completion, even if it cost twice as much or takes twice as long as the estimates orginally made. In most of the firms they had looked at, the technical risks were rather small, relative to the commercial risks. The relatively high probability of commercial failure was often due, of course, to the fact that market conditions changed, but in some cases the failure rates were due to inadequate co-ordination between the marketing part of an organisation and R & D. This interface between user and producer of new knowledge was a rocky area. In making the trade-off between technical and commercial risks it was important to remember that in project selection R & D people sometimes had an inordinate weight. Professor Mansfield said that many people had the vision of R & D being concerned with large technical advances, but this really was not true. R & D people were often risk avoiders because of the way their performance was judged. They opted for projects in which technical risks were small, but this might lead to the commercial risks often being high.

Professor Khachaturov had found the paper very interesting and provocative. He would like to make some suggestions about possible further investigations in this field. Firstly, he would like to suggest that it would be advisable to try to measure the interdependence between the speed of application of the technology and the rate of economic growth. He thought that this would be difficult over the whole economy because of lack of statistics, but in particular branches he thought it would be possible. For example, in the U.S.S.R. statistics were available for the main branches of machinery which would be useful for this purpose. Secondly, he thought it was important to look at the influence of the size of innovations on the speed of application of new technology and suggested that for this purpose it was very important to use the data of the amount of capital investment necessary for a new technology. Capital investment should be brought into the analysis not only in the macro- but also in the micro-economic work. Thirdly, on the question of the importance of size of enterprises for the speed of application of new technology, he noted that Professor Mansfield had said that the small firms tended to be quicker to substitute new technology than large firms. But no explanations were given for this. Professor Khachaturov thought that this was hardly possible because large firms had many advantages in comparison with small firms, for example in being able to finance investment or secure credits. He thought this should be further investigated. Finally, Professor Khachaturov raised the question of the relationship between the possible risk and the possible profit of new inventions. He thought that there could be measures of this and that in many cases these two items played an important role in the development of new technology. He wondered whether Professor Mansfield could take these considerations into account in further developments of his work. Dr. Sedov had suggested that standards might be a key factor in the introduction of the latest scientific and technical achievements in advanced technology. Standards in a socialist economy embodied technical, economic and legal requirements and they served as the normative scale of quality measuring, a means of regulation, systematisation and differentiation of products to meet the needs of society and a means of ensuring quality. Dr. Sedov had said that a unified system of engineering requirements for products of the engineering, instruments and automation industries was being developed in the U.S.S.R. This system would establish a uniform procedure for the development of technology, and make provision for broad application of standard manufacturing processes, standard tooling and unitised equipment. The standard was a normative requirement which had to be met by all enterprises and its task was to achieve optimal reconciliation of the interest of manufacturing and of consumers and to establish lower and upper limits of quality change within a specified period of time. Standards specified two levels of indices, lower indices that were obligatory for all and more progressive ones which were recommended. So-called prospective standards were a means of planning the application of new technological ideas in the national economy. Standards of this type had established more rigorous and stable indices which should characterise future products in some particular respect. These standards not only specified special requirements for products that were to be certified, but at the same time established higher quality indices for raw and finished materials and assembly units required for the manufacture of such certified products. Norms and requirements specified by these standards should be based on the experience of research design and experimental work. But development of the quality indices of new products and technology was a subject of the plan for standardisation which formed an integral part of the five-year national economy plan. This plan co-ordinated activity in standardising finished products and the principal raw materials, semi-finished products and components. The plan for standardisation was co-ordinated with the plans for developing new technology, capital investment and materials supply, and in the sense standards were a key factor in the development and introduction of the latest scientific and technical achievement.

Professor Gatovskiy said that the paper by Professor Mansfield threw valuable light on many of the questions about the acceleration of the application of results of research. There was, however, a tendency to confuse a technique which was completely new, based on a new principle, and which brought in its train great economic and social effects, with new technology often based on traditional principles but not yet applied. In the U.S.S.R. distinction was made between three categories of industrial capacity: firstly that which corresponded to the very highest national and international standards, secondly that which corresponded to normal modern standards and thirdly that which would soon need to be replaced by a new capacity. The policies of planning and stimulation, the policies of price, the policies of profit and of bonuses in the U.S.S.R. were based on this classification. Yet in spite of the preferences which were given to the new technology required for progress, for a variety of reasons firms and enterprises often preferred to retain a technology which was basically little modernised and to fail to adopt the technology which was essential for progress. In the U.S.S.R. in order to bring applied research into closer touch with the production stage and to accelerate application they had established the self-financing of institutes of applied research, linked with a system of contracts governing wages and bonuses in these institutes which depended on the volume and results of the profit which was brought about by the application of their researches. This had brought good results and they were now creating in the U.S.S.R. in all branches of industry firms which were devoted to science-production. These firms must include a scientific institute and the director of the institute was at the same time the general director of the firm. Thus the institute had at its disposal factories and could build prototypes and carry out experimentation and the next. This meant that all the liaison connections between science and production were carried right through from the research right up to the industrial production of the first batches. Such an institute had the task of opening the way for essentially new and economically effective technology. All the organisms of the science-production firm had a common plan. The representatives of the factory participated in the research in the preparation of the design of the new product, and the researchers participated in the application. Professor Gatovskiy said that the experience to date had shown results which were very favourable. He thought that this experience would be of general interest.

Dr. Teubal returned to the distinction between the optimal speed of diffusion and the actual speed which he had found most interesting. He asked Professor Mansfield whether he envisaged two extreme forms of capital embodiment of R & D results and whether it was worth making a distinction between these in studies of innovation. The first form was where the new technology could be embodied in capital equipment replacing existing equipment. In this case, the previous technology was not displaced and in fact the new technology could very well add to the profitability of the existing capital stock. This could occur for example with new systems of production control, with the improvement of existing machines, with better methods of conveying materials, and also to some of the inventions of Swedish industry mentioned by Dr. Nabseth in his paper. At the other extreme R & D needed to be embodied in capital which made present equipment obsolete; an example was provided by more efficient electrical generating plant. In the real world most R & D would be intermediate between these two. But he wondered whether this conceptual difference had a relevance in discussing diffusion and whether the same model of diffusion was likely to be applicable to both these kinds of innovation, with the difference appearing only in the profit variable of the diffusion function. He thought it might be necessary first to formulate alternative models. Dr. Teubal had a second question on the measurement of the rate of diffusion by the rate of imitation. He agreed that this was important but suggested that it might not be the most appropriate measure. He thought that for resource allocation it was more important to measure the total output which was produced by the new technology. He did not want to argue that no one measure was likely to be appropriate to all situations or that the rate of imitation could not tell us something valuable about the attitudes of management.

Professor Mujzel said that the comments he would like to make had been evoked not only by the stimulating paper of Professor Mansfield, but also to some extent by the questions raised by Professor Khachaturov and Professor Gatovskiy. Some decades ago there had existed in Marxist social thought a broad hypothesis that the centrally planned socialist economy should be an extremely simple and transparent system and that the efficient running of it by any society would create no problems and difficulties. He said that now on the basis of experience we were becoming aware of the *naiveté* and deceptiveness of this view. They had a highly developed educational and research system and the number of engineers employed in their industries was proportionately one of the biggest in the world. The country was spending substantial and very rapidly growing funds on science in the basic and applied research fields. Nevertheless, technological progress in products as well as in processes had been slow, insufficient, in fact, to meet the rapidly expanding expectations based on it. Professor Mujzel suggested that the managers at various levels and the representatives of workers councils had not generally been strongly motivated to accept the tasks and responsibilities involved in innovations. In most cases they preferred to do routine production, to fulfil short-run production plans and to apply familiar strategies to achieve the planned objectives. This meant that new innovations, sometimes even small improvements, had to be introduced by applying administrative and political pressures. A number of studies has been undertaken into the problem of how to improve this situation and in the field of innovation management two approaches and two views had already become evident. The first of these, which might be called the more traditional view, was based on the profit oriented system. The second approach started with the questions which seemed to be really of first importance: was it possible to construct for socially owned enterprises such quantitative measures of performance, based on profit, value added or other micro-economic magnitudes, that they would be so future oriented as to create in these enterprises an uninhibited willingness to tackle the risky, difficult and costly business of the innovation. All the measures used so far in their economy assessed results achieved in the past, but not those reasonably to be expected in the future. What they would like to know was how they were to change the character of the measures concerned or design others which would be more appropriate.

Professor Mansfield, in replying to the question that had been put to him, proposed to start with the most fundamental – that concerning the optimal speed of diffusion. The purpose of his paper had not been to try to answer this. He felt that it entailed more than economics, and he endorsed Professor Lundberg's emphasis on its importance. On the question of risks and incentives, he agreed with Professor Lundberg that in very profitable times there might be be less incentive to innovate. However, on the question of risk capital he was more concerned with the problem of the small firms, the importance of which in the innovation had often been stressed, and he felt that they had considerable problems of raising the money from sources which were quite ignorant of the technical aspects of the projects which were submitted to them. On the question of measures of profitability, Professor Mansfield said this depended upon the nature of the innovation. In some areas there were good forecasts which were generally available. In other cases it was necessary to go direct to the firms where the information was in some cases already available. However, one must not expect that because one went to a big company and was provided with printed sheets containing figures and estimates, that they were, in fact, very accurate. They had had good co-operation from firms in the United States on this, but one had to be careful about how to interpret the data. In most cases small firms had got simple back-of-the-envelope calculations. However, it must be remembered that reasonable first approximations were good enough for the work that they were doing. One did not need to worry about the last five per cent of accuracy. Professor Mansfield said that they had used the reciprocals of pay-off period as proxy for profitability, but he wanted to point out that it was really necessary to look at the profitability of doing something now versus doing it later. He agreed with Professor Lundberg that the avoidance of loss was an important criterion in undertaking an innovation and, in fact, the speed of innovations might very well be higher in situations in which loss was likely to be avoided, than when it was a question of increasing future profits. He also drew attention to the kind of situations which had arisen in the case of continuous annealing, which had really been introduced to meet the requirements of customers, and containers, which had been introduced for marketing reasons and not in order to reduce costs. But he stressed that these factors were really part of, and subsumed under, the profitability of the innovation, properly defined.

Professor Mansfield agreed that it was not a question of the faster the introduction of innovation the better. Slowness was not necessarily a bad thing. He was interested in Professor Lundberg's comments on brightness and admitted that this was one of the sections in his paper which did not have any good empirical backing. He hoped that this had been made clear in the paper. He would like to follow this up and to look at the psychological characteristics of managers if he had the time and the money to do it.

On the questions from Professor Khachaturov, Professor Mansfield said that some studies had been made on the interdependence between the rate of productivity growth and the rate of use of new technology, and mentioned Enos' study of petroleum refining, and that carried out by Hollander on the rayon industry. He also mentioned a well-known fact that many major technical changes relied very little on science. On the quickness of large firms, Professor Mansfield said that small firms tended to be slower to begin but then moved at a faster rate. He was dubious about many of the supposed advantages to be derived from giant firms. It was clear that there was some threshold size above which one had to be in certain industries. For example, one could have a three-man shop in steel. But he felt that one did not necessarily need something the size of United States Steel. In many industries it appeared that firms that were a fraction of the size of the biggest firms contributed more, on a relative basis, than the biggest firms. He certainly felt that risk was an extremely important item and that the variance of the rate of return was a partial measure of the riskiness of innovation. Referring to Dr. Sedov's point on standards, Professor Mansfield agreed that these could make a useful contribution to the speed of introduction of new technology and on Professor Gatovskiy's point he agreed that similar problems occurred in the United States. In fact he had been interested to find that many of the Soviet problems were very similar to their own. On the points raised by Dr. Teubal he agreed that a lot could depend upon whether new technology displaced previous capital or not, and that had been recognised to some extent in the model he had used. He added that he had tried to look at this more closely, but so far had not been able to identify any significant effects of this kind. He also agreed that using the rate of output figures as an alternative measure of the diffusion of innovation was useful and, in fact, they had done this. Finally, Professor Mansfield thought it was better to think of two processes and look at the rate of imitation and then the rate of intra-firm diffusion. He thought this was probably more illuminating if one had the data and the patience rather than just using rates of output.

9   A Study of Success and Failure in Industrial Innovation

C. Freeman
UNIVERSITY OF SUSSEX

INTRODUCTION

This paper discusses the main results of a recent project carried out at the Science Policy Research Unit of the University of Sussex.[1] The project (SAPPHO) was an attempt to measure a large number of characteristics of successful attempts to innovate and to compare them with corresponding unsuccessful attempts.

Previous literature on industrial innovation falls largely into two categories:

(i) Case histories of particular innovations, concentrating on a variety of different aspects, but lacking comparability of coverage.
(ii) Theoretical analysis of the innovative process, sometimes citing individual instances, but lacking systematic empirical foundation.

As a result of these characteristics, there are many half-tested hypotheses and many interesting conjectures in innovation theory, but rather little firm evidence to support them. This project was designed to contribute in a small way to establishing a firmer basis for useful generalisation in this field.

The difficulties of making such generalisations are well known. There is no regular flow of comparable statistics relating to innovation. There are no systematic archives. There is considerable resistance in private industry to providing the necessary information. Even where there is no resistance, there are genuine difficulties in assembling accurate information, as innovation is often an untidy process extending over many years. Such statistics as do exist relate not so much to innovation as to invention (patent statistics) and to the wider category of research and development (expenditure and manpower).

Our work was concerned with the measurement of innovation, not invention. The distinction is of critical importance and has been

[1] This report is based on the work of a research team which consisted of B. Achilladelis, R. C. Curnow, C. Freeman, Mrs. J. K. Fuller, A. Horsley, P. Jervis, A. B. Robertson and C. Tudway.

widely recognised by economists since Schumpeter. From the standpoint of our analysis industrial research, development and invention are all activities which may contribute to the process of innovation, which is defined as: 'the technical, industrial and commercial steps which lead to the marketing of new manufactured products and to the commercial use of new technical processes and equipment'. Although innovation is a complex social process, the crucial step is the commercial launch of a new product or system. It is this step from experimental to commercial application, which enables us both to define innovation more precisely than invention and to date it fairly accurately.

From the standpoint of economic theory it is innovation which is of central interest rather than invention, and in this context it is somewhat strange that some of the most influential works in economics literature have been studies of invention rather than innovation. The two most systematic studies (Jewkes and Schmookler)[1] were in fact both studies of invention. In the case of Schmookler this may be attributed to the availability of patent statistics which were his life-long interest. In the case of Jewkes, he and his colleagues were actually obliged to extend the focus of their study from invention to innovation, since the recognition of the 'importance' of an invention depends upon subsequent innovation. If the case studies are interpreted in terms of innovation theory, rather different conclusions emerge. This point is important since some of our findings on project SAPPHO relating to innovation have been wrongly compared with earlier work relating to invention.

I. METHOD OF PROJECT SAPPHO

This project, which was supported by the Science Research Council, was a study of fifty-eight attempted innovations in chemicals and scientific instruments. Those in the chemical industry were mainly process innovations (listed in Appendix A) whilst the instrument innovations were all product innovations. The study was deliberately concentrated on only two industries, since it was assumed that there would be some variation in each industry, as well as some common factors. Future studies will test these generalisations in relation to other industries.

'Pairing' of success and failure

A search of the previous literature had demonstrated that scarcely any of it dealt with failure and what little emerged was anecdotal

[1] J. Jewkes *et al.*, *The Sources of Invention*, revised ed. (Macmillan 1970); J. Schmookler, *Invention and Economic Growth* (Harvard, 1966).

rather than analytical.[1] Yet it was generally known that attempted innovation frequently failed, so that it seemed essential to explore the characteristic pattern of both success and failure. Our own previous work at N.I.E.S.R.[2] had shown that, typically, several firms attempted to launch new products or processes roughly contemporaneously on the world market in the chemical and electrical industries. It seemed possible to 'pair' attempted innovations and thereby to discriminate between the respective characteristics of failure and success. The technique had, of course, been widely used in the natural sciences,[3] especially in biology. When the two halves of a pair differ with respect to a particular characteristic or set of characteristics, this indicates a possible 'explanation' of innovative success or failure. Where there is a significant and repeated variation between the pattern of 'success' and 'failure', across a large number of pairs, this provides systematic evidence for the validity of particular hypotheses or groups of hypotheses. Such explanations as appear to have a significant statistical foundation may then be tested again on a new sample of innovations, either by ourselves or other teams working in this field. In this way a structured and tested foundation for theoretical work may be built up. This report analyses twenty-nine successes and twenty-nine failures but we intend in future work greatly to enlarge this sample.

Resemblances within 'Pairs'

We expected to find that the Success and Failures halves of a pair would resemble each other fairly closely in many respects, and this proved in fact to be the case. We assumed from previous experience that firms attempting to develop a particular new process would often have many characteristics in common. The analysis of similarity is complementary to the analysis of divergence, for two reasons. First, it enables us to identify some characteristics which are shared by all firms attempting innovation in particular industries. These may be necessary conditions for entry into the race. Whether they may be regarded as such, will depend partly upon whether we are able to find 'success' cases which disprove such tentative generalisations. But, secondly and more important, they enable us to focus attention on those characteristics in which the pattern does diverge between success and failure. In future stages of SAPPHO work, we shall be able to concentrate in greater depth on these significant

[1] See for example the bibliography in K. Pavitt, *The conditions for Success in Technological Innovation* (O.E.C.D., 1971).

[2] *National Institute Economic Review*, Nos. 34 and 45.

[3] A. L. Mackay and J. D. Bernal, 'Towards a science of science', *The Technologist*, II, No. 4 (1966), 319–29.

differences. Thus the analysis of 'similarity' enables us to clear the ground, by a process of elimination of unnecessary hypotheses.

Classification of 'Pairs'

Our pairs are not 'identical twins'. Their similarity is defined in terms of their market, not necessarily in terms of their technology. For example, two firms may both be seeking a new, cheaper and better way to produce phenol or urea. They may adopt somewhat different technical solutions. It is an assumption of this project that this very choice constitutes part of the 'success', and the 'wrong' choice part of the 'failure'. In a few cases the resemblance is very close as where several licensees share access to the same basic technical knowledge. But even here the design varies when two manufacturers attempt to satisfy the same demand. Success depends partly on developing the 'right' design, having regard to the available scientific and technical knowledge and to the potential uses of the innovation.

Concentration on innovation rather than invention has many consequences in terms of method. The most important of these is that the marketing aspects of the process assume much greater importance, whilst the role of that individual usually described as *the* inventor recedes into a wider social context. We did not include in our comparisons those numerous experiments by inventors and would-be innovators which are discarded or shelved long before they reach the point of commercial introduction. Such studies are undoubtedly of interest in the management of R & D, but our focus was on the wider problem of the management of innovation. Our failures were products or processes which were brought to the point of commercial introduction, and usually were in fact on the market for some years. We were therefore interested in the early stages of development work and invention work, but also in the preparations for production and sale and in the experience of marketing the innovation.

Definition of 'success and failure'

Since the project is concerned with innovation in industry, the criterion of success is a commercial one. A 'failure' is an attempted innovation which fails to establish a worth-while market and/or make any profit, even if it 'works' in a technical sense. A 'success' is an innovation which attains significant market penetration and/or makes a profit. This report analyses twenty-nine 'successes' and twenty-nine 'failures'. Often a failure is relatively clear-cut, e.g. a firm goes bankrupt, or closes a plant down, or withdraws a product or fails to sell it. Success is not always so self-evident. A product may achieve a world-wide market, but take a long time to show a profit.

One case which we originally expected to be a 'success' was withdrawn from the market on these grounds. Even with failures it is not always simple to make an assessment. There are varying shades of grey between the extremes of success and failure. We deliberately tried to investigate the fairly clear-cut 'black and white' cases of failure and success. In two cases in the chemical industry, and one in instruments, it proved to be feasible to complete two pairs, as there were several commercial successes and several less successful attempts in each case.

Definitions of 'Technical Innovator', 'Business Innovator' and 'Chief Executive'

In the innovation literature there is a great deal about the critical role of key individuals. Not only 'heroic' theories of innovation but also more prosaic accounts emphasise the part played by entrepreneurs, managers and inventors. For the purposes of our investigation we therefore attempted to distinguish between various key 'roles' in the conduct of innovation. Although these roles have been recognised in much of the earlier innovation literature, they are not always identifiable from formal titles used in firms. The job title may vary a good deal, but it was the role which we attempted to identify, defined as follows:

(i) 'Technical innovator.' The individual who made the major contributions on the technical side to the development and/or design of the innovation. He would normally, but not necessarily, be a member of the innovating organisation. He would sometimes, but not always, be the 'inventor' of the new product or process.

(ii) 'Business innovator.' That individual who was actually responsible within the management structure for the overall progress of this project. He might sometimes be the Technical Director or the Research Director. He might sometimes be the same man as the 'Technical innovator'. He could be the Sales Director, or Chief Engineer. Occasionally, especially in smaller firms, he could be the chief executive for the organisation as a whole.

(iii) 'Chief executive.' The individual who is formally the head of the executive structure of the innovating organisation, usually but not necessarily with the job title of 'managing director'. In every case in our project there was an identifiable 'chief executive', but there was not always an identifiable 'business innovator', and quite often there was no identifiable 'technical innovator'. We did not attempt to force individuals to assume these roles if they were not readily identifiable, since one of the objects of our inquiry was to assess the contribution of outstanding individuals. In order to clarify this we also distinguished one other category of 'role' – product 'champion' – which might sometimes be performed by the same individual as the technical innovator, or 'chief executive'.

(iv) 'Product champion.' Any individual who made a decisive contribution to the innovation by activity and enthusiastically promoting its progress through critical stages.

Timing of pairs

The project is concentrated entirely on post-war innovations, although it includes some work on the early background. Since the circumstances affecting successful innovations are constantly changing we deliberately chose the most recent period. But too recent an innovation may raise the difficulty of measuring its success. Although the method could be applied to earlier periods, there would be very great data collection problems. About two-thirds of the innovations came to the market in the 1960s and about one-third in the 1950s. Each pair is not, of course, exactly simultaneous. Indeed one of the hypotheses which the project was designed to test was whether it is better to be 'first', or whether as is often alleged, it is better to come in later. In most of our innovations the 'gap' between 'launching' the success and the failure is only a few years, and in half of our instrument innovations, where lead times are shorter it is only a few months. But in three of the chemical innovation pairs it was more than five years. We attempted to guard against the possibility that 'success' might simply be due to extraneous scientific or technical developments not available at the first attempt. Such a pair would be excluded from our analysis. In two of the three chemical pairs, where the 'gap' was more than five years, the successful innovation was the first one.

Interviews and Data Collection for Each Pair

The project involved collecting a great deal of information about each pair to test a large number of hypotheses. From the beginning it was recognised that most of this information would not be available from published sources, and that most of the data would have to be collected by interviews and correspondence. However, in each case we searched the technical and patent literature, and in some cases were able to obtain valuable published papers. As in the Manchester innovation studies,[1] it was found that this preliminary

[1] J. Langrish, M. Gibbons W, G. Evans and F. R. Jevons, *Wealth and Knowledge* (Macmillan, October/November 1971).

technical study was a useful preparation for the interviewers themselves. Ideally, we attempted to interview several individuals in each company involved, as well as independent sources, such as customers who had detailed knowledge of the innovation. We attempted to interview individuals who might correspond to the 'technical innovator' or 'business innovator' as defined above, whatever their formal job titles, but we also attempted to identify and interview other executives with intimate knowledge of the innovation, including sales directors, managing directors and research directors. We were not always able to achieve the ideal because of response difficulties and geographical and time limitations. The pattern of interviewing and sources of information naturally varied a great deal with each pair. Where more information was available from published sources, this sometimes reduced the need for additional interviews.

Pattern-finding

Our earlier work and our literature survey had shown that there were many possible explanations of success and failure. Moreover, it was inherently improbable in a complex social process, that single-factor explanations would be sufficient. We therefore designed the project to test a large number of single hypotheses and simultaneously to test a large number of possible combinations of factors. Our aim was to identify a characteristic pattern of failure or success. Altogether we attempted about 200 measurements of each case of 'success' or 'failure'. Some of the measures are comparative, some are absolute. Thus, for example, it is possible to test the hypothesis that large size is generally advantageous for innovation, both by testing in how many pairs the smaller firms failed, and by checking what proportion of firms with fewer than 500 employees succeeded (or 100 employees, or 1000, or 10,000 employees).

Most of our measures were comparative between the success and failure halves of the pair, enabling us to make statements such as 'successful attempts were characterised by greater . . . or less . . . or smaller . . . or shorter . . . or more . . . than attempts which failed, but our aim was to link all these comparisons together. The main hypotheses which we attempted to test related to various measures of size (employment, R & D Department, and project team); measures of market research, publicity, education of users, involvement of users; modification of the innovation at various stages; checks on the innovation at various stages; the role of engineers and scientists and of various key individuals, their previous experience, education and background; the management control and planning system in the firm; the communication network with the outside world; degree of dependence on outside technology and familiarity of the firm with the innovation; methods of organising R & D work, patent policy, competitive pressures, speed in development work, and data of launch.

A General Model of Innovation

A basic assumption of the whole project was that introspective single discipline or single factor explanations were unlikely to prove satisfactory (although the method permits the test of this possibility). We expected that a satisfactory explanation of innovative performance would have to take into account the essentially 'two-faced' or 'coupling' nature of this activity. On the one hand innovation involves the recognition of a need, or more precisely a potential market for a new product or process, and an endeavour to satisfy this market. On the other hand it involves technical knowledge which may be generally available, but may also incorporate the results of original research activity. Experimental design and development, trial production and marketing, represent the process of 'matching' the technical with the market possibilities. In the literature of innovation there are many attempts to build a theory predominantly on one or other of these two aspects. Broadly speaking, social scientists, and especially economists, have stressed the market or demand side of the process, whilst natural scientists and engineers have tended to stress more strongly the research and technical side and to neglect the market. These one-sided approaches may be designated briefly as 'demand-pull' and 'science-push' theories of innovation. Their echoes are apparent in much of the earlier discussion. By our mode of work we hoped to overcome any one-sidedness in these questions in the belief that a satisfactory explanatory model must take both these elements into account. Carter and Williams in their early model of the 'technically progressive firm' came very close to our line of thinking.[1]

III. PRINCIPAL RESULTS OF SAPPHO

The results of our analysis may be classified under three headings:

(a) Factors which were common to almost all attempts to innovate, whether successful or not.

(b) Factors which varied between innovative attempts, but in which the variation was not systematically related to success or failure.

(c) Measures which discriminated between success and failure.

[1] C. F. Carter and B. R. Williams, 'The Technically Progressive Firm', *Journal of Industrial Economics* (1959), pp. 87–103.

Statistically our sample is still too small to make any firm generalisations. Our results should therefore be regarded as provisional findings which may be validated in further work.

Here it is possible only to summarise these findings. The full results will be published shortly.

(a) *Resemblances between Pairs*

In our fifty-eight attempts to innovate we found a great many resemblances between both halves of our pairs (Table 9.1). Almost all attempts in these two industries took place within a formal R & D structure which was used to develop the innovation. Only in the instrument industry were there cases of attempted innovation without such a structure. Most of these were designs for a new product brought from an outside environment.

Since almost all of the firms involved in attempts to innovate had this formal R & D structure, it might be expected that critical differences would exist in the way in which such Departments were organised, or R & D was planned, or projects evaluated, or incentives provided for engineers and scientists. A great deal of the management and sociological literature has concentrated on these aspects of the efficiency of industrial research.

Our inquiry did not uncover systematic differences of this kind with respect to R & D organisation or incentives. As with previous empirical studies we found that many supposedly 'best practice' techniques in long-term planning and project assessment were honoured more in the breach than in the observance. But the successful innovators did not differ from the failures in this respect. There was no evidence that successful innovators expected rewards or penalties differing from the less successful. Nor was there any evidence of unusual incentive schemes for R & D personnel, or greater freedom in successful cases.

One possible explanation of more successful attempts to innovate might lie in patent priority, but again we were not able to identify differences here. Almost all our innovators, both successful and unsuccessful, took out patents and regarded them as important. But we did not find that the failures attributed their lack of success to the patent position of their rivals, except in one case.

Nor did we find that successful innovators differed from unsuccessful in the way in which they organised their project teams. One hypothesis had suggested that the less successful attempts might be characterised by departmental organisation on disciplinary lines. But this was not the case. Where firms had a large R & D organisation, they sometimes had laboratories working on conventional sub-disciplinary lines, but this did not really affect the project

TABLE 9.1

PART 1: SOME MEASURES WHICH DID NOT DIFFERENTIATE BETWEEN SUCCESS AND FAILURE

Variable Name (Mnemonic for print-out)	Question	Chemicals S>F	S=F	S<F[a]	Instruments S>F	S=F	S<F[a]	Both Industries S>F	S=F	S<F[a]
RADFIRMS	Was the innovation more or less radical for the firms concerned?	5	7	5	4	3	5	9	10	10
DECLEVEL	At what level was the decision to proceed with the innovation made?	2	11	4	1	10	1	3	21	5
TMELIMIT	Was a time limit set?	4	12	1	1	8	3	5	20	4
PATENTS	Were patents taken out for this innovation by the organisation?	—	17	—	3	8	1	3	25	1
NATURAL	Did one organisation accept the innovation as being more in its natural business than the other?	5	8	4	6	1	5	11	9	9
SERIOUS	Did one organisation have a more serious approach to planning than the other?	6	7	4	2	7	3	8	14	7
RDSYSTMC	Was there a systematic and periodically reconsidered R & D programme?	6	7	4	—	10	2	6	17	6
PUBLPLCY	What was the company's publishing policy?	5	7	5	1	11	—	6	18	5
INCENTIVE	Were there any incentive schemes to encourage innovative effort?	2	15	—	—	12	—	2	27	—
REGSTRCT	Was the innovation carried out within the regular R & D structure?	2	14	—	1	11	—	3	25	1
SUCCEFCT	What outcome was the project expected to	3	11	3	1	11	—	4	22	3

GENCCPLG	What was the degree of coupling with the outside scientific and technological community in general?	2	12	3	2	8	2	4	20	3
MOREQSES	Would the firm have recruited more Q.S.E.s if it could have done so at the time of the innovation?	—	17	—	1	10	1	1	27	1
WHENFRML	In each case, when was the decision to innovate formalised on paper?	5	5	7	1	10	1	6	15	8
TMETOSLE	How many months elapsed from prototype or pilot plant to first commercial sale?	7	3	7	5	3	4	12	6	11
FORMALRD	Was there a formal R & D department in the organisation?	1	16	—	2	8	2	3	24	2
GROWRATE	What was the scale of growth of the organisation up to the time of marketing (measured by annual growth of turnover in the five years prior to the marketing of the innovation)?	1	13	3	4	4	4	5	17	7
BEYRSEDN	How many years did the business executive spend in the educational system?	5	7	5	2	5	5	7	12	10
PRFTCNTR	Was the R & D department regarded as a profit centre?	6	8	3	1	10	1	7	18	4
NEWMTRLS	Was there any need to find or use new materials?	—	17	—	—	11	—	—	26	1

a S > F : Success more than Failure, greater than Failure etc. Or In Success but not in Failure.
   S < F : Success less than Failure, smaller than Failure etc. Or In Failure but not in Success.
   S = F : No measurable difference between Success and Failure.

TABLE 9.1 (continued)

PART 2: SOME MEASURES WHICH DIFFERENTIATE BETWEEN SUCCESS AND FAILURE

Variable Name (Mnemonic for print-out)	Question	Chemicals			Instruments			Both Industries		
		S>F	S=F	S<Fᵃ	S>F	S=F	S<Fᵃ	S>F	S=F	S<Fᵃ
RADWORLD	Was the innovation more or less radical for world technology?	10	6	1	2	9	1	12	15	2
HOWDELIB	How deliberately was the innovation sought, comparatively?	7	8	2	6	4	—	13	14	2
COMMOPPN	Was there opposition to the project within the total organisation on commercial grounds?	1	9	7	1	7	4	2	16	11
DEVELENG	Was more use made of development engineers in planning and costing for production in one case than in the other?	5	9	3	4	8	—	9	17	3
EXTCOMMS	Did one organisation have a more satisfactory communication network than the other externally?	5	10	2	5	7	—	10	17	2
CHIEFSNR	Was the R & D chief more senior by accepted status in one case than the other?	9	5	3	2	8	2	11	13	5
SLESEFRT	Was the sales effort a major factor in the success or failure of the innovation?	7	10	—	9	3	—	16	13	—
MFCNSLES	Were any modifications introduced after commercial sales as a result of user experience?	1	8	8	2	6	4	3	14	12
AFTRSLES	Were there any after-sales problems?	—	4	13	1	2	9	1	6	22
USEREDUC	Were any steps taken to educate users?	8	9	—	6	5	1	14	14	1
EQPREPRD	If new tools or equipment were needed for commercial production, were any ordered before the decision to launch full-scale production?	8	7	2	2	10	—	10	17	2
SPECCPLG	What was the degree of coupling with the outside scientific and technological community in the specialised field involved?	8	9	—	5	6	1	13	15	1

Variable	Question									
UNEXADJS	Were there unexpected production adjustments?	1	7	9	1	7	4	2	14	13
BUGSPROD	Did any 'bugs' have to be dealt with in the early production stage?	1	6	10	1	5	6	2	11	16
MKTFRCST	Was any systematic forecasting by the marketing (or sales) department involved in the decision to add the innovation to product lines or to existing processes?	5	7	5	6	5	1	11	12	6
NDSUNDSD	Were user needs more fully understood by the innovators in one case than in the other?	15	2	—	9	3	—	24	5	—
BEDVRSEX	Did the business executive have a more diverse experience in one case than in the other?	8	8	1	8	2	2	16	10	3
BESTATUS	Did the business executive have a higher status in one case than in the other?	8	8	1	5	4	3	1	12	4
BEPOWER	Did the business executive have more or less authority (power) in one case than in the other?	9	7	1	6	4	2	15	11	3
EFCTHELP	To what extent was dependence on outside technology a help or a hindrance in production?	10	6	1	6	4	2	16	10	3
LNCHDATE	What was the date of the first commercial launch?	4	3	10	9	1	2	13	4	12
TEAMBEG	How large a team was put to work on the innovation at the beginning of the project?	12	2	3	4	4	4	16	6	7
TEAMPEAK	How large a team was put to work on the innovation at the peak of the project?	9	4	4	7	4	1	16	8	5
BEYRSIND	How many years had the business executive spent in industry?	9	7	1	3	4	5	12	11	6
BEFRGNEX	Had the business executive had any overseas experience?	3	14	—	5	6	1	8	20	1
BERSBLTY	Did the business executive have a greater degree of management responsibility in one case than in the other?	10	7	—	4	5	3	14	12	3

development team which was set up in a similar way in both successes and failures.

Another hypothesis for which we found no supporting evidence was the view that business innovators might be less well qualified academically in unsuccessful attempts (or better qualified). We did in fact find important differences between business innovators which did distinguish between success and failure, but this was not one of these. Most of the business innovators were qualified scientists or engineers in both halves of the pair. Obviously in these two industries amateurs are rarely chosen to manage innovations. In the two cases when accountants were the business innovators, both failed, but this would be too small a number on which to construct any general theory. It would probably not be possible to find a sufficient number of cases in these industries where innovators were not technically qualified, to test any hypotheses relating to the supposed merits or de-merits of amateurism.

(b) *Variation Unrelated to Success or Failure*

Many other measures which we made did show considerable variation between attempts to innovate, but the variance was not closely related to success. Among them were measures relating to size of firm, size of R & D department and numbers of qualified engineers and scientists in R & D. These results need considerable care in interpretation. There was no strong systematic evidence that larger or smaller firms or R & D departments were more or less successful. For example, of the cases involving firms employing more than 10,000 people, six were successes and seven were failures. Where large firms were in competition with smaller firms, there was a tendency for them to be more successful but it was by no means clear-cut.

However, this should not be interpreted as implying that size of firm is completely irrelevant in relation to innovation in these two industries. Comparative size measures did not differentiate between success and failure clearly. But in chemicals only four out of thirty-four attempts were made by firms employing fewer than 1000, whilst in instruments over half the attempts (both successes and failures) were made by firms employing fewer than 1000. Clearly size is relevant to the type of innovations which are attempted at all, and inter-industry differences are very important. Another project which we completed this year showed fairly conclusively that inter-industry differences are critical in this respect.[1]

[1] 'The role of small firms in industrial innovation 1945–1970', Report to the Bolton Committee of Inquiry into Small Firms (Science Policy Research Unit, 1971).

We did not find any relationship between success and the number of scientists and engineers on the main board of the innovating company, although this proportion varied considerably. However, in almost all cases there were some engineers or scientists on the main board and it may be that this is the critical threshold factor.

Perhaps surprisingly, for those who believe in the amenability of innovation to planning techniques, we found no relationship between success and the capacity to set and fulfil target dates for particular stages of the project plan, nor in the general approach to planning of the innovators.

Contrary to some theories, we did not find any association between failure and the attempt to innovate in areas unfamiliar to the firm. Where firms differed significantly in their familiarity with the field the outcome was evenly distributed between success and failure.

Another set of measures which did not discriminate between success and failure related to the growth rate of the firm and its competitive environment. There were of course considerable variations between firms in the growth which they had experienced before the innovation, and in the competitive pressures to which they were subject. But these differences apparently did not affect their degree of success in attempting to innovate. Again, it is important not to over-state this finding. This does not mean that competitive pressures or declining growth may not be important in stimulating attempts to innovate, only that they do not ensure success.

A finding which rather surprised us was that development lead-time was not strongly correlated with success. We had expected that the more successful innovators would be those who found ways of shortening the development phase and telescoping the stages from prototype or pilot plant to commercial launch. But support for the hypothesis came only at the earlier stage of applied research. In the chemical industry successful firms were quicker to get through this early stage. The absence of any evidence of a shorter development stage associated with success provides support for those who have argued that hardware development is a gestation process akin to animal reproduction in that it cannot usually be artificially shortened. It may also indicate that successful firms take more trouble at the development stage to get rid of all the bugs, so that later stages are trouble-free. There is considerable indirect support for this interpretation from those measures which do discriminate between success and failure.

(c) *The Pattern of Success*

Of the 200 measures which we attempted, only a small number differentiated strongly between success and failure, and these varied a little between the two industries. The principal measures are shown in Table 9.1.

The measures which came through most strongly were those directly related to marketing. In some cases they might be regarded as so obvious that they need not have been measured. However, this is not our view, since the case studies showed that even the most obvious requirements were sometimes ignored. Successful attempts were distinguished frequently from failure by greater attention to the education of users, to publicity, to market forecasting and selling (particularly in the case of instruments where it was most relevant) and to the understanding of user requirements. The single measure which discriminated most clearly between success and failure was 'user needs understood'. In our view this should not be interpreted as simply, or even mainly, an indicator of efficient market research. It reflects just as much on R & D and design as well as on the management of the innovation. The product or process had to be designed and developed and freed of bugs to meet the specific requirements of the future users, so that 'understanding' of the market had to be present at a very early stage.

This interpretation is confirmed by the strong evidence on the occurrence of 'unexpected adjustments' and bugs after development in the failure cases, and of the need for user adaptations in nearly half the failures. About three-quarters of the cases of failure showed greater 'after sales' problems. Thus, 'user needs understood' is just as much a discriminating measure of efficiency in R & D performance as of marketing and overall management.

The fact that size of project team emerged as a clear-cut difference, while other size measures did not differentiate, is also of considerable interest. Since in a number of cases the smaller firms deployed a larger team this implies a greater concentration on the specific project. This consideration is important in considering the relative advantages of the small firm in innovation. Another measure which strongly suggests the advantages of specialisation in R & D is that relating to 'coupling' with the outside scientific community.

Much has been written about the importance of communicating in R & D. Carter and Williams[1] in particular have emphasised good communications with the outside world as one of the most important characteristics of the technically progressive firm. The most backward firms would not of course be found among those attempting to innovate. But among those who were making such attempts we were able to observe significant differences in their general pattern of communications. Better external communications were associated

[1] C. F. Carter and B. R. Williams, *Industry and Technical Progress* (Oxford, 1957).

with success, but the strongest difference emerged with respect to communication with that specialised part of the outside scientific community which had knowledge of the work closely related to the innovation. General contact with the outside scientific world did not discriminate between success and failure.

All of these differences may of course be related to the quality and type of management, so that our measures relating to the business innovator are perhaps the most interesting. First of all it should be noted that the business innovator was hardly ever the same man as the chief executive in the chemical industry, but was frequently so in the instrument industry. The most interesting difference between successful and unsuccessful business innovators, and one which was unexpected, was that greater seniority was associated with success. The successful man (they were all men) had greater power, higher formal status, and more responsibility than the unsuccessful. He was also older and had more diverse experience. Some of these differences were not so clear-cut in the instrument industry which may reflect the greater mobility and smaller size of firm, together with more hierarchical structure of management in the chemical industry. Usually the successful chemical innovator had been longer with the innovating firm and in the industry, whereas this was not true in the instrument pairs.

The higher status and greater power of the more successful innovators may be associated with their readiness to take greater risks and to recruit larger teams for their projects. In the chemical industry there was a strong association between success and a more radical technical solution, and a more deliberate search for the innovation.

The fact that the measures which discriminated between success and failure included some which reflected mainly on the competence of R & D, others which reflected mainly on efficient marketing, and some which measured characteristics of the business innovator confirms the view of industrial innovation as essentially a coupling process. One-sided emphasis on either R & D or sales does violence to the real complexity of the process.

The fascination of innovation lies in the fact that both the market and the technology are continually changing and consequently a kaleidoscopic succession of new possible combinations is always emerging. Innovation is a coupling process at this interface, but this coupling is not just the sudden flash of inspiration, or the transcendental genius of the innovator, or the idea that 'clicks'. Rather it is a continuous creative dialogue over the whole period of experimental development and design which may involve many elements of 'invention' and take many years. The critical role of the 'entrepreneur' (whatever individual or combination of individuals fulfil this role) is to 'match' the technology with the market, i.e. to understand the user requirements better than competitive attempts, and to ensure that adequate resources are available for development and launch.

In the large firm this will mean he must be high enough in the hierarchy to command resources and get things done. He must have enough knowledge of the way the firm works to know how to get things done. In the small firm it frequently means that it will be the chief executive himself, or a man sufficiently close to the chief executive to ensure the necessary concentration of effort. In either case he must be sufficiently powerful and clear about marketing objectives to ensure that the various screening and testing procedures during the course of development and trial production prevent an unsatisfactory product or process coming onto the market. Premature launch may be more dangerous than slowness. In the instrument industry the successful innovation was usually the second to market. In chemicals it was usually first, but this was sometimes after a longer development period albeit a shorter laboratory scale research period.

These conclusions will not necessarily be valid for consumer goods innovations where some different mechanisms are at work. In capital goods it is essential to satisfy certain minimal performance criteria.

APPENDIX A

LIST OF SAPPHO PAIRS

Scientific Instruments:

1. Amlec Eddy-Current Crack Detector
2. Milk Analysers
3. Foreign Bodies in Bottles Detector
4. Roundness Measurement
5. Scanning Electron Microscope
6. X-Ray Microanalyser
7. Digital Voltmeters
8. Optical Character Recognition
9. Atomic Absorption Spectrometer
10. Electromagnetic Blood Flow Meter
11. Electronic Checkweighing I
12. Electronic Checkweighing II

Chemicals:

1. Acrylonitrile I
2. Acrylonitrile II
3. Caprolactam I
4. Caprolactam II
5. Ammonia Synthesis
6. Ductile Titanium
7. Extraction of Aromatics
8. Steam Naphtha Reforming
9. Extraction of n.Paraffins
10. Urea Manufacture
11. Oxidation of Cyclohexane
12. Hydrogenation of Benzene to Cyclohexane
13. Phenol
14. Accelerated Freeze-Drying of food (solid)
15. Methanol
16. Acetic Acid
17. Acetylene from Natural Gas

Discussion of the Paper by Professor Freeman

Professor Paunio, opening the discussion, said that this study concentrated on innovations in chemicals and scientific instruments, although it had been pointed out it was hoped to lead on to other industries in future research. He said that the methods of analysis which had been used were extremely interesting and unconventional. He thought that the method of pairing which had been used to analyse the characteristics of success and failure was a very fruitful approach and he noted that this method had clearly indicated that single factor explanations were not sufficient to explain success. It was clear that both elements of demand and of research had to be taken into account in explaining the difference between success and failure. This study had made use of two hundred measurements for each of the pairs, and the provisional results of the first part of the analysis had been set out in terms of the factors common to all attempts to innovate, factors not systematically related to success or failure and factors which were peculiar to success alone.

Professor Paunio noted with interest that there was a strong correlation between success and marketing effort, and also with the size of project team. However, as the aim of the project was to establish step by step generalisations about the innovation process he felt that the methodology of the project was very important and on this question he wanted to raise a few points. He was particularly interested in how the industries, the innovations and the pairs had been selected. He was also interested in the definitions of success and failure and noted that in this particular case they did not appear to raise particular problems, and that significant market penetration had been one of the factors which had been chosen. However he felt that in future clear-cut cases might not be so easily found, and he was interested to know what other factors of success and failure might be used. He was also interested in how the hypotheses were selected. Was the availability of finance tested and was the supply of savings? Professor Paunio noted that the results on the marketing aspects appeared to go against basic assumptions that demand was a decisive factor of innovation. He also noted that the business innovator was a key personality, but that there was little discussion of the fate of the other people involved. These points, he felt, would be usefully brought up in the discussion. Finally, he said that he felt this study had been a very important one, both as regards its scientific approach and as regards the persistent systematic search for relevant information about the innovation process.

Professor Nussbaumer said that he had found this a very stimulating paper and an extremely valuable attempt to indicate the reasons for failure and success, even if we still remained doubtful about the detailed validity of the statements which had been made. He raised three points, the first of which was related to that already made by Professor Paunio about the definition of success and failure and whether this would be as clear-cut in future studies. On the question of market penetration he suggested that for some firms an increase of market share by 5 per cent might be worth while, and for another 15 per cent would not be enough. He was also a little worried about the measurements of profitability and suggested that these might be dubious, especially in the case of multiproduct firms and of international corporations where it was not easy to ascribe a specific part of the total profits to individual projects. He also thought there were difficulties in giving adequate weights to the short and long term profitability and also pointed out that in some cases organisations would be very happy to undertake projects which did not give them increased market share but rather kept them in their present position. His second question was concerned with the dispersion of results of R & D. It might well be that a research project would not look to be worth while because it neither increased market share or increased profits, but it might well be profitable to the firm because it would strengthen its whole position in the field it controlled. For example, it might be one of a bundle of products which it marketed, or again it might simply be developing background expertise which could be useful for future work in this field. He felt that this was often the case with large corporations which were active in sidelines so that they could offer complete blocks of products. He therefore suggested that it might be useful to look at the profitability of the general operations of the firm and not just at individual projects within them. His third point was the question of patents. He was not surprised that patents taken out did not provide a clear indication of success of failure or even of the volume of R & D. It seemed to him that an increasing number of firms were sceptical about the benefits which could be obtained by having their innovations registered. The dangers of imitation and circumvention might even be increased with increased publicity and it might be difficult to obtain a court verdict against the violators.

Professor Khachaturov had found the paper very interesting not only from the scientific but also from the practical point of view, and not only for a market economy but for planned economies too. He had not met similar research before but he wished to put forward some suggestions which might be considered in the future. In his opinion the authors of this work had adopted a method which was a very difficult one, because it required the identity of given pairs in every respect except the one which was studied. He said he believed it was very difficult to find many such pairs and as a result the number was bound to be comparatively small, and as such it was worth getting as much information as possible. Professor Freeman had made inquiry of many firms with successes and failures and he thought that they should do so in their economy too. A second point was that he expected more generalisation of the material which was collected and he thought this was possible. This was important for a market economy because it was possible to give recommendations to firms to achieve success and not failure and such generalisations could be used for research purposes. They could collect similar material from other branches of industry there might be quite different cases, and such research material would be of great interest, not only to practising managers but also to researchers.

Professor Oshima was very interested in this paper, and said that he rated it very highly. He would like to make some comments on the chemical innovations in Japan which they had studied, and which had produced some interesting correlations, although the study had not been so systematically conducted as that of Professor Freeman. There were however some similarities in that in most of the Japanese inventions in the chemical industry there had been a very clear responsibility taken by a senior vice-president and that high priority had been given by the company to the work. However in many cases when the product first went onto the market it was not highly successful. But it was not given up at this stage, and considerable after-sales effort produced a successful end result. Professor Oshima thought this contrasted with the findings of Professor Freeman and that this might be due to differences of market structure. But it also might depend upon the stage at which one said that success had been reached. In this respect he instanced the case of colour film, where there had been trouble in the initial stages and the process had had to be much more highly developed before it went to the production line, and success only came after this stage. Finally, Professor Oshima asked whether, in the information given in Appendix A, each of the items mentioned had a pair, so that it really represented two cases.

Professor Freeman answered Professor Oshima's point by confirming that each of the items contained two cases. He went on to say that the study under discussion related to cases where competitve activity was present and therefore there was always after-sales effort and they were only making a comparison after that effort. This meant that even successful firms had to de-bug problems in the development stage, but that the people who had failures had more problems. The question was one of comparative rather than absolute performance. Problems were not absent in successful firms but more technical problems were de-bugged before the product stage.

Professor Oshima said that he would have liked to have seen cases where the failure occurred earlier, and *Professor Freeman* replied that the Centre for the Study of Industrial Innovation in the United Kingdom had done a study of shelved projects, but pointed out that the present project had been concerned only with projects which had already been launched.

Dr. Sedov said that he had got great pleasure in reading the report of Professor Freeman, and agreed with him and others that much of the theoretical analysis of the innovative process had lacked systematic empirical foundation. He thought that the method of investigation used in this study was very foolproof and it was even now possible to make generalisations. In particular he noted that success demanded a certain relationship between the producer and the customer of a new product. He suggested that this was not the same as the simple sales relationship but that it probably demanded a new form of organisation and the establishment of new channels of information and communication. He also noted that success was based on the mobilisation of adequate resources, which was why the person in charge must be high enough in the hierarchy to command such resources and to get things done. It seemed to him that the man had to be on friendly terms with governmental officials and financial circles as well, and that he must be influential enough not only inside but outside the company. Success depended on the novelty of the product and novelty was a barrier to competition, not only for homogeneous products but for substitutes as well. Dr. Sedov thought that one could derive many useful conclusions from the questions included in the list, but he suggested that additional useful conclusions might be derived by adding some additional questions. He suggested the following: Firstly, the contacts of the business innovator or product champion with the state and financial institutions. Secondly, the size and character of the system for educating the customer and the funds available for the information and sales systems of the investigative company. Thirdly, the degree of competitiveness of the substitute products. Fourthly, the dynamics of the price and quality of the products and their correlation with the prices of the substitutes.

Professor Williams said that he had read the paper with great pleasure and an acute sense of nostalgia. This type of research was very detailed, very time-consuming, and very difficult to get adequately supported. It was easier to get money for research into the fruit-fly than for adequate study of fruitful or fruitless innovations, and he hoped that Freeman's work would make clear the worth of greater financial support.

Professor Williams went on to say that he would like to raise a number of methodological questions: Firstly he thought that the approach of using pairs was an interesting and useful one, but he had doubts concerning the conclusions Professor Freeman was drawing from these studies. He asked about the context in which he drew the conclusions. Was it directly from the particular studies or related to more general achievements of the firm? All firms had successes and failures but some firms were better in this respect than others. It would be possible to find failures associated with organisational procedures which had also been associated with great successes, and he wanted to ask whether this had been taken into account. He thought it important to relate the studies to a more detailed study of the firms concerned. On the question of the hypotheses, he assumed this was a short-list only (*Professor Freeman* replied yes) but even so he thought it was the list of hypotheses one might expect from an organisation theorist rather than from an economist. He was particularly interested in the hypotheses which were not there, and asked that Professor Freeman might throw some light on this. Finally, he commented on the conclusions about the importance of people and management hierarchy for success and failure. This was very like the study 'Investment Proposals and Decisions' that he and Scott had published six years ago, and which developed this aspect in some detail. Some innovations which they studied had depended on the ability of thrusting potential innovators to get a powerful sponsor in top management, with the thrusting innovator moving up rapidly in the management hierarchy if the innovation prospered. He wondered whether Professor Freeman had found anything similar in his study? He and Scott had also placed emphasis in their conclusions on the importance of concentrating general management attention on a project which was not going very well. Some projects they had studied which in the end turned out very well did very badly in the beginning, and others failed simply because it was not possible to get sufficient management attention focussed upon them. Professor Freeman had spoken of solving problems at the R & D stage. They had been impressed with the way problems were solved after the R & D stage, and the extent to which things had often gone wrong and been put right by an unexpected route – sometimes. though not often, by further crash development work, more often by changing production and marketing to cope with situations which had not been foreseen.

Professor Griliches said that he was very much impressed by this paper, particularly because we rarely got new data in this field. He would like to make some observations on the interpretation of the findings. As he read it, most of the findings were in some sense negative, and of the things which did differentiate between success and failure a significant number were a measure of the success itself. For example, recognising the needs of the customer, getting the bugs out, and so on, and he thought it would not be easy to put a finger on these in advance in order to predict which firms would succeed five years ahead. If this was the case, he said, we could not in fact really predict what was going to be a success in advance, except perhaps from the single item about the need for a good leader. Professor Griliches said that like most important ideas it suddenly became obvious. We were analysing people at the frontier and they were already doing their best to be successful. If anything mechanical would have ensured success they would have done it. *Professor Williams* intervened to challenge this statement, *Professor Griliches* agreed to disagree, and went on to say that a parallel occured to him in the uncomfortable theory of market valuation of firms. If the market was reasonably good economists could tell which firm would appreciate on the market. There was nothing mechanical in the statistics of the stock market or published observations on the firms which would allow one to predict whether Dupont or Dow would be more successful five years hence. He thought the same thing might be true here when it came down to assessing the likely success and failure of innovations. We could figure this out in retrospect. But if one could figure it out in prospect at the micro level more of us would be very rich. *Dr. Sedov* wanted to remind Professor Griliches about the question of who would be the first to innovate, and added that the investigation under discussion provided a good deal of information on this.

Professor Matthews said that he had had similar thoughts to Professor Griliches, in particular that some of these hypotheses were attributes and not causes, and he suggested that Professor Freeman might consider sifting out these hypotheses into those things which could and those which could not be predicted in advance. On a separate point Professor Matthews said that these industries were very different — chemicals and scientific instruments — and that other studies might again lead to different results. He also questioned whether the study related to success/failure in the British economy only, and pointed out that there was also a general difference between these industries apart from scale, in that the chemical industry had been more successful on a world standard than had the scientific instruments industry. *Professor Freeman* intervened to say that the sample was not all based on British firms but had been chosen as successes and failures in particular technologies, and some of these had obviously come from other countries, since in the type of technology under investigation competition usually took place at international levels. Less than half of the cases were exclusively British.

Professor Robinson said that this paper was an extraordinarily exciting piece of research and opened up a whole new series of questions. He was interested to see that the differences between success and failure companies were surprisingly small. He said that he had found the most interesting differences in Part 2 of Table 91. On the question of getting down to making the innovation work, he said that developments with bugs in them would always occur, but that he could not find from Professor Freeman's work whether the successes never contained bugs or whether these particular firms were good at eliminating them; he would like more information on this point. Professor Robinson went on to say that, if they would forgive research by gossip, he would like to recall that during the war years they had been concerned with the slowness in elimination of bugs from British designs compared to those of the United States in the field of aircraft prototypes. It was found that the British elimination of bugs was in the hands of pragmatic engineers who were remarkably well able to deal with minor items but who got lost when they had to go back to first principles. It was in these areas that United States firms had more scientific talent and could put more effort into going back to basic causes. Professor Robinson went on to say that he was not sure what was meant by success and failure, and pointed out that failures in wartime would often come about from trying to make big jumps, contrasted with successes in making small jumps. He therefore wanted to ask whether there was any evidence of a greater failure in making big jumps than in making smaller ones. Did Professor Freeman think that part of the difference between success and failure was a matter of strategy in seeing the scale of the jump you could make successfully?

Professor Triantis had read with admiration Professor Freeman's work because of the heroic attempt to tackle difficult factors. He said he noted that there had been found to be no association between failure and attempts to innovate in unfamiliar areas, and that he was interested to know how this unfamiliarity was defined. He thought that this was an extremely difficult problem and mentioned the case for example of conglomerates in the United States. He thought that the type of management of firms is an important distinction to make here. In some cases the strategy of management was to acquire new firms and leave them to be run by their existing management rather than to pursue the development of new technology in areas familiar to the firm. In examining then the causes of success or failure one should ask whether the management was of the kind that could manage the new area rather than being one which was only prepared to innovate in areas in which it was familiar with the technology itself. Another point which Professor Triantis raised was the question of a firm's more or less 'serious' approach to planning. He was interested to know how this had been measured. He said that if by this you meant more formal planning it might be possible to count the number of people employed in planning, but if you meant more serious he didn't know how you could do this, because planning could be informal and yet effective. On the other point above the number of engineers and scientists who were on the board of management of a firm he pointed out that these people were not generally the same animals as Q.S.E.s. They had been in management for many years. He could see the point of taking into account the background of the senior management but suggested it was more important to look at the length of experience of this management and other factors in addition to their academic training. Stock market analysts placed little emphasis on management's formal academic training.

Dr. Nabseth said that a small question puzzled him about the relation between success and failure. Professor Freeman's paper appeared to suggest that the research people had tried to find a market, a need to be solved, and then had looked for pairs in order to find a success or a failure. He questioned whether in fact there were many cases in which you would find that the firms were both successful. *Professor Freeman* intervened on this to say that there were situations where more than one firm could succeed; in fact he pointed out that it was an oligopolistic situation as far as the world market was concerned, with several firms usually innovating close behind each other. The likelihood of finding a world monopoly situation was conceivable but rare.

Professor Freeman then went on to reply to the many questions which had been put to his paper.

On Professor Paunio's point on the choice of industries he said that there had been no special significance about this and that they would like in the future to investigate other branches. In the initial stages they had obviously had to make some choice and this was partly influenced by the existence of previous innovation studies in these industries, for example the work of Enos and Shimshoi and also the capabilities of the people that were engaged on the study. On the question of the definition of success and failure he said that they had puzzled over this for a very long time and they did not claim the solution was the most satisfactory one. He said that if one took only profitability one ran up against the case where the project might be a good strategy for a firm and yet it might not appear to be extremely profitable. He said they added the criterion of penetration and that in fact in most cases their successes had both been profitable and had also achieved a considerable share of the world market. He pointed out however that in the case of Corfam they had had originally two cases for shoe leather uppers and they expected that Corfam would be a success on the basis of market penetration. It had appeared very successful with 90 per cent of the world market and greater than 20 million pairs of shoes being produced using this method, but that when they came to make their investigations more thoroughly they concluded that it was not profitable and that Dupont did not expect it to be profitable, and eventually of course the plant was closed down.

Professor Freeman said he agreed that they were fallible in making judgements and they had introduced the dual method to overcome as much of this as possible. He thought that most economists would agree with their choice and would be satisfied with their decision in about 90 per cent of the cases, but he did admit that there were some borderline ones. On the question raised by Professor Williams concerning the hypotheses generated, Professor Freeman said that they did go back to the economics literature as well as to organisation literature and also the engineering literature. They had taken six to nine months in analysing the bibliography of all explanations of success and had scanned in fact 400 items in this time. He must admit that they might have missed some of the economics literature, but they had not found it terribly fruitful, and would welcome suggestions on this point, as in the next stage they wanted to refine the original hypotheses and also to add new ones if any were put forward. On the question of patents raised by Professor Nussbaumer, Professor Freeman said he wouldn't quite agree with the way he had put his statement – he said that their findings suggested that taking out patents did not differentiate between success and failure but that they did find that almost all firms typically took out patents although they were not decisive in separating success from failure. He interpreted this as part of the game, that people had to take them out because of the danger of being caught without patents. On Dr. Sedov's point Professor Freeman said they were interested in the relationship between the firm and the government and in measuring the characteristics of the customers, although this latter was a difficult one, and in actual fact in their study the customers were the same or similar for both potential innovations. On the relationship between firm and government he admitted that this did have an effect and one of the firms which was a failure in this study completely forgot to take into account the fact that the market was really a British government monopoly, whereas the success, which happened to be a Danish firm, did take this point into serious consideration, although he added that it was interesting to see that the British firm which failed did make an attempt to interest other potential customers outside the United Kingdom. Professor Freeman thought that it was difficult to introduce systematic measures where there were only a few cases which had been the situation on the firm/government relationship side. He said this also applied to Professor Khachaturov's question – they did not have enough data for statistically-based judgement. He said they were not able to make as many generalisations as hypotheses which they wished to test, and said that they might need different codings so that they could make more valid generalisations. On the question of the competition of substitutes raised by Dr. Sedov, Professor Freeman said that you could look at the case of Corfam which in their terms was a competition between two firms introducing synthetic. leather. In this case the competition was with natural leather. *Mr. Pearson* said that this choice was an interesting one because in the paper Professor Freeman had pointed out that the similarities of the pairs was defined in terms of their market, not necessarily in terms of their technology, and he suggested that in this particular case one might be more usefully looking at poromerics as compared with plastics rather than simply firms competing

in the synthetic field using poromeric technology. *Professor Freeman* said
that they were, in fact, now looking at another firm using poromeric
materials and if this did succeed they would have one success and a choice
of two failures. In replying to Professor Williams' question Professor
Freeman said that any one firm certainly had successes and failures and said
that he would be very interested to compare firms with a run of success
with those with a run of failures. He said they had considered this and
had rejected it because firms' characteristics, their methods of management, education and so on, changed over time and it would be impossible to standardise the variables. *Professor Williams* intervened to say that
he was not questioning the validity of the pairing method but was interested in how you move from the conclusions about particulars to general
conclusions. On the question of does it work, and is this an attribute of success or is it predictable, he agreed that many of the most important characteristics which distinguish between success and failure can only be measured
ex post and not *ex ante*. On the question of technical uncertainty and the
number of bugs which are likely to arise, *Professor Freeman* said it is
very hard to see, *ex ante*, which are going to be the bugs, but he added
that although some characteristics are only identifiable *ex post* he did not
think they are necessarily tautological – it is still important from the point
of strategy of firms that they know that they should look at these things
in advance, for example if firms know customers' requirements are important this becomes an important part of their strategy. He pointed out that
the innovating firms do not always take account of customer requirements,
even ignoring their comments on prototypes as potential customers. He
also added that the divorce between R & D and marketing often prevents
a useful interplay. Professor Freeman agreed with Professor Robinson
that bugs occur in all innovations, but he added that the interesting thing
is that successful attempts were distinguished by the earlier elimination
of bugs and that successful firms took far greater pains in screening and
pilot test to ensure that those bugs which were most likely to prove to be
problems of customers were eliminated at an early stage. On the question
of big jumps raised by Professor Robinson, Professor Freeman said that
they tried to measure this in terms of the development in world technology
and also in relation to the firm's technology and this had been possible to
some extent by having scientists engaged on the study. He pointed out
that in the chemical industry more successful innovations were those which
made the biggest jump forward, whereas in instrumentation it was the
converse. They had not, however, found an explanation for this. On the
question of the relevance of the findings for practical use, Professor Freeman agreed with Professor Griliches that they had not got a cookery-
book recipe, that uncertainty in innovation would always be present.
He said, however, that they believed that one should be more aware of the
factors which are likely to cause problems and be able to eliminate them,
but one would always be faced with other firms which might do better.
He said that if our and other studies were pursued perhaps they would
result in an increase in the general standard of planning and also perhaps
reduce the cost of failures. Our results had so far had no effects whatever,

but he hoped that they might lead to the elimination of a certain amount of waste. On Dr. Sedov's point about the findings being of use to some socialist economies, Professor Freeman said that they did look at the formality of fixing dates, of evaluation, the use of planning and control techniques and so on, but he agreed that these techniques did not necessarily mean a serious approach, and that they did not ask firms about their seriousness. He said that this did not mean that planning techniques were not useful, in fact he believed that planning was important, and some innovations would not succeed without it, but the conclusions of the study were that the present procedures for planning did not have a major effect on success or failure.

10 The Diffusion of Innovations in Swedish Industry[1]

Lars Nabseth

THE INDUSTRIAL INSTITUTE FOR ECONOMIC AND SOCIAL
RESEARCH, STOCKHOLM

I. THE DIFFUSION OF A NEW TECHNIQUE

The present interest of the Stockholm Institute in problems related
to the diffusion of new technology goes back to 1967, when we were
invited by the National Institute of Economic and Social Research
in London to take part in a joint study of these problems. Today
six institutes, representing Britain, Austria, Germany, Italy, Sweden
and the United States are engaged in the project. A first preliminary
report has already been published[2] and a full report will be published
in 1972 according to the present plans. The Institute's interest in the
problem was further increased when, during the course of work with
the long-term forecast for the period 1970–5, we realised that the
annual increase of total productivity in Swedish industry during the
sixties has been unusually high. Using the well-known Cobb-
Douglas technique, we obtained values for the 'trend-factor' or the
residual of about 4·5 which seems to be out of line with international
standards.[3] Can rapid diffusion of new technique explain part of
this phenomenon?[4]

In the international study altogether ten new processes, introduced
during the post-war period, were selected for study. These were
basic oxygen processes and continuous casting (in steelmaking);
special presses (in papermaking); numerically controlled machine
tools (in metalworking); shuttleless looms (in weaving of cotton and
man-made fibres); float glass (in glassmaking); tunnel kilns (in
brickmaking); modern methods of plate making and cutting (in
shipbuilding); automatic transfer lines (in car engine production);
and Gibberellic Acid (in brewing and malting). In Sweden we
consider it of interest to collect data from firms covering not only the
ten mentioned processes but also a couple of others as well. Thus,
similar information was collected on other processes in the paper and

[1] In preparing this paper I have in different ways received much help from three
members of our research staff: Anita Lignell, Harry Lütjohann and Birgitta
Swedenborg, to each of whom I am very grateful.
[2] G. F. Ray, 'The diffusion of New Technology. A study of ten processes in
nine industries', *National Institute Economic Review*, No. 48 (May 1969).
[3] See Lars Nabseth, *Svensk industri under 70-talet med utblick mot 80-talet*,
chap. 1 (Stockholm, 1971) (Swedish Industry during the seventies looking forward to the eighties).   [4] Or avoidance of introducing bad new techniques?

pulp industry, the iron and steel industry and the brick industry.

In the present paper I shall discuss the diffusion of new techniques in Swedish industry for the following six processes: special presses, wet suction boxes, continuous pulp cooking and outside chip storage (all in the pulp and paper industry), tunnel kilns and trucks (brick-making). The processes in question were all first used in Swedish industry during the post-war period and the diffusion processes have not yet ended. Since we have data on only six processes here we will not try to make extensive comparisons between processes. Our aim is rather to discuss and analyse company behaviour as this is what our data most readily lends itself to.

II. MEASURING DIFFUSION

Diffusion of new technique can, of course, be measured in a number of different ways, for instance, as the percentage of total output in an industry produced by using the new technique, the number of companies in an industry using the technique, number of plants, number of machines or equipment using the technique in question. Our data shows when a company or a plant within the industries selected for study has first used the new technique for commercial production. Consequently in the present paper diffusion will be measured by the number of companies or plants using the new technique at different points in time. That means that in this paper I shall not discuss diffusion from the point of view of the proportion of total output which is produced by using the technique. We are primarily interested in inter-firm diffusion patterns and I shall not discuss intra-firm diffusion patterns. Although quite a lot has been written about diffusion of new technique empirical investigations are nevertheless rather few and are mostly related to conditions in the United States. The following are the authors who have mostly influenced this paper: Carter and Williams,[1] Gold, Peirce and Rosegger,[2] Griliches,[3] Håkanson,[4] Mansfield,[5] Ray[6] and Rogers.[7]

---

[1] C. F. Carter and B. R. Williams, *Industry and Technical Progress* (Oxford, 1957), especially chap. 16, *Investment in Innovation* (Oxford, 1958).

[2] Gold, Peirce and Rosegger, 'Diffusion of Major Technological Innovations in U.S. Iron and Steel Manufacturing', *The Journal of Industrial Economics* (July 1970).

[3] I. Griliches, 'Hybrid Corn: An Exploration in the Economics of Technological Change, *Econometrica* 1957, 25:501–22.

[4] S. Håkanson, 'Diffusion of special-presses'. Forthcoming article in the book about diffusion of new techniques in different countries.

[5] E. Mansfield, *Industrial Research and Technological Innovation* (New York, 1968), especially chap. 8.

[6] G. Ray, *National Institute Economic Review* (May 1969).

[7] E. Rogers, *Diffusion of Innovations* (New York, 1962).

III. THE TECHNIQUES

In order to understand the discussion of the diffusion of new techniques the reader needs to have some understanding of their significance to the firms concerned. A brief description of the six techniques investigated will therefore be given here.

Special presses in paper-making speed up removal of water from the paper web. Compared with older systems the water removal capacity of special presses is higher, which means that the capacity of the paper machine can be increased. If you can sell the increased output on the market at about the going price your pay-off period on the rather small investment will often be very short. If you cannot sell an increased output you will just save steam on your unchanged production, which makes the investment less profitable but still often worth while doing. The technique has been developed in Sweden and a patent was applied for in 1957.

Wet suction boxes also increase the water removal capacity of a paper machine. Just as with special presses, the effect of this technique on some machines is to increase the production capacity of the machine. The main advantage, however, is to increase the quality of the paper produced. The technique in question was developed in the United States and came from there to Sweden in the late fifties. It represents a comparatively small investment.

Continuous pulp-cooking has gone through big changes during the post-war period. It was greatly improved in the second half of the fifties and started subsequently to diffuse in the United States, thus replacing the conventional discontinuous batch-cooking. In the sixties the technique started to diffuse in Sweden. For a pulp-producing company, building a continuous-digester represents a rather big investment. Such investments seem also to be strongly related to increases in production capacity. The advantages with continuous cooking are lower capital costs per ton pulp when your annual output is not very small, lower labour costs and lower steam and energy costs. A disadvantage with continuous cooking compared with discontinuous, is that in the latter case you can do repair work on your digesters without stopping the whole plant.[1]

Instead of piling wood in woodyards the wood can directly after delivery be barked and cut into chip, which is then blown into piles of so-called *outside chip storage*. This method spread to Sweden from the United States in the beginning of the fifties, but did not really begin to diffuse until the sixties. Through this storage, chemical

---

[1] A great problem with this technique is that a conveniently discontinuous batch-cooking has also been improved over time partly as a consequence of the continuous technique. This makes profitability calculations very difficult.

processes start which are favourable for sulphite-pulp-production but may be unfavourable for sulphate-pulp-production. The advantages with the method are reduced handling-costs and lower interest costs on your inventories. A risk involved is that the total pile may be destroyed by chemical processes. Investment costs are very low for the method.

The tunnel kiln was invented in 1840 by a Dane. The Hoffman kiln, however, patented in Germany in 1858, conquered the market due to lower construction costs and technical difficulties with the tunnel kiln. In recent decades mechanical progress has, however, favoured the tunnel kiln. The tunnel kiln is more complex and expensive than the Hoffman kiln. On the other hand labour and fuel costs are considerably lower and the risk of damage to capital equipment is smaller with the tunnel kiln. Only one type of brick can be made at any one time in a tunnel kiln, which might be a disadvantage in smaller companies.

Already in the late forties some brick-making companies in Sweden introduced *trucks*, but it was not until the sixties that diffusion started on a broader scale. Loading and unloading bricks by hand is not only labour consuming but it also causes damage to the bricks. By using trucks labour costs as well as storage space are reduced. The purchase of a truck represents, of course, a rather small investment for a medium-sized brick producer.

IV. THE DIFFUSION DIAGRAMS

Most of the data in this paper have been obtained through direct inquiries from companies. Questionnaires have been sent to all companies in the pulp and paper industries and the brick industries in Sweden, which could possibly be assumed to be users of the techniques in question. Thus companies producing special types of paper, pulp or bricks for which the techniques in question are definitely unsuited, have been omitted from the investigation. The response rates have been very high indeed. Both users and potential users of the techniques have been investigated, which means that the diffusion process is not yet over and that further diffusion might occur during the seventies. Many non-using firms have reported plans of introducing the technique in question in the future. This will cause some problems in the econometric calculations that follow. In Table 10.1 the total number of observations is shown.

Whether the company or the plant is the relevant observation unit depends for our purposes on which of these is the decision-making unit for adopting the techniques. For SPs, WSBs and OCSs it is not certain whether it is a top management decision or whether

TABLE 10.1

NUMBERS OF OBSERVATIONS

	Companies	Plants
Special presses (SP)	25	43
Wet suction boxes (WSB)	25	44
Continuous pulp-cooking (CPC)	19	23
Outside chip storage (OCS)	21	38
Tunnel kilns (TK)	38	
Trucks (T)		44

plant managers can decide about introduction. For CPCs it is probable that top management must decide, because of the size of the investment. With TKs it is quite certain that it is a top management question, whereas the contrary holds for trucks.

In diagrams 1–6 we show the diffusion-curves for the different techniques. The curves are based on three alternative measures of diffusion, namely the number of plants and companies respectively using the technique at different dates, and the percentage of output produced by using the technique at different dates. The latter measure will not be used in the present study, as has already been said. In the diagrams the dates refer to years, whereas quarters are used in the analysis. In the case of the first two measures the dates refer to the year when the techniques in question first started to be used on a commercial scale. From the diagrams it will be seen that SPs started to diffuse in Sweden in the sixties, TKs and Ts in the forties and all the others in the fifties. Use of trucks is the only technique of which it can be said that diffusion at the plant level was nearly complete during the period considered.

As was stated above, I have no intention of trying to explain differences in imitation-rates for different types of techniques. Six observations form too small a sample for that. Nevertheless we may keep in mind hypotheses put forward in other studies[1] when looking at the diffusion-curves in diagrams 1–6. No doubt diffusion-patterns differ between different techniques. For some techniques diffusion is a very slow process (for instance in CPC, TK and T) while SP is an example of very rapid diffusion. The 'logistic curve' and the bandwagon effect is clearly visible for some techniques, but in the case of TK and T it is difficult to find any such tendencies. Since we are not in a position to compare the profitability of the techniques we cannot say anything about the hypothesis that on the average more profitable innovations diffuse more quickly than less profitable ones. As to size of investment, CPC and TK seem, relatively speaking,

[1] For instance, Mansfield, *Industrial Research and Technological Innovation*, chap. 8.

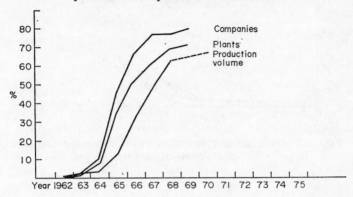

FIG. 10.1 Diffusion of special presses among plants, companies and on a production-volume basis

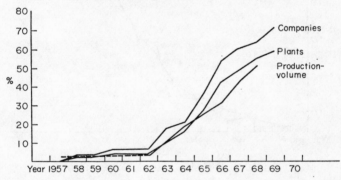

FIG. 10.2 Diffusion of wet suction boxes among plants, companies and on a production-volume basis

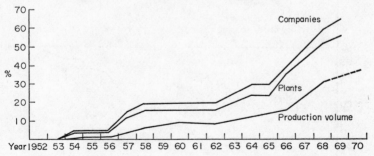

FIG. 10.3 Diffusion of continuous-pulp-cooking among plants, companies and on a production-volume basis

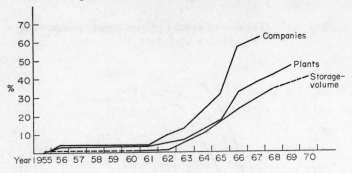

FIG. 10.4 Diffusion of outside chip storage among plants, companies and on a storage-volume basis

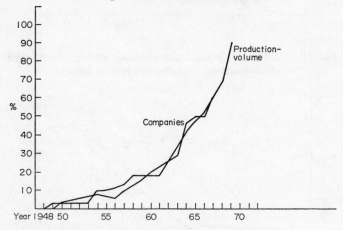

FIG. 10.5 Diffusion of tunnel kilns among companies and on a production-volume basis

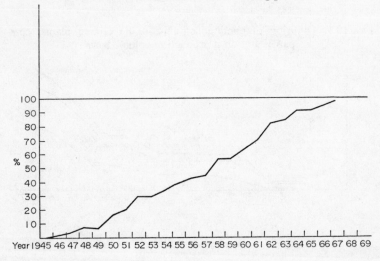

FIG. 10.6 Diffusion of trucks among plants

to involve the biggest investments for the companies and they also show rather slow diffusion rates, which is in accordance with results from elsewhere. Still the diffusion-pattern for trucks does not fit into that picture at all.

V. THEORIES AND HYPOTHESES

When trying to explain diffusion rates among companies and among plants there are a few things that should be remembered, but which sometimes tend to disappear in the analysis. First, many techniques improve over time, which means that profitability calculations of companies will not be the same at different points in time. This is often very difficult to take into account in the analysis. In this paper this applies especially to CPC during the fifties and to some extent perhaps also to OCS. The profitability of TK depends among other things on the relationship between coal and oil prices, which has changed during the post-war period. Secondly, one must be careful not to interpret early introduction of a new technique as a sign of a progressive and successful management. It may be so, but it may also be more profitable for a company to wait a short time before introducing a new technique in order to be certain that it really works and is well developed.

When trying to set up testable hypotheses about the diffusion of new techniques one can rely on different types of theories. Our hypotheses are basically drawn from economic theory but may be said to incorporate elements of sociology as well.[1] I shall put forward four types of hypotheses, which have in various forms been advanced in the studies referred to earlier.

A. My first set of hypotheses relates to the time when information about a new technique is received and the ability to evaluate this information.[2] Economic models are often based on the assumption that information about new processes spreads instantly among companies or plants. This is probably a highly misleading assumption as well as incompatible with profit maximising behaviour by firms since obtaining information is a cost consuming activity.[3] Interestingly, we have found that very often information about new technique is spread by producers of the equipment which incorporates the technique in question (the change agent in Rogers' terminology).

[1] Especially through the influence of Rogers, *Diffusion of Innovations*.
[2] Rogers, *op. cit.*, speaks of different stages in the adoption process. Here we are concerned especially with the awareness stage and the evaluation stage.
[3] Carter and Williams, *Industry and Technical Progress*, state that the technically progressive firm has a high quality of incoming communication. It is quality rather than quantity that matters. 'In consequence, the backward firm may not hear of an idea for several years after it is first made known.'

Thus, our hypothesis will be that the earlier information about a new process is received and the greater the ability of the decision-making unit to evaluate the information, the sooner the process will be introduced in the company or plant.

Since we have information on when companies or plants first got to know about the techniques, the first part of the hypothesis is testable. However, I am not convinced that our data are very reliable on this point, since it is possible that the questionnaire answers here involve a subjective estimate on the part of the responding company official. Alternatively, it could be tested indirectly by using the share of exports in production as a proxy for information data, on the assumption that a large export share means more contacts with foreign producers, foreign markets, and so on and will be positively correlated with receiving information about new techniques early.[1] The second part of the hypothesis has been tested by using the amount of resources, which may be used by a company to evaluate information, as a measure of the company's or plant's ability to evaluate information. In this paper I shall use expenditures for R & D as a proxy for evaluation capacity. Alternatively, company or plant size could serve as a proxy, assuming that size is positively related to the amount of resources which may be used to evaluate a new technique. On the other hand, size may be a proxy for a number of other things as well.

B. The second type of hypothesis is of a more purely economic character. Basically, the hypothesis is that the more profitable it is for a company or plant to use technique the earlier it will be introduced. Chances of high profitability thus take over the risks of early introduction. If capital markets are less than perfect (i.e. if marginal yields on invested capital are not everywhere the same) the hypothesis is not as evident as it looks to be and the statement will not necessarily hold in a comparison between companies. Companies are likely to have different opportunity costs and a pay-off period or internal rate of return which is satisfactory for one company may be too high or too low for another company. Furthermore, this hypothesis raises a controversy between economists and sociologists as to the relative importance of different variables in explaining diffusion rates. Sociologists stress much more the so-called interaction effect – that is the process through which individuals who have adopted an innovation influence those who have not yet adopted it. Being an economist, I prefer the economic hypothesis here, but we must of course be aware of the fact that what we are seeking are the *ex ante*

---

[1] It has been found in many studies that early adopters of innovations seem to be more cosmopolite than late adopters. Rogers, *Diffusion of Innovations*, p. 183.

calculations on profitability, not the *ex post* ones. In view of the uncertain problems involved in introducing a new technique these *ex ante* calculations may to quite an extent be influenced by experience by others.

We have encountered difficulties in finding adequate measures of profitability for our analysis. For SP we have a pay-off period for each machine in the various companies as a consequence of Håkanson's work for the international study.[1] For WSB we have used speed of machines and type of paper produced as proxies.[2] For CPC we have used age of cooking equipment as a proxy since the installation of CPC mostly means scrapping of old equipment. Consequently the older the equipment in a company the earlier CPC should be introduced. In OCS we have used handling equipment in the woodyard and type of wood used as proxies.[3] For TK and T we have not succeeded in finding acceptable proxies for profitability, a question to which we will have reason to return in our calculations later. Alternatively, increase in output could be used to measure profitability. The use of this variable is based on the assumption that new technique is introduced in connection with investment made in order to increase capacity.[4] Then, the more output has increased, the more profitable it has been to use the technique in question. We have figures on the increase of production for all companies and plants during the relevant period for all processes except OCS.

C. It is often said that the financial structure of a company influences its investment activity and its readiness to take risks.[5] The higher the degree of internal financing a company can afford the more prepared it is to invest in new uncertain technique. The theory

[1] The following formula has been used where r = Pay-off period in years ; I = Investment cost ; a = Capacity increase (%), which is a function of the age of a machine ; Q = Annual capacity of the paper machine ; P_{pa} = Price per ton of paper ; P_{pu} = Price per ton of pulp.

$$r = \frac{I}{a \cdot Q(P_{pa} - 1 \cdot 05 \, P_{pu})}$$

[2] WSB is best suited on machines with high speed and producing kraft paper, newsprint paper or journal paper.

[3] OCS has been considered to be a less suitable method for companies having installed cranes in the woodyard. Furthermore fir and pine have advantages over other types of wood when using OCS.

[4] See for instance, Salter, *Productivity and Technical Change* (Cambridge, 1960), or Wohlin, *Skogsindustrins strukturomvandling och expansionsmöjligheter*, (Forest based Industries: Structural Change and Growth Potentials), (IUI, 1970). Gold, Peirce and Rosegger, *Journal of Industrial Economics* (July 1970) are more sceptical towards this hypothesis.

[5] Rogers, *Diffusion of Innovations*, p. 175, finds that in general early adopters of innovations have a more favourable financial position than late adopters.

is based on an assumed unwillingness on the part of management and owners to borrow above a certain limit as that might increase the creditors' position within the firm. Furthermore, credit-institutions very often ask for certain relationships between owned and borrowed money in a company's financial structure. We have tried to obtain information on the financial structures of the companies during the period 1955-64. We have related total net revenue generated within a company during that period (net profit + depreciation + other savings-dividends) to total investment during the period. The hypothesis we are testing is that the larger the ratio between net income and investment expenditure the easier it has been to finance investment in new technique and the earlier such techniques should have been introduced. Size may also be considered a proxy for financial strength in the sense that the bigger a company is the easier it can take a loss on introduction of a given new technique.

D. It has been suggested that there are differences in attitudes towards new processes and that these may affect the rate of diffusion of new techniques (see, for instance, Carter & Williams, *Industry and Technical Progress*, chapters 10 and 16). Our discussions with managements also indicate that there may be such differences. Producers and sellers of equipment involving new processes (for instance the Kamyr Company in Sweden for CPC), also stress this point. In the present study we have tried to consider the role of differences in attitudes in explaining diffusion in various ways. For SP and WSB we have information about the behaviour of the companies and the plants as to four other new techniques introduced during the post-war period. From this information we have produced a scale ranging from 16 to 0 depending on how quick the companies have been to introduce these four techniques. For TK and T the same procedure has been used although we have here only had information for one other technique (TK for T and *vice versa*). As to CPC and OCS, companies have given us information as to whether (a) they aim to be the first to introduce a new technique in Sweden, (b) they wish first to know that it works in some other company, (c) they wish first to be sure that the technique works in a number of companies, (d) they wish first to be sure that the technique has worked satisfactorily for some time in a couple of companies. From this we have formed a scale ranging from 4 to 1 which has been used as a proxy for attitude towards new technique.

R & D activity can also be used as a proxy for attitude towards new technology, the hypothesis being that the more interested a company is in R & D activity the earlier it will introduce new technique. In this investigation we only have information on whether the companies have R & D activity or not. We have used this information as a dummy variable giving the value 1 to companies with and 0 to those without R & D.

In Table 10.2 we have summarised the independent variables used for the different techniques in the investigation. The dependent variable is the first date at which a company or a plant has used the technique in question for commercial purposes. Time has been measured in quarters from the first quarter in which the technique was used anywhere in Sweden (giving value 1 to that company or plant). For those companies that have not yet introduced the technique we have assumed that they will introduce it during the first quarter of 1975. This assumption will be more fully explained later on. From a statistical point of view, however, it seems to be preferable to try to use all information available than to discard part of the diffusion-curve.

TABLE 10.2

LIST OF INDEPENDENT VARIABLES

	SP	WSB	CPC	OCS	TK	T
Size of company or plant (s)	x	x	x	x	x	x
Increase in production (p)	x	x	x		x	x
Attitude towards new technique (b)	x	x	x	x	x	x
R & D activity (fou)	x	x	x	x	x	x
Export share (exp)	x	x	x	x		
First information (inf)	x	x	x	x		
Financial capacity (fi)	x	x	x	x		
Pay-off period (r)	x					
Type of paper produced (pgr)		x				
Speed of machine (hast)		x				
Age of capital equipment (k)			x			
Handling equipment (hant)				x		
Type of wood (ved)				x		

There are, of course, a number of problems involved in our choice of independent variables which can only be mentioned briefly here. One is the validity problem. Do the variables chosen really measure what we want them to measure? A second and related problem is that the variables chosen can sometimes be used for testing different hypotheses. This is for instance the case with size, R & D, and increase in production. We must remember this when interpreting our results.

VI. ANALYSIS AND RESULTS

The analysis has been divided into two parts. In Part 1 we have merely grouped the material into different subgroups, the grouping being made according to the values of the dependent variable, companies or plants which introduced the technique relatively early, somewhat later and so on in a descending scale. Such a grouping can of course be made in a number of ways (see Rogers, *Diffusion of Innovations*, chapter 6). We have tried to get about the same number of users in each group. All non-users have, however, been grouped together. In Part 2 we have tried a formal econometric approach. One advantage with the presentation in Part 1 is that it gives us some information about the functional relationships between the dependent and the independent variables. In Part 2 we must, of course, make assumptions on this point.

(1) *Grouping of the Material*

In Tables 10.3–8 are shown the groupings for the different techniques and the corresponding average values of the independent variables. Inspection of the tables reveals that the explanatory variables perform quite well in many cases given the hypotheses presented earlier. Taking them in order of presentation we start with the information- and evaluation-capacity variables. Information about the techniques considered seems to have diffused at about the same time in Swedish industry, although time-lags between the first and the last are sometimes astonishingly long. Differences in average awareness-times between the groups have been rather small. One explanation for this might be that companies in the Swedish pulp and paper industry work rather closely together in different forms of industry organisations and in some cases they have common research laboratories working in special fields. Another reason might be that sellers of the new technique have approached the companies at about the same time. This was, anyhow, the case with CPC. In spite of this the average-values mostly agree with the hypothesis that the earlier a company or a plant receives the information, the earlier it has introduced the technique in question. As to size, export share and R & D activity the average-group values act in the expected way in most cases, both on the company and plant level. CPC and T constitute the exceptions as far as the size variable is concerned. We know from producers of equipment that the big pulp producers were a little hesitant towards continuous-pulp-cooking from the beginning. CPC also gives a mixed picture of the role of export shares, although the most export oriented plants and companies appear to have introduced CPC first. The R & D variable performs well on a company level except for TK. The data on the plant level are rather questionable here, though.

The proxy-variables for profitability work astonishingly well in most cases. Pay-off periods, type of paper produced, speed of

TABLE 10.3

SPECIAL PRESSES

(a) Companies

Group (No. of Companies in brackets)	Date of Introduction	Number of Employees in 1968	Export Share of Turnover as Percentage	R & D Activity[a]	Time of First Information	Growth of Production 1962-8 (1962 = 100)	Pay-off Period in Years	Self-Financing Ability on Average between 1955-64	Attitude Variable Ranging from 16-0
I (5)	1964:2	1865	68·0	0·6	1960	139	0·30	1·035	9·8
II (6)	1965:3	1623	68·2	0·8	1961	152	0·28	1·291 (5)	8·8
III (5)	1966:1	1552	47·6	0·8	1961	133	0·92	1·052	9·4
IV (5)	1967:3	584	68·6	0·2	1961	131	0·60	0·786 (4)	7·4
V (4)	(1975:1)	330	55·5	0·25	1963	132	1·90	0·902	2·5

(b) Plants

(No. of Plants in brackets)	Date of Introduction	Number of Employees in 1968	Export Share of Turnover as Percentage	R & D Activity[a]	Time of First Information	Growth of Production 1962-8 (1962 = 100)	Pay-off Period in Years	Self-Financing Ability on Average between 1955-64	Attitude Variable Ranging from 16-0
I (8)	1964:3	1004	59·4	0·5	1961	142	0·31		8·5
II (9)	1965:3	922	69·3	0·8	1961	154	0·36		6·0
III (8)	1966:2	716	62·9	0·75	1961	162	0·16		6·2
IV (9)	1968:1	581	61·6	0·4	1961	132	0·63		5·9
V (9)	(1975:1)	314	54·5	0·7	1963	124	1·83		4·0

[a] As a dummy variable R & D activity = 1; no R & D activity = 0.

TABLE 10.4

WET SUCTION BOXES

(a) *Companies*

Group (No. of Companies in brackets)	Date of Introduction	Number of Employees in 1968	Export Share of Turnover as Percentage	R & D Activity	Time of First Information	Growth of Production 1962–8 (1962 = 100)	Type of Paper Produced[a]	Speed[b] of Machines	Self-Financing Ability on Average between 1955–64	Attitude Variable Ranging from 16–0
I (5)	1960:2	1864	76·8	1·0	1956	157	0·8	352	1·145 (4)	12·0
II (5)	1965:1	1396	68·6	0·4	1961	149	0·4	232	1·216 (4)	10·6
III (5)	1966:2	987	68·2	0·4	1961	131	0·8	286	0·926	7·8
IV (5)	1968:2	1094	42·0	0·4	1963	138	0·4	204	0·978	9·2
V (5)	(1975:1)	731	50·4	0·4	1961	127	0·2	122	0·997	5·4

(b) *Plants*

(No. of Plants in brackets)	Date of Introduction	Number of Employees in 1968	Export Share of Turnover as Percentage	R & D Activity	Time of First Information	Growth of Production 1962–8 (1962 = 100)	Type of Paper Produced[a]	Speed[b] of Machines	Self-Financing Ability on Average between 1955–64	Attitude Variable Ranging from 16–0
I (10)	1962:4	914	75·2	0·8	1957	157	0·7	291		7·4
II (9)	1966:1	652	66·3	0·6	1961	155	0·6	292		6·7
III (10)	1967:4	524	54·0	0·5	1962	131	0·5	281		7·3
IV (15)	(1975:1)	541	55·5	0·7	1961	130	0·3	174		5·5

[a] As a dummy variable kraft, newsprint and journal paper = 1; other types of paper = 0.
[b] In metres/per minute.

TABLE 10.5

CONTINUOUS PULP COOKING

(a) Companies

Group (No. of Companies in brackets)	Date of Introduction	Number of Employees in 1968	Export Share of Turnover as Percentage	R & D Activity	Time of First Information	Growth of Production 1954-68 (1954 = 100)	Average Introduction-Time of Cooking Equipment	Self-Financing Ability on Average between 1955-64	Attitude Variable Ranging from 4-1
I (7)	1962:1	686	81·3	0·9	1949	330	1936	0·650 (5)	3·3
II (6)	1968:1	1355	61·5	0·8	1950	261	1947	1·182	3·0
III (6)	(1975:1)	656	77·2	0·7	1952	208	1950	0·987	2·3

(b) Plants

(No. of Plants in brackets)	Date of Introduction	Number of Employees in 1968	Export Share of Turnover as Percentage	R & D Activity	Time of First Information	Growth of Production 1954-68 (1954 = 100)	Average Introduction-Time of Cooking Equipment	Self-Financing Ability on Average between 1955-64	Attitude Variable Ranging from 4-1
I (7)	1962:1	686	81·3	0·9	1949	330	1936		3·3
II (6)	1968:1	1355	61·5	0·8	1950	261	1947		3·0
III (10)	(1975:1)	512	78·0	0·7	1952	219	1945		2·4

TABLE 10.6

OUTSIDE CHIP STORAGE

(a) Companies

Group (No. of Companies in brackets)	Date of Introduction	Number of Employees in 1968	Export Share of Turnover as Percentage	R & D Activity	Time of First Information	Type of[a] Handling Equipment	Type of[b] wood Used	Self-Financing Ability on Average between 1955–65	Attitude Variable Ranging from 4–1
I (7)	1963:1	1712	69·6	0·9	1959	0·4	0·9	1·236 (6)	3·1
II (7)	1966:2	1308	77·0	0·9	1958	0·1	1·0	0·731	3·3
III (8)	(1975:1)	1028	56·8	0·25	1960	0·5	0·5	0·882 (7)	2·4

(b) Plants

(No. of Plants in brackets)	Date of Introduction	Number of Employees in 1968	Export Share of Turnover as Percentage	R & D Activity	Time of First Information	Type of[a] Handling Equipment	Type of[b] wood Used	Self-Financing Ability on Average between 1955–65	Attitude Variable Ranging from 4–1
I (9)	1963:4	618	72·1	0·9	1958	0·3	0·9		3·1
II (10)	1967:3	696	71·3	0·6	1960	0·4	0·8		2·8
III (19)	(1975:1)	495	70·3	0·6	1961	0·6	0·5		2·4

[a] As a dummy variable good handling equipment for introduction = 0; bad for introduction = 1

[b] As a dummy variable favourable type of wood and pulp = 1; unfavourable type of wood and pulp = 0

TABLE 10.7

TUNNEL KILNS, COMPANIES

Group (No. of Companies in brackets)	Date of Introduction	Number of Employees in 1968	R & D Activity	Growth of Production 1958-68 (1958 = 100)	Attitude Variable Ranging from 5-1
I (6)	1955:2	54	0·3	255	4·0
II (6)	1957:4	29	0·2	155	3·3
III (6)	1964:3	24	0·5	220	4·3
IV (5)	1966:4	27	0·4	196	2·6
V (15)	(1975:1)	26	0·2	95	1·5

TABLE 10.8

TRUCKS, PLANTS

(No. of Plants in brackets)					
I (9)	1949:2	38	0·3	171·1	3·6
II (8)	1953:4	39	0·4	180·6	2·6
III (8)	1960:4	31	0·5	143·4	3·5
IV (9)	1964:2	42	0·3	164·3	3·0
V (9)	1967:3	34	0·1	129·2	1·4
VI (1)	(1975:1)	10	0	79	1·0

machine, age of equipment, and type of wood used, all show that high profitability on a new technique and early introduction coincide. The opposite holds for low profitability and late introduction. Early introducers of a technique have also had a more rapid expansion of production than late introducers or non-introducers.

The financial variable performs very poorly. (Here, of course, we only use values on a company level.) At most you might say that there is a slight tendency for early introducers of SP, WSB and OCS to show a somewhat better self-financing ability that late introducers or non-users, but the tendency is very weak. Before drawing any conclusions regarding the relationship it should be remembered that our proxy-variable is very crude and perhaps only vaguely related to the hypothesis. It deserves mention that a number of companies have stated that they have not had money to finance the necessary investment. This is the case with 5 companies in SP, 6 in WSB, 3 in CPC, 3 in TK and 2 in T.

As for our last hypothesis, that is the role about the attitude of management towards new technique and new processes, the proxies in our groupings stand out much better than expected, both on a company and on a plant level. Companies or plants which are early introducers of one technique also seem to be early introducers of other techniques. Furthermore, if they say they want to be early introducers in the country, they also seem to act accordingly when new processes become available. The opposite also seems to be true. (Håkanson's results, however, are not quite in accordance with this last statement.)

Of course, this grouping of our data has not been a test of our hypotheses in any real sense. It just shows that the correlations between the independent and the dependent variables work mostly in the expected direction. The most outstanding exception from this is the data on self-financing ability. Here the crude nature of our data must be remembered, however.

(2) *Econometric Approach*

In the econometric analysis we have used the standard programme of step-wise regression (BMDO2R). The results for all techniques both on a company and on a plant level are shown in Table 10.9. Some problems, however, must be mentioned before discussing the results.

As mentioned earlier we have used information pertaining to not only users but also non-users of the techniques in the analysis. Having removed all companies that cannot possibly be expected to introduce the technique in the future because of wrong type of paper produced and so on, the remainder is supposed to consist of potential users. In fact, many of the non-users have explicitly stated in the questionnaire returns that they are going to introduce the particular technique in the future. Since we did not like to discard this information we have been faced with the problem of finding an appropriate introduction date for these future users. Assuming a logistic diffusion curve diagram 1–6 shows possible introduction dates. The years 1972, 1975 and 1980 have then come out as possible candidates. Trials with these different assumed introduction years have not changed the R^2-values very much and so we decided to fix the first quarter of 1975 as introduction time for all non-users in all processes. This is, of course, somewhat arbitrary.

The choice of functional form of the regression equations has from a theoretical point of view been rather tricky. From the tables presenting the material it appears that the relationships between the dependent and the independent variables often seem to be of a difficult curve-linear-type. We have experimented with three kinds of forms.

In the first alternative we assume a linear relationship of the type:

$$y = a + b_1 x_1 + b_2 x_2 \ldots \tag{1}$$

Table 10.9

REGRESSION EQUATIONS FOR SP, WSB, CPC, OCS, TK AND T
t-ratio in brackets

SP; companies ($n = 25$)

$$\log y = 1.719 - 0.043\ b - 0.070\ \text{fou} - 0.026\ \log s - 0.233\ \log p$$
$$(3.078) \quad (0.647) \quad (0.155) \quad (0.442)$$
$$+\ 0.283\ \log r - 0.047\ \log \exp + 0.639\ \log \inf \qquad R^2 = 0.767$$
$$(2.532) \quad (0.238) \quad (2.829)$$

SP; plants ($n = 43$)

$$\log y = 3.289 - 0.023\ b + 0.046\ \text{fou} - 0.490\ \log s - 0.395\ \log p + 0.153\ \log r$$
$$(2.106) \quad (0.568) \quad (3.388) \quad (0.949) \quad (1.483)$$
$$+\ 0.028\ \log \exp + 0.281\ \log \inf \qquad R^2 = 0.572$$
$$(0.144) \quad (1.493)$$

WSB; companies ($n = 25$)

$$\log y = 3.424 + 0.036\ \text{pgr} - 0.007\ b + 0.045\ \text{fou} - 0.158\ \log s - 0.198\ \log \exp$$
$$(0.280) \quad (0.369) \quad (0.380) \quad (0.934) \quad (0.839)$$
$$-\ 1.188\ \log p + 0.103\ \log \text{hast} + 1.139\ \log \inf \qquad R^2 = 0.736$$
$$(1.579) \quad (0.435) \quad (5.256)$$

WSB; plants ($n = 44$)

$$\log y = 2.903 - 0.023\ b - 0.072\ \text{pgr} + 0.082\ \text{fou} - 0.291\ \log s - 0.737\ \log p$$
$$(1.707) \quad (0.809) \quad (1.006) \quad (2.573) \quad (1.728)$$
$$+\ 0.181\ \log \text{hast} + 0.803\ \log \inf - 0.051\ \log \exp \qquad R^2 = 0.565$$
$$(1.048) \quad (4.810) \quad (0.270)$$

CPC; companies ($n = 19$)

$$\log y = 0.073 + 0.076\ \text{fou} + 0.280\ \log s - 0.371\ \log p + 1.138\ \log k - 0.387\ \log \exp$$
$$(0.300) \quad (0.662) \quad (0.521) \quad (1.306) \quad (0.621)$$
$$+\ 0.699\ \log \inf \qquad R^2 = 0.397$$
$$(1.775)$$

CPC; plants ($n = 23$)

$$\log y = 0.489 + 0.034\ b + 0.410\ \log s - 0.487\ \log p + 0.768\ \log k$$
$$(0.335) \quad (0.873) \quad (0.810) \quad (1.147)$$
$$-\ 0.977\ \log \exp + 0.913\ \log \inf \qquad R^2 = 0.349$$
$$(0.837) \quad (2.334)$$

OCS; companies ($n = 21$)

$$\log y = 2.051 - 0.187\ \text{fou} - 0.022\ b + 0.130\ \text{hant} + 0.020\ \text{ved} - 0.062\ \log s$$
$$(1.323) \quad (0.311) \quad (1.043) \quad (0.127) \quad (0.332)$$
$$-\ 0.266\ \log \exp + 0.378\ \log \inf \qquad R^2 = 0.511$$
$$(1.082) \quad (2.694)$$

OCS; plants ($n = 38$)

$$\log y = 1.792 + 0.104\ \text{hant} - 0.056\ \text{ved} - 0.025\ b - 0.033\ \text{fou} - 0.043\ \log s$$
$$(1.379) \quad (0.666) \quad (0.617) \quad (0.385) \quad (0.410)$$
$$-\ 0.163\ \log \exp + 0.398\ \log \inf \qquad R^2 = 0.413$$
$$(0.893) \quad (3.486)$$

TK; companies ($n = 38$)

$$y = 103.630 - 0.043\ s + 3.668\ \text{fou} - 0.058\ p - 9.599\ b \qquad R^2 = 0.419$$
$$(0.358) \quad (0.441) \quad (0.905) \quad (3.364)$$

T; plants ($n = 44$)

$$y = 93.156 - 0.375\ s - 0.093\ p + 1.898\ \text{fou} - 6.379\ b \qquad R^2 = 0.253$$
$$(1.793) \quad (1.680) \quad (0.238) \quad (2.707)$$

In the second alternative we assume a multiplicative relationship of the type:

$$y = ax_1^{b_1} \cdot x_2^{b_2} \ldots$$

which in its logarithmic form is transformed to

$$\log y = \log a + b_1 \log x_1 + b_2 \log x_2 \qquad (2)$$

It must be remembered, however, that attitude and dummy variables such as R & D etc are not in a logarithmic form.

In the third alternative the dependent variable and one independent variable (in this case information time) (besides the attitude and the dummy variables) have not been logarithmised. The lin-log-equation has then got the following form (only tried on SP, WSB and CPC):[1]

$$y = a + b_1 x_1 + b_2 \log x_2 + b_3 \log x_3 \ldots \qquad (3)$$

According to Mansfield (*Industrial Research and Technological Innovation*, p. 158) the multiplicative relationship ought to make most sense because the effect of each of the independent variables is likely to depend on the level of the other. We do not feel certain that this argument applies to all the independent variables we have tried. Actually the logarithmic form worked out best for the pulp and paper techniques, whereas the linear form did the same thing for the brick techniques.[2]

Another problem has been the treatment of the financial variable. Lack of information meant that we had to throw out data from two companies when testing that variable. It then turned out that the financial variable only became significant (according to a *t*-test) for CPC on company level. We then included the two companies and excluded the financial variable. That did not change the picture very much for all techniques except CPC. In Table 10.9 the financial variable is excluded in all calculations.

Looking at the results in Table 10.9 we find that the R^2s are always higher on the company than on the plant level. This is not really surprising. Our hypotheses relate more to company than to plant behaviour. For instance, variables as R & D activity and information time seem to make much more sense on company than on plant level. On the other hand we do explain quite a lot of the

[1] The relationship being of the following type:

$$e^{y-b_1 x_1} = a \cdot x_2^{b_2} \cdot x_3^{b_3}$$

[2] In Table 10·9 we have only shown the log-equations for the pulp and paper techniques and the lin-equations for the brick techniques.

diffusion on plant level too, which means that the same kind of model can be used in both cases. This is in accordance with Mansfield's thesis (*Industrial Research and Technological Innovation*, p. 190) that the same kind of model can be used for explaining interfirm and intra-firm diffusion rates.

Another feature of our regression equations is that the coefficients mostly show the right sign. It is only for WSB that many of them have the wrong sign, especially as to R & D activity and speed of the machines. For SP (plants), CPC (companies), TK (plants) the R & D variable also has the wrong sign. In some cases export share also has the wrong sign. But otherwise the agreement of the expected sign-values and the actual ones must be looked upon as quite encouraging.

Furthermore, the R^2- and the t-values show that our variables have a rather low explanatory value for the brick-techniques and for CPC. This might be due to the bad proxies used for profitability in these cases. For TK and T we have only used growth of production as a proxy. We have got the highest R^1 for SP, which is not surprising as we have been able to use a well-calculated pay-off period for that technique. In Table 10.10 we have numbered those

TABLE 10.10

SIGNIFICANT EXPLAINING VARIABLES

	lin	log	lin-log
SP			
Companies	b[x] r[xx] s[xxx]	b[x] r[x] inf[x]	b[x] r[xx] s[xx]
Plants	r[x] s[x] p[xx] b[xx]	s[x] b[xx] r[xxx] inf[xxx]	s[x] r[xx] b[xxx] inf[xxx]
WSB			
Companies	p[xx]	inf[x] p[xxx]	p[xxx]
Plants	p[x] pgr[xx]	inf[x] s[x] b[xx] p[xx]	p[x] pgr[xxx] s[xxx]
CPC			
Companies	k[xx]	inf[xx] k[xxx]	k[x]
Plants	inf[xx]	inf[xx]	k[xx]
OCS			
Companies	inf[xx] fou[xx] exp[xxx]	inf[x] fou[xxx]	
Plants	inf[x] hant[xxx] ved[xxx]	inf[x] hant[xxx]	
TK			
Companies	b[x]	b[xxx]	
T			
Plants	b[x] s[xx] p[xx]	s[x] b[xxx]	

x = significant on 1 per cent level
xx = significant on 5 per cent level
xxx = significant on 10 per cent level

variables which are significant on 1, 5 and 10 per cent level for different techniques and different functional forms. Relating the symbols in Table 10.10 to our hypotheses we find that the information-hypothesis does better than expected from our groupings. First time of information rather often turns up as a significant variable (the problem being of course what type of data we have got here). The evaluation-capacity hypothesis does less well, however. Size turns out to be a significant variable for three techniques on plant level, which is less than expected while the hypothesis seems to be less appropriate on company level. The caveat that size may be a proxy for other things as well should be kept in mind. Our data rather tends to support the hypothesis that when companies introduce a new technique they start by using it in their big plants. The profitability-hypothesis gets rather strong support from our data as many of the proxies for profitability are significant on the 10 per cent level. Rapid increase in production, however, is not as powerful an explanation as we had thought (which is in accordance with the results of Gold, Peirce and Rosegger, *Diffusion of Major Technological Innovations*). The attitude-hypothesis, finally, gets strong support in four of our techniques, which means that we have to take into account differences in management attitudes when trying to explain and forecast diffusion rates. This will be somewhat more elaborated in the next section.[1]

VII. COMPANY BEHAVIOUR AS TO DIFFERENT TECHNIQUES

As already mentioned we have in this study had the possibility to compare company behaviour as to different new techniques introduced during the post-war period. Thus, it is interesting to try to ascertain whether some companies always tend to be leaders and others laggers of new techniques. For the four pulp and paper techniques we have got six rank correlations for the four techniques in question, whereas we have got one rank correlation for the brick techniques. These values are shown in Table 10.11. From the figures in the table we can draw the conclusion that there seems to have been some form of technical leadership in Sweden in the industries concerned during the post-war period. From interviews with people in the industries and with sellers of equipment this hypothesis has also been confirmed for the pulp and paper industry. Some companies have always wanted to be in the lead in installing new technology. On this point we seem to have got stronger evidence

[1] We have investigated for multi-colinearity among our independent variables and this does not seem to be a great problem in our calculations.

TABLE 10.11

RANK CORRELATIONS BETWEEN THE
DIFFERENT TECHNIQUES

rSP-WSB $= 0.492^{xx}$
rSP-CPC $= 0.026$
rSP-OCS $= 0.611^{xx}$
rWSB-CPC $= 0.606^{xx}$
rWSB-OCS $= 0.495$
rCPC-OCS $= 0.309$
rTK-T $= 0.434^{xx}$

^{xx} = significant on 5 per cent level.

than, for instance, Mansfield (*Industrial Research and Technological Innovations*) did.

VIII. SUMMARY AND CONCLUSIONS

In this paper certain theories have been presented by which we have tried to explain the diffusion of new technology. The data used in the empirical analysis have been obtained from the Swedish pulp, paper and brick industries. Naturally, the quality of available data must be taken into account when interpreting the results, and this caveat applies especially to the measures of profitability which we have used. Still, the hypotheses stand up fairly well when confronted with the data, although in general it explains company behaviour better than plant behaviour. The analysis shows that profitability seems to be an important variable in explaining the pattern of diffusion between companies and plants. This proposition had probably been further strengthened if our measures of profitability had been more accurate (especially in the brick industry). Our results are in accordance with results found elsewhere, for instance by Mansfield (*Industrial Research*) and Griliches (*Econometrica* 1957), although Griliches' analysis is based on aggregated data and not on data regarding individual decision-making units. Thus, our results give some support for the economist's point of view in the debate between profitability and interaction effects on the diffusion of innovations.

Our data also support the hypothesis that there are marked differences in the attitude of different managements towards new technology. Some managements more readily take the lead and the related risks, whereas others prefer to wait. This is in accordance with results found by, for instance, Carter & Williams (*Industry and Technical Progress*). On the other hand we have not been able to test the hypothesis that managements of firms introducing new technology early must also be considered as technically progressive and more successful than managements doing so relatively later. Our data permits us, however, to compare productivity growth in different types of companies and plants which might throw some light on the question of technical progressiveness. It has not been possible for us to find out if technical leadership is constant or is changing with time as Mansfield's results (*Industrial Research*) seem to suggest. In Sweden, however, it seems to have been somewhat more steady than in the United States.

According to our analysis awareness of a new technology also leads to early introduction and *vice versa*. Since receiving and evaluating information on new processes is a cost consuming activity, this relationship may very well be interpreted as another way of showing attitude differences between managements as to new technology. Managements wanting to be in the lead also make sure that they get early information on new things happening in their field.

Company or plant size also has explanatory value in our calculations although less frequently than expected and found in other studies (for instance Mansfield *Industrial Research*). Whether big firms introduce new processes early or late often seems to depend on the technique in question. For instance, they were rather reluctant in the beginning to accept continuous pulp cooking and also basic oxygen steel (Ray, *National Institute Economic Review* (May 1969)).

Rate of increase in production and financing capacity have not been of essential importance in explaining diffusion among the techniques studied. As to financing ability, however, this might be a problem of validity; our proxy has not measured what we wanted it to measure.

In this paper it has not been possible to analyse whether rapid diffusion of new technology contributed to the extraordinary high total productivity increase in Swedish industry during the sixties. This problem will be dealt with in the international study mentioned in the beginning of this paper.

Discussion of the Paper by Dr. Nabseth

Mme Fardeau, in introducing Dr. Nabseth's paper, said that she had found it very interesting, not only because of the data it contained but also because of the possibilities it opened up of international comparison. The paper was principally concerned with the study of the behaviour of firms in respect of new processes. The rate of diffusion had been measured in a number of different ways, using the percentage of product and also the number of plants using the techniques at different dates, and the calculations were based on quarterly census data. Dr. Nabseth had also gathered information by use of direct questionnaires, and had had a very good rate of response to this. It was clear, however, that he did not have enough cases to explain the differences in diffusion rates. However, it was possible to compare his hypotheses with those in other studies, for example those of Professor Mansfield. The study had revealed that in some cases the innovation went very fast, for example in special presses, and in others rather slow, for example in continuous cooking of pulp. It would appear that the logistic curve was more appropriate to some techniques than others and that where an innovation called for high investment the rate of diffusion tended to be slower.

Mme Fardeau noted that there were four groups of hypotheses in Dr. Nabseth's paper. The first related to the time when information about a new technique was received and the ability to evaluate the information. Dr. Nabseth had said that the data might not be very reliable on this point, and suggested an alternative of measuring the proportion of product exported. The second group of hypotheses was that the more profitable was an innovation the sooner was it likely to be taken up. On this point it was very difficult to get *ex ante* information and there were, of course, difficulties even in measuring profitability *ex post.* Dr. Nabseth had used a number of different methods for different innovations. In one case he used the pay-off period and in others he used such things as the speed of machines and types of paper as proxies. The third group of hypotheses was concerned with the structure of the finances of the company, and he argued that that the greater the share of self-financing the easier it was to innovate as managers were unwilling to borrow beyond a certain limit. This hypothesis was tested in the paper by relating total net income generated within a company to total investment during the period. It was hypothesised that the larger was this ratio the easier it was to finance investment in new techniques, and the earlier such techniques were likely to be introduced. In addition the author took into consideration size as a proxy for financial strength and suggested that the bigger was a company the more easily it could accept a loss on introduction of a given new technique. Mme Fardeau suggested that might be some contradiction with the argument in Dr. Nabseth's paper that the speed of invention might not be a sign of good management – it might be better to allow the bugs to be eliminated first. The fourth hypothesis related to the attitude of firms to new processes. On this question Dr Nabseth collected information and gave a set of marks to organisations, which were higher the more rapidly they introduced innovations. It was interesting to note here that Professor Freeman had said in the previous session that in fact there were often rapid changes in the structure of firms over time, and this suggested that Dr. Nabseth's current approach might not be the most appropriate. His approach of also looking at novelty and the existence of R & D in a firm seemed very close to the first hypothesis and she wondered why it was necessary to include this. She went on to say that Dr. Nabseth, having set out his hypotheses, raised two problems which were not yet fully solved. The first was related to the validity of the choice of methods proposed and the second was on the size of sample required to test different hypotheses.

Having discussed the explanatory variables Dr. Nabseth then went on to look at the results, using an econometric approach, and presented the analysis in a series of tables. On the basis of these data his major conclusions were that new techniques had been promulgated relatively quickly in Sweden and he attributed this to the availability of information. He suggested that salesmen proposed new technology simultaneously to all clients and that for example in the paper industries enterprises were very close to each other. On profitability the measures had given good results but were inconclusive on the question of financing. There was, however, a high correlation between the attitude of managers and the date of adoption of new technology. In the econometric analysis various regression equations were attempted, with the log form being found better for paper and the linear for bricks. The regression coefficients were relatively high and were higher for enterprises than for plants, which was perhaps not surprising as the variables concerned enterprises. In these results one found confirmation of Professor Mansfield's thesis. On the behaviour of a given firm, one found leaders and followers in Swedish industries. This was interesting and international comparison would no doubt follow and throw new light on the subject.

Dr. Nabseth's findings were again in accordance with Professor Mansfield's on the profitability variable and with Professors Carter and Williams' earlier study on the attitude of management. The role of profitability was not so easy, however, and one needed to look at this on a long-term basis. The problem was linked with that of the rate of profit sought by a firm. Industrial strategy might be more important than psychology and the choice of defensive or offensive strategies was important. She thought it was necessary to distinguish major and minor innovations and to ask whether we could expect to use the same model for each type of situation. We had not in this paper enough cases for satisfactory analysis of this factor. Finally, on the question of the independence between the independent variables it appeared that some of these were interconnecting. This was a delicate problem we had already met, whether *ex post* or *ex ante*.

Professor Oshima wanted to put a question to both Professor Mansfield and Dr. Nabseth relating to major and minor innovation differences. He pointed out that the intentions of the innovator were not included in the model. When the innovation was a core innovation then he would probably not want to have his technology diffused but would want a monopoly. Therefore Professor Oshima suggested that for peripheral innovations diffusion might be well explained by these models, but that it might be interesting and important to discriminate between the two types of innovation, since technical monopoly or oligopoly could become relevant.

Professor Rasmussen wanted to draw the attention to the consequences on the macro-level, when the diffusion of technology was described by logistic curves as assumed by Professor Mansfield and well illustrated by the graphs of Dr. Nabseth. If new technologies were utilised over time in different industries according to a logistic curve, the result of an aggregation might be, more or less, a linear development if one related the diffusion to the number of companies or plants. If, however, one operated in terms of output, the convergency level of the logistic curve would, in a growth economy, steadily move upward. This would on the aggregate level result in total output increasing along a curve, the second derivative of which was positive. In specific cases one might get the exponential curve dominating growth theory. It was, however, interesting that the speed of diffusion did not matter if one went to the limit, that is if one had a steady state. It was easy to show that what mattered was the convergency level and its upward movement. However the speed of diffusion (that is the steepness of the logistic curve) greatly affected the speed of approaching a steady state – if you were outside it – and presumably you always were.

Professor Mansfield said that he regarded Dr. Nabseth's paper as a very fine one, and he thought that many of the results were in agreement with his own. For example, with respect to the financial variables mentioned, they had both found no effect. He added that he thought Professor Oshima was right, and the distinction he had made was an important one on which they were trying to do some work at the present time. On Professor Rasmussen's point Professor Mansfield said that he did not believe that the rate of diffusion was unimportant.

Professor Matthews wanted to make a point related to one made previously by Dr. Teubal on Professor Mansfield's paper. The introduction of trucks in brick-making was hardly an innovation in the normal sense, but simply an increase in capital intensity, that is to say a movement along a given production function. He suggested that if this was correct one would expect some of the factors affecting diffusion of innovation to be absent – notably Schumpeterian imitation. He asked whether this was perhaps the reason why a logistic pattern was not found in this case.

Professor Triantis returned to the question of whether big firms introduced new processes early or late. Dr. Nabseth had suggested that this would depend on the nature of the process. He wanted to ask whether he had followed this up systematically. On the possibility of internal financing facilitating innovation he asked how this might relate to management attitudes. He suggested that retention of profits might reflect conservative management and lead to expansion with traditional technology. Or, the profits might be used for conglomerate expansion. He wanted to ask why Dr. Nabseth spoke of borrowing money as the only outside financing. Would sale of treasury stock be included in internal financing or external?

Dr. Sedov suggested that some additional economic hypotheses might be used in the future investigation of the problem. For example it would be valuable to show the factors affecting profitability. This might be due to a decrease of the prices of products diffused or to an increase of the quality of these products. Knowledge was not a direct factor. It set in action the economic interest which predetermined the objectives of the economic behaviour of the companies. These objectives might be the increase of the profitability or survival in face of the competitive necessity to decrease cost in order to meet a change of prices. The diffusion was connected with the attitudes of the companies and users of the technology. Dr. Sedov thought that their economic characteristics should be given in more detail and be classified with greater regard to policies at the different stages of diffusion. *Dr. Nabseth* said that the answer to the first part of Dr. Sedov's remarks was very simply 'yes'. As to the second question, this was what they had been trying to achieve.

Professor Williams raised a question which was relevant to the papers of both Professor Mansfield and Dr. Nabseth. Dr. Nabseth referred to the effects of attitudes to new processes on the rate of diffusion. Professor Williams wondered how far he had attempted to explain these differences in attitude. Was he taking attitude as a very general thing, or more related to the nature of the innovation? Some innovations were very easily introduced without disturbance to the existing process and structure of management while others involved fundamental changes. The tunnel oven for instance took a very long time to become established in the pottery industry, because the new process required continuous rather than batch production and thus required a change in the whole structure/process of management and far-reaching changes in social relationship in the factory, which were not readily accepted by the craft workers. By contrast a new machine tool might be slotted in with great ease. As an intermediate case he mentioned computer controlled machine tools that were likely to require more production planning and significant change in the process of management. Professor Williams suggested that people were often reluctant to go into new technology if it was 'not their type of business', Finally *ex ante* calculations of profitability varied enormously. How far had Dr. Nabseth examined this? In particular, with reference to calculations made by different centres? He was not happy with the average figures and suggested that we needed to get down to company level.

Professor Robinson said that when technological progress was very rapid it was often difficult to know when one should jump in. Would next year's equipment be better than this year's was an important question to ask. Should investment be made in new equipment now, when the bugs might not have been ironed out, or should one wait a year or two? The more optimistic one was about the future course of innovation the more cautious one might be about jumping in prematurely. Slow jumping in on progressive innovation did not necessarily show conservatism. It might reflect a better knowledge and appraisal of the trend of development.

Dr. Nabseth said he was happy to have an opportunity to comment on what had been said on both his own and Professor Mansfield's papers. Professor Mansfield had tried to give a history of the world in his paper. In their international group they had leaned heavily on his earlier work. Professor Mansfield was the innovator and they were the laggards. *Professor Mansfield* interjected 'you are just missing Professor Robinson's point', to which *Dr. Nabseth* replied 'maybe', and continued by saying that when they started work they thought it would be easy to find new, well-defined processes which had been introduced in the post-war period. This however had not been so easy – improvements were often of a small character and difficult to define. In addition the problem of profitability measurement cropped up the whole time, and they had had difficulty with using Professor Mansfield's average profitability concept.

Dr. Nabseth said he could not find what average profitability really meant in different companies. Profitability calculations were always very difficult, as Mme Fardeau had pointed out. The problems was often that if you had scarce resources on the capacity or the management side, then profitability could be high for one company but not for others, because their opportunity costs might be different. On top of this, the capital structure was likely to vary between different industries and different countries. Dr Nabseth said that the problem raised by Professor Robinson was a difficult one. The technique was not defined at once, or only in comparison with older ones. Change was often continuous and when one was intending to use a particular innovation one had to be careful not to jump in too early. A second point which Professor Robinson had not mentioned was the change in the old techniques which was a consequence of the new techniques coming along. He cited continuous pulp cooking which brought rapid development in the old discontinuous method, so that the essential thing was to look at the difference in profitability between the old and new techniques.

Dr. Nabseth said that in the international study it had been very difficult to get estimates of expected profitability. For example in continuous casting engineers were now very unwilling to give estimates of this variable. They had tried to get proxies of different types as Professor Mansfield had done. But it could be argued that they did not adequately represent profitability. He agreed with this. But one had to try, and one could not just throw this variable away. In processes where they had their best measures of profitability they seemed to have their best results. On the point raised by Professor Oshima concerning the problem of major and minor innovations and licences and patents Dr. Nabseth agreed that this must be taken into account. With respect to licences it was possible to get this into the profitability calculations. The patents position was very important, for example in the float glass process for which Pilkingtons gave licences. This sort of innovation probably needed a different model to explain the diffusion. On the problems of finance and R & D, Professor Mansfield had said that they had got the same results. Dr. Nabseth would like to be careful here. He was not certain how to test the hypotheses. One could put it the other way round. It was difficult to test, and hard to define the real time period, and they had been wrong in taking only single time periods. on the question of external financing and new equity capital, in the case of Sweden it had not been very important. Equity capital was rather small in the total financing, and he would like to treat it as a third way of financing. On the point about multi-colinearity made by Mme Fardeau and Dr. Sedov he thought it was not a big problem, but they had been trying to do something about it. To Professor Matthews on the question of whether in some cases it was just a shift along the production function and not a change in the production function, Dr. Nabseth said this was very interesting, but knowing the brick manufacturers he was not certain that they knew so much about the truck that it was just a matter of asking questions about the labour and capital costs. On Professor Williams' point about the different attitudes to different innovations, Dr. Nabseth said that they had not yet found it profitable to do anything with this, although they had looked at general attitudes – particularly in the pulp and paper industry. Here the producers of equipment for the pulp and paper industries maintained that there were big differences between companies for all equipment. Some tried everything, whether a small or big innovation, rapidly while others were always very slow.

ADDENDUM

In private discussions during the conference, *Professor Matthews* pointed to the analogy in growth theory: the rate of saving does not influence the rate of growth, but the level. *Professor Mansfield.* though accepting the logic, opposed the relevance of the statement partly by pointing to the fact that a slow process of adjustment makes the 'steady state' less relevant, and secondly and more important, because there were many reasons to expect an inter-relationship between the speed of diffusion and the convergency limit on its upward movement. (In growth theory the analogy could be an inter-relation between the rate of savings and the growth of population or between the rate of savings and technological progress.)

11 Science Policy and Development Strategy in Developing Countries

Vidosav Trickōvić[1]

UNIVERSITY OF BELGRADE

I. THE FRAMEWORK SET BY TECHNOLOGICALLY ADVANCED COUNTRIES

In recent years problems of science policy have acquired a prominent place and have led to growing interest among social scientists in discovering the relationships between scientific research, technical progress and economic growth. The marriage of scientific research with physical production, which characterises present economic development and current changes in world economic relationships, has clearly shown the short-comings of the tools of economic growth analysis. The introduction of a residual term 'technical progress' in production functions and the magnitude of this residual term, which has been found to be greater than that of both capital and labour together, was merely the recognition of the fact that the predictive power of equations derived from past performance of national economies and their various sectors, has in fact been unpredictable.

Two main groups of factors combine in producing a given residual trend in a particular phase of development of a given individual country. One group comprises those factors which are concerned with changes in the level and structure of economic and social activities, and reflect the basic features of national development policies. These include the emphasis on expansion of particular industries, and within branches of industry the emphasis on particular techno-economic and organisational production methods. These in turn can substantially change the production pattern in respect of product-mix, technical coefficients and the research content of the whole branch of industry. In developing countries these factors may be considered 'controllable' in the sense that development plan

[1] The author is Associate Professor of the Faculty of Economics, University of Belgrade. At the time of the conference he was temporarily occupying the post of Director of Planning in Zambia under the U.N. OPAS Programme. This paper draws much on the material presented in Vol. I of *A Study of the Needs for Scientific Research and Advancement of Technology in relation to the Economic Development of Yugoslavia* (1969), undertaken in co-operation with the Department of Scientific Affairs, O.E.C.D. The author was Project Director of that Study.

targets can take account of desirable production standards already existing in technically more advanced countries.

The second group of factors reflects the influence which is exerted by technically advanced countries on the pattern of world trade and consequently on the degree of competitiveness of industries in developing countries. Every new technical breakthrough imposes a new constraint on the production efforts of developing countries.

A technical, or better, a techno-managerial, gap has been recognised as existing between the United States on the one side and other already developed countries in Western Europe. To explain the causes of this gap the symptoms of which appear in such forms as new products, new cost-reducing processes, and so on, international comparative analysis has almost always relied on comparative figures of the numbers of R & D personnel and R & D expenditures in relation to population or G.N.P. But the very nature of scientific R & D effort, the product of which is the discovery of something new in the field of industrial production, requires that absolute rather than comparative figures are considered in measuring the dimensions of the technical gap. R & D effort, as a creative activity, differs fundamentally from other productive processes in that it cannot be regarded as reversible. It is a process of growth and accumulation of knowledge, the output of which is invention. The product-mix at every stage of the growth curve differs from that of the past. In the production of a given set of goods, since the indicators will show higher *per capita* stocks of capital goods, higher proportions of engineers, technicians and top-level managerial staff in the total labour force, and other similar factors, inevitably there will be higher productive capacity of the national economy as a whole and in individual branches of industry. It is not surprising that these indicators are closely correlated with G.N.P. *per capita*.

On the other hand, the results of scientific research successfully applied to production processes – that is inventions which have been translated into product or process innovations – represent in fact broadly defined production technologies whose absolute and not relative scale measures the productive potential of particular national economies, branches of industry or individual businesses. A successful innovation can be turned into a theoretically unlimited number of higher quality or lower priced goods, which can improve the competitive position of those who succeed in exploiting the technological advances.

We can admit that in general the chances of an absolutely greater number of technological discoveries are greater, the greater is the concentration of research efforts in terms of both personnel and resources. Yet the idea, the invention, is not the only essential. The history of various inventions and their technical application shows the important role that has been played by the financial resources of an enterprise, and particularly by its management, which has often proved decisive in the transformation of the invention into new process technology or into new products.

The size of enterprises and their financial strength is thus becoming an increasingly important condition of further break-throughs in technical production. This is shown by figures for the distribution of total expenditure on starting new lines of production. In the process of transforming an invention into an innovation in production, the relative expenditures on the various phases have been estimated as follows: research in the narrow sense, including the basic idea and its elaboration for application to production, 5–10 per cent; technological elaboration and development work on preparing the product for the market, 10–20 per cent; setting up the production process and the manufacture of equipment, 40–60 per cent; costs of starting up production, 5–18 per cent; initial costs of market research and sales, 10–25 per cent.[1]

Research work alone, *sensu stricto*, thus accounts for less than 10 per cent of the total effort that enters into the creation of a new technology, while the remaining 90 per cent is accounted for by a variety of other activities, among which the organisation of production, the selection of an optimal economic solution among a set of technical alternatives and the related research all represent very important elements in an innovation process, the scales of which are often little known and thus tend to be underestimated.

Although major scientific discoveries leading to the advance of technology have been made by individuals, the scale of the organisational and financial resources necessary for a full economic exploitation of discoveries is such that it requires companies of a considerable size to undertake the risks involved in these ventures. The necessity for a certain absolute level of resources in respect of R & D and other personnel is an exogenous factor that sets practical limits on the capacity of medium and small companies to take part in the competition to make technical innovations.

All these considerations point to an important corollary. Countries with larger G.N.P. *per capita* and larger populations are in a position to exploit the benefits of higher concentration of R & D resources and of the larger market which permits a higher concentration of production and consequently a better management and organisation of production.

[1] *Technological Innovation: Its Environment and Management* (U.S. Department of Commerce, 1967); material submitted at the E.C.E. Conference of Senior Economic Advisers, Geneva, October 1967.

The impact of this 'size factor' is best demonstrated by comparison between the United States and other developed countries. Thus it is only by taking twelve European countries together with Japan and Canada that it is possible to equal the 'scientific capacity' of the United States in 1965 measured in terms of research personnel (674,000 scientists, engineers and technicians as against 696,000 in the U.S.A.) while in terms of financial resources the United States was even then deploying 2·8 times more funds for R & D. Even when the comparison is confined to the so-called 'economically motivated' R & D expenditures, the total funds expended in the United States for R & D amounted to $billions 6·9 against $billions 3·2, or about twice what the six most developed countries of Western Europe spent in real terms.[1]

It is clear that the size of the real G.N.P. of a nation is a principal determinant of the size of its research and development potential and its output. The fact that the research effort is very unevenly spread over different economic sectors and industries, and that within research intensive sectors there is a further concentration of R & D capacity in bigger enterprises only adds to the disadvantages of small national enterprises when struggling to survive in competition with the giant corporations on the world market.

The 'technological-organisational gap' is a term which seems more adequately to express the elements that lie at the root of the differences in economic efficiency between the United States and the most advanced countries of Western Europe. This situation, which further scientific and technological progress will make progressively more difficult for developing countries, and particularly the smaller developing countries, imposes upon them an exceptionally important task of working out an appropriate strategy for the science policy and its integration into the plans and programmes of long-term national economic development.

II. IMPLICATIONS FOR THE STRATEGIES OF DEVELOPING COUNTRIES

It is obvious that the selection of a strategy for the development and structuring of research capacity, designed to achieve optimal effects on total economic development, is a very complex task. On the basis of the results of various studies, some of them referred to in Section I, the following conclusions may be drawn.

(i) There are very big differences between the countries that are

[1] United Kingdom, Netherlands, France, Sweden, Germany, Belgium: Data from *The Overall Level and Structure of R & D Efforts* (O.E.C.D. 1969).

technically advanced and the remaining countries, particularly the less-developed countries, in regard to the absolute level of technological and research capacity in various branches of production. These require serious study of priorities in the science policies of developing countries. In absolute terms, the human and financial capacities of the smaller countries should be concentrated on those economic activities in which there are natural and other advantages for efficient economic development (natural resources, existing development, manpower resources and so on). The faster is economic development, the less the advantage of 'cheap labour' in international trade, for which technical progress can be substituted.

(ii) The less-developed countries have an advantage in that they can make use of the general fund of knowledge accumulated in the technically advanced countries. To make wide use of this advantage is one of the most effective strategies of science policy. In this respect, Japan and its implementation illustrates a strategy of development which produced results in technical progress and economic expansion that are well known.[1]

(iii) In the case of the less-developed countries, the desirable level of expenditure on research, measured as a proportion of national income cannot be judged merely by comparison with advanced countries. It would appear that the need to absorb the existing higher technology in the world requires an even greater proportionate effort on the part of developing countries in the provision of funds for research and development. This is shown primarily by the great difference in the absolute levels of total research expenditures between the developed and under-developed economies. In order to ensure a sufficient absorptive capacity for the transfer of technology, it is necessary to organise, on a wider scale, sufficient research to make it possible to keep abreast of modern trends and developments in world science and technology. Without this knowledge, the education of high-level personnel for economic and scientific work becomes seriously handicapped through lack of first rate up-to-date teaching. Moreover, a country with low research capacity will find difficulties in the successful transfer of technology from advanced countries. Nor will it be able to adapt the imported technology efficiently to local conditions.

(iv) International co-operation with enterprises in developed countries, on the basis of bilateral arrangements and exchange of interests between the parties concerned, provides a possible means of speeding up the transfer of technology in its broadest sense. In this respect, some international organisations (for example, O.E.C.D. and the European Economic Commission) have already taken a lead.

[1] *Review of National Science Policy*, Japan (O.E.C.D. 1967), p. 156.

(v) The integration of the development efforts of a group of developing countries appears to be one of the most effective ways of overcoming the limitations of narrow national markets and of making possible production units of a size which permits the adoption of technical advances and higher management efficiency. These opportunities which can materialise in various forms range from multinational trade agreements facilitating specialisation and trade between developing countries to more advanced forms of co-operation such as joint development projects. But these opportunities have hitherto been little exploited. As a consequence, the developing countries continue to sell to developed countries their precious, unprocessed mineral wealth, while importing products manufactured from the same primary products that they have exported, although the equipment, technologies and necessary technical manpower could easily be imported. To free themselves from this basically colonial pattern of exploitation of their national wealth, developing countries should look more for the benefits that can be reaped within their own family.

(vi) The 'gap in the quality of management' is no less important than the 'technological gap' in its strict sense. Without the creation of an adequate institutional base in the enterprises themselves, in the sense of a very considerable improvement in the methods and quality of management, the 'absorptive capacity' of the economy for the results of research cannot be created. A science policy, no matter how well conceived, cannot create demand for production-orientated research, which must emanate to the greatest possible extent from business organisations themselves. The presence of high quality management in enterprises combined with large plants and/or large integrated or co-operative units, is becoming a *sine qua non* of an efficient national science policy.

III. RESEARCH INTO THE FACTORS DETERMINING STRUCTURE AND DEVELOPMENT STRATEGY IN INDIVIDUAL PRODUCTION SECTORS

Research and development work related to economic development constitutes only one link in the complex chain of socio-economic, organisational and technological cells in a living national-economic organism, which itself is a part of a broader system of international relations and ties. The efficiency of R & D, from the point of view of the needs of any economy or society therefore presupposes the simultaneous presence of other elements in the system. Thus, in addition to scientific capacity there is an essential need for an adequate educational infrastructure, for efficient financing, for planning and co-ordination of research work, for successful economic planning, and especially for an efficient organisation of the economy at a micro-level, particularly in respect of the size of enterprises and their methods of organisation and management, which basically determine the extent of demand for technological innovations and consequently for the output of research and development work.

The long-run general strategy of economic development cannot be worked out without a preliminary study of the role played by technology in the growth of various branches or product groups, that is to say, without an appraisal of the necessary conditions for the achievement of a definite rate of technological progress in a given sector.

Thus, it becomes essential to make systematic studies of the principal features of the main branches of industry, keeping in mind not only the existing 'technology gap', but also the potential gap which is likely to develop in the near future on the basis of the trends in the growth of applied research in developed countries. In particular, these studies should provide the basis and criteria for a national science policy designed to establish priorities and to co-ordinate a national research and development effort that can be matched with an allocation of funds and with educational targets for R & D personnel. In this way the studies should directly contribute to a closer linking of the science policy with the requirements and needs of the general and particular development trends of the country.

The following subjects need to be examined in the studies of selected branches of production:

  (i) the place and importance of this productive activity in the national economic development;
 (ii) its level of technology and organisation of production;
(iii) the research and development work currently being conducted;
 (iv) the use of manpower and the manpower resources;
  (v) the institutional set-up of the production branch concerned.

A brief outline of the desirable analysis of each of these five subjects follows:

(i) *The Place and Importance of this Productive Activity in the National Economic Development*

The study of this should in the first place examine what are the natural resources and comparative advantages on which the growth of the activity basically depends. Against this background the level of development of the activity has to be examined in terms of a number of indicators, including: the contribution of the branch to the total G.N.P. and to the total of the country's employment; the import and export ratios; the market demand in the past and forecasts for the future. It is often true that in developing countries a natural resource is far from being properly assessed, but there are always indicators of the presence of some particularly valuable natural resource. It is in this context that serious thought has to be given to the best possible use of the existing resources, by diversifying utilisation of primary commodities, further manufacturing and processing, and to the possibilities not only of adding to the national income and employment but also to much-needed diversification of the economy and the stability of growth.

(ii) *Its Level of Technology and Organisation of Production*

In almost all developing countries there is an urgent need to assess the 'techno-organisational level' in various activities. No uniform methodology exists for tackling this problem. However, the recent pilot-team studies organised by O.E.C.D. have already made some progress in this direction. Under this head the following subjects need to be examined:

(a) The characteristics of the equipment used in production, particularly with regard to its functional suitability. These findings need to be compared with more modern and even with the latest known equipment in the activity concerned in the technically most advanced countries in the world. Analysis of the economics of introduction of more modern equipment should then follow, to provide economic justification, and the phasing of the introduction of more modern equipment should be considered.

(b) The extent of the use of modern technological processes and industrial production methods is a further logical step in defining the level of the technology applied. Here again the comparisons with more recent and advanced technologies, as well as the conditions under which they can be supplied (in terms of the demand for new equipment, specialised technical and other personnel, licensing possibilities, and so on) have to be worked out.

(c) The third aspect of the techno-organisational level should concern the scope of the application of the methods of modern organisation of work and management, including: organisation of staffing; degree of automation; application of quantitative methods in programming and planning (simulation of production processes and use of computers; PERT analysis and other operational research techniques).

(d) The productivity levels and costs per unit of output of selected commodities compared with the results obtained in technologically advanced countries in similar branches of production will serve as aggregate indicators of the level of technology, management and organisation of production achieved in the sector of national economy under consideration.

(iii) *Research and Development Work*

It is desirable and interesting to examine the following aspects of R & D:

(a) The research capacities and their organisation as related to an individual branch of production: the independent institutes, university institutes, and R & D units within individual enterprises and their industrial affiliations, together with information on the number of research units, their personnel and its composition, their equipment, their financial resources. The type of research work undertaken (fundamental, applied, development). The comparison of the research capacity with that in countries technically more advanced in the activity concerned.

(b) Evaluation of the R & D results and their contribution to the improvement of technology in the light of recent research trends in the world. Their contribution of original research results as compared with imported innovations. Successful and potential achievements on the world science front.

(c) Evaluation of the national science policy affecting the activity under consideration; the adequacy of co-ordination, finance and other incentives, with particular attention to experience in countries which have achieved significant results in this field.

(iv) *Manpower*

(a) The composition of research staff in terms of qualifications and education; the ratio of engineers to the total number of employees; the occupational distribution and qualifications of executive staff.

(b) Comparative analysis with technically advanced countries and the evaluation of the needs for improving the level and composition of the staff.

(c) Evaluation of the capacity of the educational system to educate top-level engineers, economists and research staff needed for the advancement of technology and the creation of a modern organisation of management and production in the country's industrial enterprises.

(d) The remuneration and other conditions necessary to stimulate the creative human factors in productive enterprises and research institutions.

(v) *Institutional Factors*

Under this heading the factors regarding the institutional set-up at both macro and micro level should be analysed, so as to focus attention on the legal, financial, administrative, organisational and other bottle-necks which may hamper the fuller and more efficient utilisation of the available human and physical resources for production. Clearly the depth at which these predominantly 'economic aspects' can be analysed depends on the existing socio-political set-up and the possibility of an objective analysis and ranking of the various alternatives within the framework of the system of values and development objectives that reflect the priorities of the country's social and economic needs.

The results of the studies under (i)–(iv) should be presented in such form as to give a detailed picture of the technological-economic structure of the activity concerned, in terms of a distribution of enterprises, total production, employment and fixed capital in the activity concerned analysed in terms of the technological-economic levels achieved. This would involve a necessary breakdown of the averages for each activity and would show the most important problems, on the solution of which depends a more rapid transition of the activity to higher economic technological-levels and thus to more efficient production.

(iv) *Further Applications of Data on Techno-Organisational Levels*

On the basis of time-series of output, fixed capital, manpower and data on the degree of utilisation of capacity and in the techno-organisational level referred to above (the structure of fixed assets in respect to their technical level, the composition of labour employed and the extent of research and development work), an attempt can be made to determine the specific production function for each activity or product group with the objective of breaking down the residual trend of technical progress into its actual determinants. Production forecasts could then be based on a quantification of a sufficiently wide range of factors to permit a more realistic estimate of the growth rates of main activities and product groups by deliberately creating the necessary conditions to ensure the desired expansion.

For this purpose the data on different aspects of the economic-technical-organisational level need to be presented in the form of specific indicators. Various elements of the techno-organisational level of production can be quantified in terms of indicators, assigning, for example, the number 10 to the highest known level in the world. By assigning indicators to each particular element (for example,

the efficiency of production equipment and its functional composition, the degree of application of modern technological processes,
the use of modern management techniques), and by weighting these
individual indicators according to the relative importance of each
separate element, one can arrive at an aggregate index of the techno-
organisational level of the enterprise selected for analysis. Similarly,
one can evaluate the level of research and development work and the
quality of staff in a particular enterprise by way of calculating
indices for a number of components selected to indicate the nature,
characteristics and performance of the R & D activities or of the
staff engaged in production.

Finally, by weighting the indices worked out in the sample of
enterprises by a weight reflecting the share of the output of the
particular group of enterprises in the total output of the activity,
one can calculate both separate indices for each particular element
of techno-organisational level in the activity concerned and an index
which would measure the over-all techno-organisational level of the
branch.

Selection of the weights to be attached to individual indicators
requires the team work of experts in particular fields. It is obvious
that any weighing system would always express the value judgements
of the members of the team and thus would contain a degree of
subjectiveness and bias. However, this bias is more likely to affect
the absolute level of individual or aggregate indices rather than the
relative achievement and performance, as reflected in the aggregate
indices representing the general characteristics of the equipment, the
technological processes, or the organisational standards and productivity levels. There is, moreover, some cancelling out of errors
contained in individual indices, as a result of their aggregation first
into group indices and then into an over-all index for the activity
as a whole.

Special consideration must be given to the fact that for each
relevant technology group it is necessary to acquire sufficient data
for purposes of comparison relating to higher levels of technology,
organisation of work, and R & D activities in the more advanced
economies. The results already available or soon to be published
of the work of various international organisations and specialised
institutes in the field (for instance the work of the O.E.C.D. Direc-
torate for Scientific Affairs on the 'Technology Gap') will considerably facilitate international comparison. Nevertheless, it will remain
necessary to organise other methods of acquiring the essential
information and data for international comparison.

In spite of their limitations, such techniques as have been described above will continue to represent the best way of bringing

together in systematic form the principal characteristics indicating the technological, managerial and organisational levels of a given industry or activity. The value of these indices – particularly those which are in more aggregated form – is to be found in the possibilities of comparing the techno-organisational levels of various industries or activities, and particularly those which, for various reasons, play a prominent part in the economic development of the country concerned.

From the outline of the material required under sub-headings (i)–(iv), it can be seen that most of the data need to be collected by special surveys of enterprises and of R & D institutions and units, through interviews and with the full co-operation of enterprises and institutes. By taking a sample of a sufficient number of enterprises in each industry or activity, the aggregate indices showing the quality of various productive factors, such as equipment, organisation and management, R & D personnel, can be used as explicit variables in a cross-section production model. One technique might take the form of using these indices as correction factors (in percentage terms) to arrive at a level which more adequately expresses the productive effect of the variable in question. For example, the stock of capital expressed in monetary terms can be multiplied by say a factor 0·80 which specifies a measure of efficiency of capital in relation to a highest level marked as 1·00. In a similar way, R & D effort measured in money terms can be corrected appropriately. Another technique might take the form of using the index of techno-organisational level as an explicit variable in addition to capital, labour and R & D expenditures.

As a result of this cross-section analysis, a multi-dimensional function can be obtained, specifying the relationship between output and the various determining variables by way of the elasticity coefficients. Once these elasticities have been established, they can – if the data on the determining variables such as capital, labour, R & D expenditures and their 'efficiency indices' can be compiled for a sufficiently long period – be utilised as the given parameters which will then make it possible to discover what may be called the real, unexplained, residual trend in the production functions of various activities based on time-series data.

This procedure resembles the well-known technique of demand analysis, whereby income elasticities of consumption are exogenously determined by cross-section analysis of family budgets, and then used as given parameters in the time-series analysis of consumption data, to help in the calculation of price elasticities. This technique was designed to overcome the considerable difficulties encountered in econometric demand analysis of time-series data, as a consequence of the high intercorrelation of incomes and prices.

It remains to be seen how such a model can best be applied in practical work. No doubt there will be difficulties in collecting the necessary data on the techno-organisational level, particularly over past periods. Yet, once these data become available, their value for the improvement of production planning and the closer inter-connection of economic, development and science strategies, will become obvious and will undoubtedly foster interest in further inquiries into the methodological problems involved.

Discussion of the Paper by
Professor Trickōvić

Professor Simai, in opening the discussion, said it was difficult to introduce the paper in the absence of the author, as he would not be able to give answers to the questions which would be raised. However, he said the issues discussed in the paper were extremely important. It was well known by all economists, particularly those working in developing countries, that the gap was widening between developed and developing in many areas, but particularly in the field of science and technology. According to O.E.C.D. statistics only 5 per cent of the world research and development expenditure was in the developing countries, and even there the distribution was unequal, with a concentration in three or four countries only. In these developing countries the import of modern technology was slow and the imported technology did not serve sufficiently to reduce the gap. The development of modern industries did not contribute sufficiently to the reduction of unemployment, and in particular urban unemployment was increasing. Professor Simai went on to say that the managerial gap was a vital issue and many developing countries could not use the available know-how within the framework of their production programmes. In some countries there was no policy for science and technology at all. In others it received mention in the plans but was rarely implemented. Scientific manpower itself was not sufficient in many cases, and their capacity was not utilised. In some countries there was intellectual unemployment and research institutes were often short of funds. These problems had been taken into account implicitly if not explicitly by Professor Trickōvić in suggesting strategies for implementing science policy. Professor Trickōvić very properly emphasised that the long-term general strategy of economic development needed a preliminary study of the role of science and technology and urged systematic studies of different production branches, keeping in mind not only the existing but also the potential technological gap which might arise in the future as the result of trends in developed countries.

In looking at possible topics for study Professor Trickōvić suggested the place and importance of production branches in national development, the level of technology and organisation of production, research and development, manpower and institutional factors. The outline on these points given in the paper was sufficiently general for the suggestions to be taken into account in any developimg country independent of its size and state of development. However, the international aspects of the problem could not be so generalised because the greater part of science and technology was imported. The research sector in developing countries was very closely related to the outside world and therefore the international aspect of science policy needed careful study. However, after studying the outlines given in the paper by Professor Trickōvić he was encouraged to ask what should be done next, after the data has been obtained. Should one look at specific production functions for every branch in general terms without going into details. Professor Simai said he had doubts on this. There had been many sceptical comments made during the conference on the quality of R & D data in developed countries and these comments were highly relevant to developing regions. Besides a static scale in a dynamic world did not give much help to the planners. He also had some doubts about our general ability to judge the soundness of the suggested approach confidently. Did we have enough results in terms of case studies, for example, from more advanced developing countries, such as Brazil, India, Mexico and Egypt? Had these countries been able to formulate and implement science policies and integrate them into development planning? We did not know enough about this. And were the recommendations of the United Nations Science and Technology Conference utilised? We seemed to know a lot more about the brain drain and the costs of transfer of science and technology. But the reader missed in this paper the practical analysis of the problems involved. To judge the value of the suggested policies, we would need to know more about practical experience.

Mr. King said the subject was of great interest to him and one of which he had some experience. Professor Trickōvić had been a colleague of his. He would like to address his few remarks to the question of lessons for the strategy of developing countries. It was true that less-developed countries had access to the totality of world knowledge. But generally speaking they were less able to use this knowledge than other countries. The example of Japan was not a good one to take, as they had been able to use science, technology and education in such a way as to contribute to economic development. In general the less-developed countries spent a higher proportion of their R & D on fundamental research and existing science was not adequately used to help the development of the country. This might perhaps be an exaggeration but it was not very far from the truth. In listening to discussions on developing parts of the world he thought that lip service was being paid to the need to reduce the gap without being much done about it. In Latin America there was a general impression both on the economic and political side and on the scientific side that if you did more research you would automatically get more economic development. This was an extremely naive view, and the concept of innovation and its complexity was just not properly understood. The chief necessity was to achieve articulation between the scientific and economic development policies. Much of the fault came from advanced countries which had exported a lot of academic snobbery.

Mr. King said that the immediate need for pure science could be said to be small in developing countries. It was very unlikely that there would be original innovation for many years. The emphasis should be on the transfer of technology. The strategy for developing countries needed the building up of scientific potential for long-term needs. There seemed to be a necessity for each country and even each firm to be above the threshold of science and of awareness of technology. Below this threshold they could not use imported technology successfully. In developing there was a need for each country to build up a degree of scientific awareness. Only then would it know with any certainty what techniques to buy. Otherwise it would simply buy itself obsolescence. This principle should be the first short-term objective in building science policy in such countries.

Professor Rasmussen intervened with a comment which, as he explained, might have been made on many earlier occasions and might be considered as a more general comment for the conference. He argued that in discussing the effect of R & D on 'economic development' we tended to forget the role of our own field, economics. That was not fair. We were taking 'science' in a far too narrow sense. In fact we could play as economists a very important role, provided we succeeded in developing the social sciences in general and economics in particular. Let us remember how far too modest we had been. When in physics the general feeling spread that high energy physics represented the world of the future, the final outcome of heavy pressures on individual governments turned out to be the building of a new (300 GEV) accelerator within the framework of CERN in Geneva. The natural sciences were very expensive, but even so they succeeded in developing at a very fast rate, even in cases in which there were no military gains to be seen. Considering how costly these fields were, and how cheap as a science economics was, Professor Rasmussen asked why we accepted this situation. Why did we not ask for an Institute in Geneva devoted to the study of the consumption function, or a huge institute in Paris devoted to the study of import elasticities? Whatever one might imagine to be the size of such institutes, it would be cheap compared with any other scientific activities.

He wished to argue further that for several reasons the problems of the social sciences were probably very much more difficult to tackle than those of physics, biology, and the other natural sciences. If so, this would only be one more reason to increase the activities. Why was it that the social sciences, and *inter alia*, economics, got such a small share of the cake? Professor Rasmussen stressed that in this context he was thinking of the social sciences as a whole, and would not claim that the problems considered by economists were more important than those of sociology, or other social sciences. But as economists, he felt we ought to make a plea for our own science and its potential impact on 'the wealth of nations'.

Professor Munthe wondered how much the paper had to do with R & D and how much with the development of under-developed countries in general. Professor Trickōvić was trying to seek the weak points in development. The question remained whether these weaknesses should be cured by research effort or rather by technical expertise and the importation of different production techniques. If it were the latter, were his other arguments really relevant? So many of the developing countries were said to have a dual economy – a small export sector with foreign capital and access to all kinds of knowledge, able to use the most advanced techniques, and another sector confined more or less to domestic production, not for export. In this agriculture was a big subsector, and here improvement in performance was a social/economic problem rather than a technical one in most cases. If we excluded agriculture, Professor Munthe asked, were Professor Trickōvić's proposals for dealing with the problem really relevant in other sectors? Here there was a parallel with all industrial countries. There was a wide range between the best and the worst firms in productivity and narrowing this gap brought a significant rise in productivity. This was what technical assistance programmes were really designed to do – to bring techniques available in other parts of the world to developing countries and to advise them on what to adopt. He was puzzled by Professor Trickōvić's argument about the size of enterprise which he suggested was crucial. Professor Munthe thought that what was important if one wanted to introduce advanced techniques was the responsiveness of the firms. This had nothing necessarily to do with size. If one restructured and created bigger firms, one might not improve production techniques at all. Finally, Professor Munthe said that Professor Austin Robinson had earlier asked the Russian members about the guidelines they used for allocating R & D resources. He was not sure that in the Western world we knew a great deal about the optimal way to allocate such resources. It seemed to him that so many of the resources for R & D had to be allocated proportionately to the existing industrial structure. And this structure did not change rapidly. Thus the big points now were generally big in ten years' time. The assumption was that, if one could improve these sectors, one would get significant gains to the economy. Perhaps then one would see what was happening in science and technology in these major industries in the rest of the world, and this might give good guidelines for developing countries.

Professor Freeman said that he would like to comment on two aspects – first on the world balance of science and technological effort, and second on the pool or stock of technical knowledge. On the first point he said he would like to endorse Professor Trickōvić's estimates of the tremendous imbalance between industrialised and developing countries in the scale of R & D and other science and technological activities. This was perhaps the biggest problem of world economic development. Professor Trickōvić did not suggest how it should be overcome, and it was probably a more difficult problem than he allowed for. In the past twenty to thirty years the world R & D system appeared to have been purely responsive to strong and effective demand, particularly in capital goods where enterprises wanting new process technologies had been able to initiate them. This also applied in the government sector, with the obvious case of military systems, where lavish expenditures on R & D had increased the effectiveness by orders of magnitude and in medicine where some diseases had been almost eliminated. More recently we had had the spectacular example of space R & D. This all indicated that the world science system was responsive to stronger effective demand created by those who had a a knowledge of scientific capacity.

If we made a comparison with the developing countries, the most urgent economic problems of the world appeared largely to have been ignored by modern science and technology. The metaphor of a world pool of knowledge could be a barrier to understanding and had only a very limited validity. For fundamental research it appeared to be true. For example the laws of physics were true everywhere. But it was a completely different picture when we turned to the application of science and technology. Because R & D was concentrated in the industrial countries it was largely concerned with market demands in those countries and therefore usually with capital intensive technologies. If we tried to transfer these without regard to the different factor endowments in developing countries there were likely to be grave consequences. In some cases it might produce high growth, but this was likely to be accompanied by rising unemployment because of the nature of the capital intensive technologies. Another reason why the concept of the world pool of science and technology was inappropriate was because a large part of world technology was proprietary knowledge and publication was not open. It was difficult for the weak enterprises to negotiate with the powerful multinational companies and some of these big corporations were doing more research than a whole continent. When it came to negotiating licensing agreements it was a case of 'to him who hath shall be given'. Thirdly, in the area of science and technology the price of transfer was participation in the process of generating science and technology. One could not understand unless one was involved, and therefore transfer was not costless, and not free. We must therefore look again more carefully at the question of availability of world science and technology. If there was to be an effective demand, then it called for social and political changes in developing countries and also for growth within developing countries of scientific and technical capacity in order to assimilate and modify and more important to generate indigenous technology. This would lead to a shift of the balance of world R & D into the developing countries themselves. Finally, in the industrialised countries we should learn to recognise the real problems, and in balancing policy to see that the problem of poverty was more important than prestige. Professor Freeman said that in this respect the new Canadian institute was a good example. Five per cent of its R & D effort was being applied to generate science and technical knowledge which was relevant to developing countries. This pattern needed to be followed by other countries and international bodies.

M. Chapdelaine said the paper was of considerable interest to UNESCO whose technical assistance programme in science policy covered about forty countries to date, and it was also of general interest for the United Nations, within the framework of the Second Development Decade. In this context it was necessary to stress the plea made by the developing countries to UNCSAT (United Nations Conference on the Application of Science and Technology to the Development of Less Developed Countries held in Geneva 1963) for bringing U.N. programmes in science and technology to bear on their own development needs. Despite this and the mandate given to UNESCO by the U.N. Economic and Social Council, the Third World had not always followed up its declared intentions or philosophy with action, for instance by allocating sufficient resources and giving sufficient priority to science policy formulation. The essential question was whether a strategy was possible? Did we make an act of faith here, or did we have serious reasons to believe that it could be put into practical operation? Developing countries devoted ten times less of

G.N.P. to R & D than developed countries and their *per capita* G.N.P. was one-tenth of that of developed countries. In absolute terms expenditure is now less than one-hundredth for equal population. But over what period of time could developing countries hope to multiply their effort by 100? To reach this by the year 2000, it would be necessary to double every five years. The experience of the last ten years had not been encouraging in this respect. Perhaps it demanded too great an effort – possibly 10 per cent of G.N.P. Was it possible to reduce the need for R & D substantially? By strategic choice? To the extent that the R & D system formed a unity, this seemed difficult. Specialisation might permit economies but it was necessary to have in mind that a system of scientific and technological education had to be set up; a significant proportion of fundamental (or non-orientated) research had to be supported (this was 30 per cent in Japan); technological transfer could sometimes cost as much as R & D on the spot; it was necessary to set up scientific services comparable to those that were already at the disposition of the developed countries; it was necessary to train a whole generation of technicians; it was a complete illusion to suppose that the under-developed countries could concentrate their R & D effort towards economically oriented activities to the extent of putting aside alll research aimed at social objectives. Finally one had to remember that only six developing countries (seven with China) had populations of more than 50m, only eighteen have more than 20m. For sixty to seventy others, scale effects operated remorselessly and virtually eliminated them from the race. This emphasised the necessity of groupings of nations, not to speak of integration. Certain encouraging signs were appearing in Asia and Latin America (Central America and the Andes region), regarding scientific co-operation. But at the level of technical and industrial co-operation, with its economic and political implications, one could see a strategy of scientific policy running up against highly adverse factors.

Professor Williams said that he suspected an inconsistency in Professor Trickōvić's paper between his suggestion that developing countries had certain advantages in that they could draw on the fund of knowledge in the technologically advanced economies and his later argument that developing countries had difficulty in absorbing existing technologies if they did not have significant indigenous R & D efforts. What evidence was there that capacity to absorb foreign technology depended on indigenous R & D? Professor Williams argued that cross-country comparisons of R & D percentages, on which many economists leaned to support argument for R & D, were not very useful, and that it was important to take into account the quality and capacity of scientific manpower in research and application, and the actual and potential constitution of production. The R & D in the United States and the U.S.S.R. was dominated by space and advanced military systems which under-developed countries were not likely to produce; in the United Kingdom R & D percentages were dominated by aircraft and atoms, and here again a small under-developed country like Zambia was not likely to get involved in such forms of production. Professor Williams said that a most important issue, stressed in Professor

Trickōvić's paper, though somewhat obscured by the idea that science was a 'good thing', was how to get into a situation where known technologies could be used. Somewhere in the world there was knowledge relevant to the problems faced by developing countries, whether in agriculture, mineral exploitations, or medicine. There had been talk of intermediate technology and there might be something in this, but there were many known technologies which were relevant to the capital and manpower resources of the under-developed countries and the cheapest and quickest way to encourage economic development was to identify and use them, and not waste resources on re-invention. An emphasis on application of known knowledge, with some adaptation for special local conditions, would need to carry with it emphasis on applied science in education. The appropriate forms of training could well be similar to those which were used inside the United States 50–100 years ago. He had noted with interest recently that one Indian university required students of agriculture in their final year to operate a plot of land as a commercial venture, and that high profits made a handsome contribution to final year marks.

Professor Matthews said that the educational system was important in bringing about the poor articulation between science and technology and the economic needs of U.D.C.s stressed by Mr. King. Lack of facilities for advanced training in under-developed countries made them send their best students for graduate study to the United States or other advanced countries. Hence these students become interested in problems studied there, often either pure research or applied research relevant to developed countries. Thus when they got home, they felt intellectually dissatisfied or their work was irrelevant or both. Conflict between relevance and intellectual interest was not unavoidable. There was no lack of problems of high scientific interest that were relevant to under-developed countries. This came out vividly at the UNESCO conference. Professor Matthews did not wholly agree with Professor Williams that solutions to all these problems were known and needed only to be transferred. Especially in primary production, climatic conditions were very diverse between countries. So were the social conditions relevant to production. There were also differences in factor prices. What was the moral? There was a need to make more and better provision for advanced training in science and technology of a kind relevant to under-developed countries, whether actually located in the latter or in developed countries. In economic education we had perhaps done rather better than in science and technology education. Students who came for postgraduate training in economics to developed countries did usually study problems relevant to their own environment.

Professor Tsuru said that he agreed with Professor Freeman that the world pool of science and technology was not freely available, that it was often proprietary and costly to transfer. In contrasting the 1870s and 1880s, which was when Japan took off, with the present period, he was struck by the cheapness of the way in which developed countries at that time transferred technology to his own country. For example, the transfer of French techniques for silk reeling and weaving. He felt that this would be impossible at the present time because similar technology was proprietary, and he wished to ask why this should be so. One reason, he suggested, was that science and technology was now a major cost item and it was a form of know-how for which a charge could be made. It was important for economists to recognise this fact and if possible to do something about making such knowledge more freely available.

Professor Mossé said that much had been said about strategy for science and technological development, but he believed that we had neglected the first rule – to know your enemy well. One must know their forces, their strength and their position. The enemy was everywhere and very powerful. In a general way most men were hostile to progress and better techniques and technology. This was certainly true of the less developed countries, but also of the developed. We could try to think of ways and means of promoting progress. But we were a minority and very few in comparison with the forces against us. We came up against the hostility of most people. And those who opposed progress might be right – new technologies might fail at first, and this reinforced the strength of the opposition. We must study the religious and philosophical reasons which made most people hostile to progress, and also the social and human repercussions of applying technology without adequate regard to its drawbacks.

Professor Trickōvić, who was unable to be present at the conference, after seeing the record of the discussion, commented on it as follows. It had been a difficult tack for Professor Simai to introduce the paper in his absence, and it was an equally difficult task to reply systematically to the remarks made in the discussion. A number of points could have been easily clarified by word of mouth. His impression from the comments was that there had been a much higher degree of agreement than disagreement on the essential issues which he had examined in his paper. Professor Simai had rightly pointed to the central idea that his paper was intended to convey that there was a need to integrate technology and science policy properly into the programmes and plans of economic development of developing countries, and that to this end he had recommended a series of sectoral studies, the content and scope of which should be sufficiently flexible to allow for the variety of situations faced by different countries in respect of a stage of development, national resources available, educational and manpower potentials and other factors. His paper drew on the experience of the methodological approach applied in the preparation of a Yugoslav study. This was concerned with the need for technological improvement and scientific research as a means of enhancing the development of various sectors of the economy. These covered, for example, forestry and timber, agriculture and agro-based industries, both based on the principal natural endowments of the country, as well as other sectors. As the conference had rightly assumed, his paper was not designed to provide the answers to particular problems of particular developing countries, but rather to provide an analytical framework for a study in any individual country. In this respect, it was to be hoped that Professor Simai's appeal for more country case studies, supported when necessary by special assistance from international organisations including UNESCO and other agencies would meet with a wide response. And the sooner this was done the better.

Most of the participants in the discussion had added contributions to particular aspects of the very broad subject of his paper. Mr. King had made a particularly important point in emphasising 'a necessity for each country and even each firm to be on the threshold of science and technology awareness' as a pre-condition for any efficiency in using imported technology. Professor Freeman had elaborated further on this and brought to their attention the important limitations to the transfer of technology and the use by developing countries of a 'world pool of knowledge'. This difficulty often forced developing countries to employ their applied research potentials to discover what had been already discovered. Although as a means of breaking this technological blockade, Professor Freeman had suggested that we should 'shift the balance of world R & D much more into the countries themselves', he himself would think that M. Chapdelaine's emphasis on the need for groupings and co-operation between developing countries was the only effective way of gradually closing the gap in technology and research between developed and developing countries, which – as M. Chapdelaine had shown – was assuming threatening proportions. This conclusion stemmed from the recognition of the importance of measuring the gap in terms of absolute and not relative level of R & D efforts – a fact which he had stressed in his paper. Again, it appeared that properly conducted sectoral studies, along the lines suggested in the paper, would be of great assistance in emphasising the necessity for regional economic and R & D co-operation between developing countries. The national sectoral case studies should also contribute substantially to understanding how improvement in advanced training in science and technology could best meet the needs of individual countries. In this regard, he could not agree more with Professor Matthews when he emphasised the role which properly planned education in advanced science and technology might play not only in regard to actual national research and technology but equally to the assimilation and transfer of known technologies.

He did not think there really was the contradiction to which Professor Williams had referred. To absorb foreign technologies, which often required adaptation, the developing countries had to have a certain minimum level of indigenous research capacity, which in turn became an important ingredient in the quality of higher education in science and technology. This supported Professor Williams' argument that the quality and capacity of scientific manpower was a significant indicator of a country's research potential. His paper had particularly stressed the necessity of building up national research and educational potentials as well as that of achieving international co-operation to correspond with the nation's needs for technological progress in the most relevant sectors of the economy. Thus Professor Williams' fears about the implied dangers to developing countries did not seem to be warranted.

Professor Munthe had referred to developing countries with dual economies, with a large sector of subsistence agriculture, the transformation of which (he very much agreed) was basically a socio-economic problem, though it also required some increase of agricultural research. But the argument that for the establishment of local processing industries technical assistance advisers would be sufficient, without the need for any research effort even for educational purposes, went too far. The right answer, he thought, should again be to conduct case studies, taking into account all the relevant features of individual economies. In fact the framework of case studies which he had suggested was intended to cover the problems facing a wide range of developing countries which differed in stage of development, in structure of their national economies, and consequently in the role of the improvement of technology and research. This in turn should have a bearing on the content and scope of the case studies, the outline of which might be worked out to fit the conditions of countries at various typical levels of development. Professor Munthe had also commented that the responsiveness of firms to the introduction of advanced techniques was not necessarily related to the size of firms. Yet the small size of firms in particular sectors was often found to be a crucial bottleneck which in turn limited the capacity of firms to support the necessary research, however much they might be aware of the need for it. Professor Munthe's proposal to allocate R & D resources in proportion to the existing industrial structure seemed to underestimate the power of R & D in particular sectors of the economy to generate differential development.

Professor Tsuru was to be commended for his appeal for freer flow of technological knowledge to less-developed countries, to which he hoped there would be a response from the international organisations. As an economist, he fully supported the ideas advanced by Professor Rasmussen. Similarly, the importance of sociological factors – stressed by Professor Mossé – undoubtedly deserved consideration. Finally he wanted to express his wholehearted thanks to all the participants in the discussion for their most valuable comments and their contributions to the better understanding of the issues with which he had been concerned.

12 Research and Development and Economic Growth in Japan

Keichi Oshima

DEPARTMENT OF NUCLEAR ENGINEERING,
UNIVERSITY OF TOKYO

I. INTRODUCTORY

There is no question that science and technology play one of the most important roles in the economic growth of advanced countries. However, when one attempts to discover the relation between the effort devoted to research and development and the economic growth of a nation, it is not easy to find a simple correlation between them.

In the case of Japan, in fact, there exist several studies on the relation between expenditure in research and development and the growth of national income or gross national product. The correlations are generally very good; for example it was shown that during the period 1953 to 1960 G.N.P. (y) could be related to investment in research (x) by an equation of $y = 6·55 + 44·3x$ with a correlation of $r = 0·980$, and a much higher correlation was found when research investment was related with G.N.P. with a lag of two to four years [1].

Once, however, one turns to international comparisons the facts become more complicated. In contrast to the continuous high rate of economic growth since World War II, the total expenditure of research and development in Japan has always been at a relatively low level compared with similarly industrialised European countries And again, an international comparison of indicators of technological innovation made by O.E.C.D. shows that the ranking of Japan is low among O.E.C.D. countries [2]. Thus a simple macroscopic comparison indicated no positive relationship between research and development and economic growth in Japan.

This apparent paradox has not been generally understood even by the Japanese themselves, because though the high rate of economic growth of Japan may be attributed to several other causes, no one can deny that the rapid growth has been based on successful introduction and development of advanced technologies in industry.

A clue to the interpretation of these facts may be found in the recent understanding of technological innovation as an integrated process in the industrial system, combining such stages as technical development, investment, production and sales. The effect of research and development effort on the economic growth of a nation should be evaluated in such context.

In the following paper the author has attempted to set out his views, though based on qualitative observations, regarding the pattern of research and development and its contribution to the growth of the economy as illustrated by the process of technological innovation in Japanese industry [3].

II. THE TRANSFORMATION OF INDUSTRIAL STRUCTURE

The rapid growth of 9 to 10 per cent per year in the Japanese economy since the war has not been a mere matter of expansion but has resulted from a substantial change in industrial structure. There has clearly been a general shift from primary to secondary industry. The most remarkable change, however, has taken place inside the sector of manufacturing industry itself.

After the war the objective of Japanese industry was to strengthen its competitive capacity in international trade and to transform the industrial structure to a more advanced type, with emphasis, that is to say, on 'heavy and chemical industries'. After complete defeat and complete loss of her pre-war colonies, everyone, the government and industry alike, was convinced that this would be the only way to reconstruct the Japanese economy.

The success of the transformation can be seen in the figures showing the relative proportion of heavy and chemical industries – the percentage, that is, of products of machinery (including electric), metals, and chemical industries, within the total of all industries. In 1950, the figure was about 50 per cent and reached 64 per cent in 1960, which is about equivalent to that of the industrialised countries in Europe and the United States. It also can be seen that the structure of exports has followed the same pattern as that of industrial production with a lag of a few years; the relative share of the heavy and chemical industries in exports was 38 per cent in 1955, 43 per cent in 1960 and reached 62 per cent in 1965.

This shows that the transformation of industrial structure started first in production for the domestic market and then extended to export. It should also be noticed that the demand in the international market has followed a similar trend of decline in raw materials, food and natural fibres and increase in advanced industrial goods such as machinery, vehicles, and chemicals, and that this has contributed to the increase in the competitiveness of Japanese industry in the international markets.

Since heavy and chemical industries are all technology-intensive, the growth of the economy and the transformation of industrial

TABLE 12.1

THE CONTRIBUTION OF TECHNOLOGICAL INNOVATION TO THE INCREASE OF PRODUCTION IN JAPAN, 1956–66

Industries	Increase of Capital (1956 = 1·00)	Increase of Labour (1956 = 1·00)	Estimated Production Increase without Technical Progress (1956 = 1·00) (a)	Actual Production Increase (1956 = 1·00) (b)	Contribution of Innovation (c) %
Whole manufacture industry	3·94	1·57	2·52	3·57	41·1
Textile	1·88	1·13	1·43	1·70	38·3
Paper, Pulp	3·93	1·54	2·51	4·74	59·7
Chemistry	5·43	1·23	2·68	7·07	72·3
Metal material	3·97	1·61	2·45	3·66	45·6
Metal products	8·67	2·31	3·62	5·88	46·3
Machinery	5·17	1·74	2·52	4·19	52·3
Electric machinery	7·43	2·59	4·29	6·32	38·2
Transportation machinery	5·37	1·82	2·74	5·37	60·2

Note: $c = \dfrac{b-a}{b-1} + 100$ (%)

See [4] for source material for this and subsequent tables.

structure have required high technological innovation, and, in general, industries of high growth rate were highly innovative in technology. An analysis of the contribution of technological innovation to the increase of production of manufacture industries (see Table 12.1) shows that 40 per cent of the increase of production in manufacturing industries as a whole during the ten-year period 1956 to 1966 could be attributed to technological innovation. Moreover a high contribution of technology is seen in the high growth industries, such as chemical (72·3 per cent), machinery (52·3 per cent), and transportation machinery (60·2 per cent).

As is well known, except in a few cases such as the shipbuilding and optical industries, the basic technologies of these innovations were in most cases derived chiefly from imported foreign technologies. It is interesting to note that in these two exceptional cases the technologies were supplied by the transfer of technologies developed for military uses during and before the war. In other cases, however, technology in Japan at the end of the war was far behind that of European countries or the United States and there was little chance to develop a competitive technology in a short time.

An analysis of the contribution of imported technology to the sales of products in several industries is shown in Table 12.2. The contribution to the sales of manufacturing industries as a whole was 33·6 per cent in 1961 and 20·6 per cent in 1967, and large contributions can be seen in major sectors, such as chemical, machinery,

TABLE 12.2

CONTRIBUTION OF IMPORTED TECHNOLOGY TO SALES
(Million dollars)

Sales (1967)

	Total Sales (A)	Sales related to Imported Technology (B)	B/A %	B/A 1961 %
Chemicals	7,850	2,111	26·8	(59·0)
Textile	1,550	258	16·9	(15·7)
Pulp, Paper	711	44	6·3	(29·5)
Rubber	706	402	56·9	(45·4)
Ceramics	878	160	18·3	(19·1)
Iron and Steel	5,508	440	8·0	(61·1)
Electric machinery	7,775	3,400	43·7	(30·8)
Precision machinery	372	87	23·5	(25·2)
Machinery	3,214	528	16·4	
Transportation machinery	5,217	333	6·4	(5·9)
Total manufacturing industry	43,542	8,994	20·6	(33·6)

iron and steel, and electrical industries, and even though there appears to be some decrease in the contribution in some industries, the trend is still continuing up to recent years.

The payment for import of technology in 1967 amounted to 90,563 million yen (U.S.$251 million) and was about 2·8 per cent of the sales based on imported technology. Of the total payment for imported technology, 32·4 per cent was paid by electric machinery, 28·7 per cent by the chemical industry and 10·1 per cent by machinery. It can be seen that technology-intensive industries are paying larger amounts. All these figures show that the transformation of the industrial structure of Japan has been achieved by full use of the import of advanced technology originating abroad. This means that the transfer of technology or invention can be used effectively in the process of technological innovation, if it is made under suitable conditions.

III. RESEARCH AND DEVELOPMENT

We may now turn to the role of the domestic efforts and investment in research and development in relation to Japanese economic growth. Figure 1 shows a comparison of expenditure of research and development of advanced countries. Japanese expenditure is comparatively low. In 1968 the total amount was 768 billion yen (U.S. $2·1 billion), about 1·45 per cent of the G.N.P. Furthermore, the proportion of government expenditure in the total is low; in 1967, the government paid for only 30·2 per cent of the total research and development expenditure. This is very different from a country like the United States or the United Kingdom, where 64 and 50 per cent respectively is paid for by the government.

In this regard the figures shown in Table 12.3 are interesting. In Japan, the greater proportion of the government expenditure on research and development (75·4 per cent) is devoted to economic purposes and very little (16·5 per cent), to national projects such as military, nuclear energy and space. This reflects the fact that, in Japan, the main objective of research and development even in the case of the government has been to contribute to and support the industrial efforts for economic growth.

When, however, one compares the figures of the ratio of expenditures of research and development to sales in various countries, as shown in Table 12.4, it can be seen that the figures of Japanese industries are also lower than those of other industrialised countries, with an exception of the iron and steel industry.

If one accepts that the industrial technology in major Japanese industries is high enough to be regarded as being at the same level

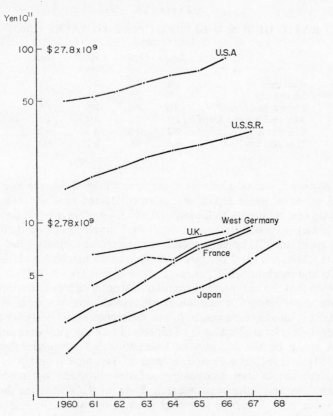

FIG. 12.1 National R & D expenditure

TABLE 12.3

PROPORTION OF GOVERNMENT R & D EXPENDITURE
BY CATEGORIES
(percentages)

	Military, Space and Nuclear Energy	Economy	Health and Welfare	Support to Universities etc.
United States	86·6	3·2	6·8	3·4
France	73·6	17·2	1·9	7·3
United Kingdom	61·4	24·9	3·1	10·6
West Germany	74·0	12·4	1·4	12·2
Japan	16·5	75·4	5·7	2·4

TABLE 12.4

RATIO OF R & D EXPENDITURE TO SALES (1967)
(percentages)

	Japan	France	United States	West Germany
Average	1·1	3·3	4·2	2·8
Machinery	1·3	2·9	4·3	3·2
Electric machinery	2·6	8·1	8·5	6·8
Automobiles and Ships	1·5	2·3	3·4	4·5
Chemicals	2·2	4·0	4·3	5·2
Iron and Steel	0·7	0·9	0·7	1·0

as in other advanced countries, these figures suggest that the research and development in Japan have been conducted in a very efficient way, in combination with the import or transfer of foreign technology.

In fact, expenditures on research and development in industry are spent mainly by such sectors as electric machinery (26·9 per cent), chemicals (21·9 per cent), transportation machinery (11·1 per cent) and machinery (8·5 per cent): these figures are for 1971. It can be seen that the heavy and chemical industries which have the largest share in the import of technology are also the major sectors of expenditure of domestic research and development. This shows that these industries are leading in all aspects of technological innovation.

A study by the Ministry of International Trade and Industry (MITI) in 1961 shows some features of Japanese industry in the development of new technology and new products by imported technology and indigenous research. Some of the results are shown in Table 12.5. It can be seen that in most industries the imported technology achieved much larger sales per item, which indicates that the core of the innovation has depended mainly on imported technology. None the less, the contribution of indigenous research is also very considerable and it exceeds 50 per cent of the total sales. This means that already in 1961 Japanese industry was reaching a stage of development by its own technology and imported technology was thoroughly absorbed and integrated into the total system of industrial technology.

At the same time, there is a good correlation between domestic research and development and import of technology. A study shows that the indigenous research and development were related to the import of technology. The percentage of the current level of research and development that was already going on at the time of the import of the new technology was asked. In the case of technologies imported by 1500 companies during the period between 1950 and 1966, 83 per cent of the companies were already doing some kind of

TABLE 12.5

SALES AND EXPENDITURE OF R & D AND
COMMERCIALISATION PER ONE NEW PRODUCT
DEVELOPED [5]
(Million dollars)

	Sales (1961)		Expenditure of R & D + Commercialisation	
	Products by Imported Technology	Products by Own Technology	Imported Technology	Own Technology
Food	5·1	0·8	1·3	0·6
Textile	1·6	1·7	0·7	0·8
Pulp, Paper	3·6	3·0	4·1	2·5
Chemicals	5·0	0·9	4·6	0·8
Petroleum, Coal	12·7	2·4	2·2	—
Rubber	1·1	0·5	0·7	—
Iron and Steel	9·9	1·1	1·9	0·6
Nonferrous metal	1·8	1·3	3·3	0·5
Metal	2·0	0·4	0·4	0·1
Machinery	0·9	0·6	0·4	0·2
Electric machinery	1·0	0·6	1·2	0·2
Transportation machinery	1·1	5·5	1·8	0·8
Average	2·5	1·3	1·9	0·5

research and development related to these technologies; 31 per cent of them were already carrying on some development work.

This also shows the fact that the import of technology was not an isolated phenomenon; parallel efforts in the form of indigenous research and development were going on in the enterprises concerned to enable them to absorb and transfer the imported technology smoothly into the integrated technological system of the enterprise or industry concerned. In other words, most of the indigenous research and development was a part of a technological innovation process which had the imported technology as a core. And, in fact, many imported technologies have been improved after introduction into Japan and in the case of some of the technologies they were commercialised for the first time in the world in Japan.

Other characteristics of research and development in Japan are her high proportion of basic research and of research manpower engaged in it. In general, the proportion of basic research decreases with the increase in the total amount of expenditure on research and development. But Japan is an exception. In Japan, the proportion of basic research is not only high (28·1 per cent in 1967) but also the absolute amount of expenditure (170 billion yen (or U.S.$474 million) in 1967) is larger than in any Western European country, and second only to that of the United States.

Since there have not been very large expenditures for development of national projects, it is understandable that government expenditure on basic research is large compared with that on development. It is to be observed, however, that even in private industry, the expenditure classified as on basic research is larger than in other countries. Japanese industry spent 10·2 per cent on basic research in 1967; this is to be compared with 3·9 per cent for the United States, 5 per cent for the United Kingdom and 3 per cent for France. This indicates that a larger proportion of research in Japan is oriented to the general raising of the technical and scientific level, and less to development of large projects.

On the other hand, the number of researchers in Japan is also very high. The figure of 157,612 in 1968 is not only high in absolute number – the third highest in the world – but in proportion to population (1·6 per thousand) it is also higher than in any European country.

In this regard, it has to be noticed that the level of general education is high in Japan. The percentage of attendance at primary school is 99·9 per cent, the highest in the world; the period of compulsory education is nine years, attendance at high school is 70 per cent, and college and university attendance is 20 per cent.

The percentage of students of natural science and engineering in higher education is about 20 per cent, lower than in European countries, but the proportion of engineering students among this group is extremely high, more than 70 per cent. This trend reflects the government policy of meeting the demand of industry for high-quality technical manpower.

These figures for basic research and manpower show that in Japan a much larger effort is devoted to improving the scientific and technological background of technological innovation rather than to the direct effort on new developments. In most cases there is little direct link between basic research in universities and public institutions on the one hand and industry on the other. And the contribution of activity in basic research has been mainly to supply a scientific and technological soil of high standard in which to cultivate active technological innovation in industry.

It should perhaps be mentioned that the contribution of universities to basic research is high in Japan; 65 per cent of all expenditure on basic research is spent at universities in Japan, 48 per cent in the United States, 45 per cent in United Kingdom and 64 per cent in France. Moreover the larger share of engineering (42 per cent) and medicine (32 per cent) is characteristic of basic research in Japanese universities. This may suggest that economic and social needs have a much stronger influence on research in universities in Japan than in other countries.

IV. GOVERNMENT POLICY

As has been outlined above, the process of technological innovation in Japan has been a combination of the import of foreign technology and of domestic research and development. One of the most powerful and influential factors guiding and promoting such behaviour in industry has been the policy of the government, and especially that concerned with industrial technology in relation to the reconstruction of industry.

After the War, it was a major objective of Japan to catch up with Western European countries – that is, to modernise her industrial structure and increase the economic standard of living. This was not only an objective for the Japanese government and industry, but it could be regarded as a target for the whole nation. Thus, all the policies of the government were naturally carried on in accordance with this national consensus to achieve the growth of the economy.

The policy regarding technology for industry was also decided and implemented under this guide line. That is to say, technology was considered as one element in the integrated economic activities designed to develop modern industry and it was treated in the same context as other economic factors, such as investment, labour, and market promotion. Thus, once the basic policy to establish a modern technology-intensive industry was decided, the import of foreign technology was actively encouraged, as a natural consequence, for the purpose of filling the technological gap between Japan and the advanced countries, and especially the United States.

One point to be emphasised is that, for this reason, the policy of technology for industry was handled and decided by MITI, even though the responsibility for general policy in science and technology was carried by the Agency for Science and Technology. Thus in the early period, some of the measures now considered important for technological development and innovation, such as market allocation, investment in the commercialisation of advanced technology, and tax incentives were not properly appreciated as necessary ingredients in a policy for science and technology, but were mainly regarded as features of a general economic policy. This indicates that these important aspects of a policy for technological innovation, which have since made a major contribution to the growth of the economy of Japan, were not properly identified, in the first stages, as an aspect of science and technology, and this obscured the analysis of the economic effects of research and development.

Now, however, after success in achieving a high rate of economic growth as a result of technological innovation, it has become clear that close co-ordination of technological and economic policies has been the major factor in successful technological innovation.

When the Japanese economy needed to start its reconstruction from a state of wholesale destruction, there were many opportunities for new enterprises to enter the market. And, naturally, the government tried by all means to encourage economic activity. But once there was a comparatively clear objective of improving the industrial structure and suitable criteria had been established to achieve this objective, the government could implement its policies with full collaboration from industry.

At this period, because of destruction and shortage, the government was obliged to control and allocate scarce resources. The two major functions of the government in this regard were to allocate investment and foreign currency. Even though it was not the official purpose of such allocations, the government could make use of these measures as a means of influencing private industries to implement government policy in promoting technology-intensive industries competitive with those abroad.

The government was also obliged to impose strict regulation on imports, including payments for foreign technology, and industry thus needed its approval. The government had set up various criteria for the approval of the import of technology and first priority was given to technologies contributing most to the development of domestic industry that would save imports of basic industrial commodities. In most cases, there was already existing demand or at least a potential demand for the product, and a large future increase of demand with the growth of economy was also forecast.

In general, the programmes of future industrial development and the fields of priority were decided through discussions between the government and industry, usually in the form of an advisory committee of the MITI composed of experts drawn from industry, the government and the universities. Estimates of future market demands, the basic policy regarding the scale and priorities of investments, including locations and technology, were discussed at these committees. In many cases, the government even issued special edicts for the promotion of important technology-intensive industries, such as electronics, petrochemistry, and machinery, and provided various incentives, but industry itself already had a strong incentive to enter these new fields of advanced technology.

In principle, the government followed a policy of securing the autonomy and initiative of Japanese enterprises in all deals with foreign partners and the policy was implemented by supervising the conditions in contracts for imported technology. Consequently, Japanese enterprises had to take the major responsibility of the commercialisation of the technology imported.

Enterprises were also asked to submit programmes, a sales plan and proposed production capacity. This information was used to balance and to control, at least through unofficial guidance, the enterprises in each area. On the one hand, new entrants were limited in order to avoid too much competition between domestic enterprises before the earlier entrants had reached a minimum economic scale and had sufficiently established their technology. On the other hand, since free enterprise was the basic principle in Japan, any enterprise which fulfilled the requirements of the foreign exchange regulations had a right to ask for the approval of import; and in fact, partly because of the rapid growth of the domestic market, approval for the import of competitive technology needed to be permitted to several enterprises in order to avoid a technological monopoly for a few.

Once production was started, the import of foreign products was severely restricted, and the domestic market was given to the Japanese industry. As a result, an enterprise which could introduce a superior foreign technology at an earlier stage could enjoy a dominant economic position until the government gave approval to another enterprise for the import of competitive technology.

As another means of promoting advanced industries, the government provided financial aid for investment. Because of the break-up of large concerns after the war and the scarcity of funds, new investment was mainly supplied by public capital through banks and security corporations. The government was in a position to influence the policies of the financial houses, both through its own direct investment through the Development Bank and also by providing risk coverage to commercial banks through the central bank. A progressive attitude in investment to advanced new technologies was one of the most important factors in the success of rapid introduction of technology-intensive industries in Japan.

These policies and measures on the part of the government formed the basis of the successful technological innovations in Japan. Since innovation is essentially an integrated process, the introduction or transfer of technology can lead to successful innovation in industry only when other factors such as investment and demand are associated with it to produce favourable conditions.

V. CONCLUSIONS

It can be understood from what has been said above that the main features of technological innovation in Japan represented what have been called 'absorptive' strategies. The core of the technology was in most cases transferred from abroad. But the importance of the indigenous efforts in research and development and the supply of high-quality research manpower should not be underestimated.

In the process of innovation, research and development played the most important role of absorbing and integrating the transferred core of technology into the total system of industrial production. This is the only way that a new technology can contribute substantially to the growth of an economy, since the chief economic effect of technology comes from the general advancement of the whole technological structure of industry rather than from the isolated improvement of a specific item of technology.

If there had not been an effective combination of research and development with the transfer of technology, Japanese industry would not have been able to make such a rapid and self-sustaining transformation to an advanced-technology-intensive industrial structure, which formed the basis for the high rate of growth of the economy.

The principal characteristics of research and development in Japan may be summarised as follows.

(1) There was general agreement regarding the objectives of economic growth through modernisation of the industrial structure to correspond with that of the highly industrialised countries of Europe and the United States. Thus the objectives and criteria of research and development were sufficiently clear.

(2) Research and development was considered as a part of the whole process of technological innovation, and much effort was given to the absorption and improvement of the core of technology transferred.

(3) Development of technology was considered simultaneously with other economic factors, such as investment, markets and labour. Consequently, a success in technological development could be exploited smoothly at the stage of commercialisation.

(4) The government itself promoted technological innovation with guide lines and protective measures, but the competition between enterprises in the domestic market was very severe and continuing research and development of technology in enterprises was required if they were to be competitive. For example, as an extreme case, seventy-three companies in the

motorcycle industry in 1955 were reduced to no more than six companies in 1965 by fierce economic competition.

(5) There was comparatively little effort devoted to national research and development projects for non-economic purposes. The major research effort was directed to economic applications. This was equally true of basic research in universities.

In short, it may be concluded that though the total amount of expenditure for research and development in Japan was not large, the combination of research and development with transferred foreign technology was achieved in a way that was extremely effective for the growth of the economy.

Finally, however, it should be made clear that new conditions have arisen which may change the present pattern. Recently with success in catching-up with the technologically advanced countries and with liberalisation of trade and capital, the transfer of foreign technology has become more difficult and less effective. Japanese industry has now to originate and develop for itself technology of high originality and competitive superiority. This must require some change in the present strategy of research and development.

An even more important factor is the change in the social attitude to the development of technology. There are now strong and widespread feelings of doubt as to whether first priority should continue to be given to economic growth and rising living standards. Much more effort in research and development needs to be devoted to social problems, such as better environment, health and welfare. A recent recommendation of the Council of Science and Technology has emphasised this point. Moreover, even in the programme for the development of industrial technologies in the seventies recently issued by MITI, emphasis is put on technology for the prevention of pollution, projects to develop technology for social welfare, and the introduction of the assessment of technology. This change of the social climate is likely to change the pattern of research and development in Japan, which has hitherto been strongly oriented towards economic growth.

REFERENCES

[1] A. Uchino, *Investment for Research in Japan* (Jitsugyokohosha).
[2] K. Pavitt, 'Performance in Technological Innovation', ch. 10 of *Technological Innovation and the Economy* (Wiley-Interscience, 1970).
[3] K. Oshima, 'Setting the Scene Three: Japan', ch. 4, *ibid.*
[4] *White Paper on Science and Technology*, 1968 and 1969 issues, Science and Technology Agency, Government Publication.
[5] *Report of Survey on Industrial Technology*, Ministry of International Trade and Industry, Jitsugyokohosha (1964).

Discussion of the Paper by Professor Oshima

Professor Triantis, in opening the discussion, said that the paper by Professor Oshima was concerned with 'research and development' in relation to economic growth in Japan. At the outset, the author pointed to studies of the Japanese economy which, for relatively short periods, showed a close correlation between investment in research and growth of G.N.P. On the other hand, inter-country comparisons showed that research and development had accounted for a smaller proportion of G.N.P. growth in Japan than in Western countries. Yet Japan's income has been increasing rapidly. Professor Oshima had not examined whether domestically the share of G.N.P. going to research had varied with the rate of growth in G.N.P. Nor did he try in his inter-country comparison to relate the share of research in G.N.P. also to the level of income of the various countries, though this might not have produced an explanation of the relatively low share of research in Japan. Any such relation might be obscured easily by the many other factors, some of them unquantifiable, which affected inter-country comparisons. And in any event, the causation would be running in both directions. The rate of growth in G.N.P. should also be related in a two-way causation to the share of research in G.N.P. Professor Oshima proceeded to explain why expenditures on research and development were low in Japan and yet were important for the rapid growth in her income. He noted that the post-war growth of Japan's income and exports had been based on the expansion of technology-intensive industries: mainly heavy and chemical industries. However, for the most part Japan relied on imported technology and she proceeded to adapt and improve it, and to absorb it as well to diffuse it and popularise it. Professor Oshima concluded that both government and industry fostered technological innovation in Japan through appropriate technology-absorbing strategies.

Professor Triantis said that we might summarise some of the fundamental relationships in a simple model. Japan had an abundant supply of labour. Of the other factors of production, the supply of land could not be increased easily. The supply of capital would rise if the rate of growth in income increased, and so it has. Technology could be imported at a reasonable cost. It served to increase productivity. Also, it induced and entailed investment in new plant and equipment, which raised incomes and consumption, thus stimulating growth. In turn, the high rate of growth in income and the concomitant increase in the supply of capital provided for a rapid increase in investment, which embodied technological improvements and increased the return on expenditures on research and development.

Professor Oshima noted that the introduction of new technology in Japan was helped by an adequate supply of technically qualified personnel and managers. Another factor that might have been mentioned is the high level of advancement of Japan in the fields of information and communication. On a *per capita* basis Japan was generally ahead of Europe in the number of highly-educated persons, the circulation of books and newspapers, the number of computers, TV sets and telephones, and other important indices of communication. Professor Triantis said that to his mind the significance of this infrastructure for the diffusion of imported, as well as native, technology could not be overemphasised.

Since reliance on imported technology would perforce diminish as Japan caught up with the West, Japan opted for a high share of basic in total research and an ample supply of research manpower. Her ratio of researchers to total population was greater that that of any European country. Even in absolute terms, her expenditure on basic research had exceeded that of any other country except the United States. Japan's indigenous research and development was advancing, and she now depended on her own development of technology very considerably. She would depend even more so as she shifted her attention from economic growth to problems of space limitation and physical planning, pollution and the environment, the quality of work and life, housing, urbanisation, health and social welfare. These were issues which the Japanese people now seemed bent on emphasising.

Professor Oshima noted also that the Government played a crucial role in guiding and promoting the import and development of industrial technology in Japan. It realised at an early stage that innovation would have to be an integral part of the process of industrial growth. Accordingly, it proceeded to couple policies for the development and introduction of new technology with policies for advancing investment and developing the labour resources and markets. Responsibility for promoting the development and application of new technology as an element of a well-rounded policy for industrial restructuring and growth was given to the Ministry of International Trade and Industry, while the Agency of Science and Technology had been concerned with the general policy in this field. Some governments in the West – Canada was an example – seemed to be missing this point. Science and technology tended to be treated in relative isolation, and the elements of economic policy required in order both to foster the transfer and development of technology and to bring it to fruition had been neglected.

Professor Oshima had presented us with a very useful review of the role of technology in post-war Japanese development. In order, however, to comprehend Japan's experience more fully and to draw lessons useful to other countries Professor Triantis suggested that we should have to answer a number of additional questions. Professor Oshima had attributed Japan's success partly to the existence of 'a national consensus' on the need to modernise the economy and advance economic growth. To quote from the paper 'there was a comparatively clear objective of improving the industrial structure and suitable criteria had been established to achieve this objective', so that 'the government could implement its policies with full collaboration from industry', and later 'the objectives and criteria of research and development were sufficiently clear'. Professor Triantis said that targets of this sort, and a consensus about them, would seem to exist in many other countries, but somehow they did not produce results.

What precisely did the Japanese targets involve? How were the goals translated into operational objectives and how did these, in turn, lead to
criteria for research and development? What were the manner and terms of
bargaining by which consensus, general as well as on specific issues, was
achieved? What were the conditions which contributed to and guaranteed
the effective co-operation of the various interests in the formulation and
application of the relevant policies? Students of Japanese development had
pointed to the relations between big business on the one hand and central
government bureaucracy and the Liberal-Democratic Party on the other.
They had pointed to the daily contact of officials of the Ministry of
International Trade and Industry with large firms, the early retirement
of many bright men from the civil service and their transfer to private
industry, and other such factors. It would seem that social scientists would
have to carry their research a good deal further before we could claim to
understand the mechanics by which the consensus and co-operation noted
by Professor Oshima were achieved, as well as the role of these factors in
the process of innovation and economic development.

Professor Triantis went on to say that at a different level, it would be
interesting to look into the factors which helped or hindered the transfer
of technology, preferably (in his view) by analysing a number of specific
cases. What were the precise technical and economic requirements for
such transfers and the various manners in which these transfers were
effected? He noted that we also needed to know about the transfer of 'core
technology into the total system of industrial production'. These questions
called for an examination of the role of education and training, labour
movements, the attitudes towards innovation, the system of information
and communication, industrial co-operation and competition, the size of the
industrial units and the investments involved, the profitability of new
techniques, and so on. The conclusions would be of considerable interest
to other countries. It might be noted here that, because Japan was quite
advanced in supply of qualified management and manpower, foreign
technology transferred as software was imported almost entirely through
licensing arrangements, purchase of patents, and technical contracts
between Japanese and foreign firms, rather than through foreign direct
investment in Japan. There was an important difference here between Japan
and many less-developed countries. It bore very significantly on the kind of
technology that could be transferred and the conditions of transfer and
diffusion of such technology in the recipient country – quite apart from
the political implications in the two cases.

A systematic comparison of Japan with the advanced Western economies,
as well as the less-developed countries, would serve us well in an effort to
validate and confirm arguments concerning the role of technology in
Japanese economic growth. It would also suggest, as well as help to check
on, the possibilities of similar developments in other countries. Professor
Triantis said that he did not have the time to proceed with this topic, but
wished to conclude with a few points relating to the less-developed countries.
Had other, less-developed, countries tried to follow the Japanese path to
technological progress? To what extent had they succeeded? What had

been the obstacles in their way? To these and other related questions we had only scant or vague answers. What we did know was that less-developed countries often pointed to Japan's economic growth as an example for imitation. Important differences between these countries and Japan were often overlooked. The size of Japan's economy was very different from that of most underdeveloped countries, while the structure of the Japanese economy was closer to that of the Western economies. Accordingly, as has been pointed out in previous discussion, Japan could import and use fairly readily Western technology, both software and in hardware. Furthermore, the Japanese had demonstrated a penchant for foreign trade. This, combined with the rapid growth in their productivity and favourable political developments, opened up large overseas markets for Japanese products. A careful study at the micro level would probably reveal some striking differences between Japanese and other salesmanship abroad, as well as in the backing which exporters had received from their respective governments. Compared with many of today's less-developed countries, Japan possessed resources critical for receiving and developing new technology. They included an educated people and labour force, and an advanced system of communication in the broadest sense. In addition, Japan had had a strong government, fairly free from foreign interference, either overt or concealed. She was also reputed to enjoy the advantage of a traditional understanding between government and industry. By comparison, in many less-developed countries, whose economic and political system was presumed not to be basically different from Japan's, government had vied with governments in the West and had often outdone them in matters of interference in and control of private enterprise. Relations between government and industry were strained. Unlike those of many other countries, the Japanese Government was until recently slow in increasing public investment and expenditure on welfare, urban development and other social overhead capital. Rather than having the fruits of growth distributed to the aged, the widowed and the needy, these were reserved for the labour force which contributed to modernisation and growth.

Professor Triantis said that he wished to conclude by saying that when he accepted the invitation to introduce Professor Oshima's paper he was somewhat concerned about problems of the interface between economics and engineering. As it turned out he had enjoyed Professor Oshima's study immensely. He wished to end his remarks by expressing a deep conviction about the need for a much closer contact between social and other scientists than we now enjoyed, and for more co-operation and mutal stimulation in analysing the complex process of economic development.

Professor Hsia said that he had found Professor Oshima's paper highly informative, and it had aroused his appetite for more information. In particular he was interested in knowing how the figures of the contribution of imported technology to sales of products and the payments for imports of technology had been arrived at. What was the coverage of such data? Did it only include embodied technology? And in Table 12.2, column B, giving figures of sales related to imported technology, how these had been obtained?

Professor Sargent said that he had found Professor Oshima's paper very illuminating as an account of the necessary conditions for Japanese success in using imported technology. He would like rather more precise information on the role of government towards enterprises which had imported technology, particularly with regard to the statement made that 'Japanese enterprises had to take major responsibility for the commercialisation of the technology imported'. He took this to mean that having imported the technology they must then stand on their own feet in using it. However, in a subsequent paragraph Professor Oshima had said that once production had started the import of foreign products might be restricted. Professor Sargent asked for what period this restriction might apply, and how was it determined. If an enterprise introducing technology could enjoy dominance until the government gave approval to another enterprise, what criteria determined the time when the import restrictions were dismantled? He suspected here a carefully judged balance between protection and exposure of enterprises to foreign competition, and he would like a little more information on this. On a second point Professor Sargent said that Professor Oshima had made a statement in the paper to the effect that 40 per cent of the increase of production of all manufactures could be attributed to technological innovation. He took this to mean 40 per cent of the increase which could not be related to growth in capital and labour stock was a residual. If so he thought it was surprisingly large.

Professor Khachaturov said that the high rate of economic growth in Japan was well known. But this could be attributed to many other factors besides R & D. They had had small military expenditures, they had had land reforms, a high level of education, and a government policy in favour of high tempo and transfer of technology. He thought it was therefore very difficult to know how to distinguish the role of R & D from all of these other factors.

Professor Oshima said he would like to explain how he had approached his problem. He had made the preliminary assumption that economic growth and R & D were connected through the innovation process. His paper started to analyse the Japanese figures in the context that economic growth and science and technology were linked in the framework of technical innovation. On the role of government he said the prime responsibilities lay with private industry, with rather strong restrictions on foreign currency and investment. Private industry would first propose a scale of investment, for example for polyethylene, including an indication of the scale they would like, and then the government examined it within the general framework of planning criteria. Market protection had two meanings. Domestic competitors could come in when the market was large enough for full operation of the minimum scale orginally proposed or just before it. This was not decided simply by the government departments concerned but in *ad hoc* meetings between industry and government, and was implemented with the approval of both parties. On Professor Hsia's point, Professor Oshima said that Japan had strong regulations governing the import of technology and the figure had been obtained from detailed government sources in this area. On the question of sales of imported technologies, this was derived from a MITI study undertaken through inquiries from all major companies. On the question of the contribution of technology, this had been estimated by the Japanese Science and Technology Agency, and was a residual factor, using a Cobb-Douglas type production function. Professor Oshima said he agreed there are many factors contributing to Japanese growth. What he had tried to do first of all was to have a framework in mind, and he now thought they would need more micro studies to follow this up. His paper showed that growth and total R & D expenditure were not easily related in macro-economic terms.

Professor Triantis said there were two possible strategies. The first was to do something that succeeds. The second was to fail to do something that would obstruct success. The Japanese followed the policy of encouragement rather than that of expenditure of public capital. He wanted to ask whether this low level of government expenditure was a deliberate plan. Was the high level of imported technology part of a similar plan, or was it simply a failure of the government to try to develop home-based technology?

Professor Kawano said that Professor Oshima had dealt with the state of Japan since 1945. He would like to say a few words regarding the attitude of governments to R & D and technology, much of which had been remarkably constant and went back over a hundred years. Already in the nineteenth century Japan had universities with all the necessary teaching of technology, and not just basic science. This was a remarkable feature of their system in comparison with the other countries in Asia and so-called developing countries. Technology had been treated as an integral part of fundamental science for a very long period of time. He did not know whether this policy had been based on fundamental assumptions or not. Rather it must have been the imitation of the Western model, he thought. However it could not be denied that the governments of the earlier periods had a far-reaching vision of the future of the country and the role of technology in it. For example the Kyoto University had been founded in 1897 with a technology department and eleven professors from the outset and today there were 130 chairs of technology, whereas the number of chairs of economics was only 16 or 17 even now. This showed the remarkable preponderance of technology throughout the modern history of Japan.

Dr. Nabseth said it was difficult for economists to explain Japanese development in any ordinary terms. Many of the common hypotheses broke down. For example, after the war wage and salary levels in Japan were lower than in Western Europe, and thus one would have expected rapid expansion in labour-intensive branches of industry. But in fact this did not happen. Growth tended to be in the capital-intensive industries, and we must ask why this was. Had the government policy attempted to diminish capital costs, so that the relation between labour and capital costs were not what we would assume? Secondly, when one had the concentration on capital-intensive branches of industry, did Japan use different techniques from those in the West? For example, did Japan use more labour-intensive production methods than, say, the United States or the

United Kingdom? For shipbuilding in Japan and Sweden this was the case, although it was not as different as one might have expected. Another problem was the diffusion of new technology. Japan had imported this to a great extent, but we had no information on the speed of diffusion, except in a few isolated cases. For example there was the basic oxygen steel-making process, which had been introduced much faster in Japan than in the majority of European countries. We might ask whether the Japanese government had tried to reduce risk in introducing new techno-logy. How did their policy differ from that of other countries? Table 12.1 suggested that the increase of capital was extremely high, and the 1965–1966 increases were far above those in other countries. Some of this might be explained by new techniques but it was interesting to ask whether there had been diminishing returns to capital.

Professor Robinson said this paper of Professor Oshima was a very fascinating one, but he would like to go back to the paradox – that the growth in Japan was highly correlated to growth in R & D expenditure in Japan, but in fact Japan was low on the world scale in R & D, though with the highest growth rate in the world. Professor Robinson said it seemed to him that the conference had been oscillating between two views, both exaggerated. Firstly that the world stock of knowledge was freely available, and secondly that there were such obstacles to access to this world stock of knowledge there was no advantage in it. Professor Robinson wanted to argue that there was a very valuable stock of knowledge but there were costs of access to it. There were costs of adaptation to local conditions. For example, in agriculture the start of the green revolution came from Americans working in Mexico. When it came to be applied to rice the principles were well known, but they still had problems first of applying the technology to rice and later of selecting improved types of rice suitable for different countries. The cost of adaptation for each country was there. But it was not necessary to invest in discovering the basic principles again. Professor Robinson suggested that we might distinguish research costs along the following lines. Firstly there were the costs of technological leadership in creating new knowledge and spearhead industries. Here one had a great deal of abortive research, for example the early research in the United Kingdom on computers, hovercraft, nuclear processes and aircraft. Such research was never likely to pay off properly. The creation of tech-nological leadership seemed to be inevitably expensive. Secondly there was the cost of following. This was very different. The broad outlines of the solutions were likely to be known, and the proportion of abortive research was likely to be less than in the spearhead-type activity. Professor Robinson wanted to argue that until lately Japan had been a follower and not a leader. Hence the cost of required research was less, and they were pro-ducing apparently far better results. But now Japan was ceasing to be one of the followers and in some areas, for example in the development of camera lenses, was right in the forefront. Thus we were likely to see the proportions of abortive research get much higher in Japan. Europe had tended to be in the middle ground for some time past. We had imported from the United States, often via multinational organisations, and had also carried out quite a lot of our own spearhead research, much of which had never paid off. As the technology of Japan caught up on the world, Japan would have to increase the ratio of R & D needed to help to remain high in the world league table. Someone must create the world stock of research. Part of the problem was to see how the followers might pay a contribution to it.

Professor Nussbaumer asked whether the Japanese situation was not quite different from that of developing countries, and also from that of the smaller industrial economies in Europe, both of which experienced considerable difficulty in carrying out R & D and deriving full benefits for their economic growth. First Japan seemed to follow an infant industry policy in controlling the influx of new methods of production and of new products. This then allowed Japanese firms to compete with each other on the basis of similar standards of production, this competition contributing to the growth process. On the other hand competition by foreign forces at a higher stage of development could be restricted, thus reducing the danger of foreign control and of incentives for growth slipping away from the national economy to international companies operating abroad. Secondly, Japan had been able to absorb the results of foreign science and technology due to the large number of scientists available both in absolute and in relative terms. In the other countries either these scientists had been lacking or where they have been trained in sufficient numbers, sometimes at considerable cost to the educational system of these countries, they have been exposed to a considerable brain drain. Japan seemed to have benefited therefore both from a large number of scientists available and from a surprisingly small amount of brain drain and had thus kept its capacity to absorb the results of foreign R & D high without incurring excessive cost of training scientists.

Professor Griliches said he would like to raise a small point for clarification. When Professor Oshima talked of licensing committees and approval of the import of technology, what did this really mean? Did one need permission to import technology? Or did one get a subsidy and therefore have a body to ration it? What in fact was the Licence Committee awarding?

Professor Simai said that he would like to revert to the question of the relation between R & D and economic growth. Professor Tsuru had raised the question of the generosity of the outside world in the export of knowledge, especially prior to the First and Second World Wars. The important question was how things were after 1945. Were the conditions for Japan more favourable compared with those of other countries in this period? It would also be interesting to know to what extent imported technology contributed to export sales, and to receive more information about the main sources of imported technology and innovations.

Professor Munthe thought there was a danger in looking at the post-war period in isolation. We should also consider the influence of the Second World War itself on the Japanese economy. It would appear that in this period Japan conducted the war effort in isolation but on a large scale, and developed new industries rapidly, including industries of an advanced technological character, for example optical machinery and transport equipment. The economic laws of wartime were very different. You put in what you could and you did not count the cost. During this period Japanese engineers acquired techniques that were important for post-war development, and they had a huge stock of technical knowledge almost freely available to them. He suggested that a good deal of the Japanese miracle was due to the events of the 1939–45 war.

Mr. King said that Professor Khachaturov had suggested the possibility of lack of raw material resources contributing to making a country a major world force. In such circumstances technological development was more likely to concentrate on skills than on exploiting natural resources. He thought the importance of education had not been sufficiently stressed by Professor Oshima in the upgrowth of Japan, since this had been a big factor which had helped their development. Mr. King said that it appeared from the paper that many of the things that happened might not have been deliberate policy, but had arisen from the tacit co-operation of government and industry. These informal policies seemed to have worked better than the very formal ones in Western countries. Finally, Mr. King wished to asked a question on the projected trend towards social goals rather than economic growth. The G.N.P. per acre of utilisable land was very high in Japan. Thus it was possibly significant to consider Japanese trends towards more social goals. He was struck by the seriousness with which they had taken their pollution problems, and their preoccupation with social problems such as student unrest. He would like to ask Professor Oshima how considerable was this reorientation, and what effect it was likely to have on the growth rate of Japan? Did, for example, a high growth rate lead to aggravation of these social problems? And would science and technological effort respond to these new objectives with the same smoothness as the orientation to economic and growth goals which occurred in the last twenty years?

Professor Williams said that one of the most interesting features of the Japanese case was the contrast with Western Europe. Japan had a lower level of technology, except in the war industries, but had a more adequate supply of technicians. Both Japan and Western Europe were involved in diffusion processes, but they absorbed foreign technology by different means. Both used licences, but Western Europe used foreign enterprise and foreign capital as well. This Western European policy seemed quite deliberate – to encourage American firms in to increase capital and to get American technology and American management. Part of the difference was probably to be found in different national attitudes. But Japan must have been confident of her ability to maintain a high savings rate and must have had a strong belief in the capacity of her managers. Japan was now a highly developed economy, was placing greater emphasis on her own technology and was moving into technologies which were more capital-intensive. There were some signs that very high savings rates would be harder to maintain in the future, and it was interesting to speculate whether there would be a shift to more European attitudes to the import of capital and foreign management.

Professor Freeman argued that it would be wrong if we let the impression of a low level of R & D dominate our thinking. A comparison in terms of conventional exchange rates put Japan at the bottom of the league. If one took account of the fact that most of Japan's R & D was growth orientated, then there was less difference. If the comparison was made in manpower terms then in fact Japan had a bigger R & D effort than European countries. Professor Freeman therefore endorsed what Professors Munthe and Kawano had said that we were not dealing with a sudden leap. During and just after World War Two there had been a high level of education and a high indigenous base. He also endorsed what Professor Robinson had said on the balance of research. Japanese policy makers had been able to combine an indigenous scientific effort with the acquisition of the most valuable science and technology for their own purposes.

Professor Oshima said he would try to answer the many questions put to him briefly. He said he must confess that Professor Triantis had hit a most important and weak point. It was not a deliberate policy at all. There had always been strong argument as to whether the government should spend much more money on national projects to catch up with the West. But it was really left to the initiative of industry with protection and guidance. On the question of capital- or labour-intensity, Professor Oshima said he was not an economist, and could not say anything about policy on capital costs. They had a picture of the Western economy which they had aimed to reproduce which was capital-intensive and technology-orientated, and they had aimed to modernise industry and make it internationally competitive. The government had many investments with the development banks. But Professor Tsuru could answer better on the real mechanism of capital supply. In the early stages there had been many arguments as to whether there should be an emphasis on light goods, for example toys and cameras, or on heavier industries, such as automobiles and petrochemicals. In the first stages the decision was to concentrate on heavier industries and they had been successful in this. But perhaps if they had followed different policies they might not have had their present pollution problems. On the question of shipbuilding, this was an interesting case study, but looking at the development of this technology it was much more capital- than labour-intensive. On Professor Robinson's point, Professor Oshima said he agreed that Japan was facing a turning point and was now starting to develop spearhead technologies. He also agreed with the point made by Mr. King on the increasing importance of social factors. These were two areas in which they must develop their R & D and they no longer had a map to guide them. If they made a bad choice it would have a big impact on the growth of the economy. On the question of the licensing of imported technology he said there were no rigid criteria. General policy decisions had been made to develop say electronics or petrochemicals, and this was given approval by government officials, not acting on written criteria, but depending on rather vague and personal guidance. He said there was no reason for a company intending to import technology not to start the next day. There were no legal restraints on this. On the sources of imported technology, of 5840 examples of such imports between 1950 and 1969, 3468 came from the United States and the next largest number from West Germany; and then came the United Kingdom and Switzerland. As to the influence of the Second World War, it was true that Japan was isolated from the outside community, and thus had to develop indigenous R & D. There were generous funds allocated to universities and industry for military purposes. Some were abortive, for instance oil from coal. But synthetic fibres developed during the war had become basic for industry after the war. In reply to Mr. King, they had a most serious problem of pollution – power stations could not be located within certain areas, with restrictions on SO_2. There were many limitations to growth imposed by pollution problems. Most graduate students did not want to work for the companies which did most to pollute the environment, and there were strong feelings among the young on this topic. The technological policy which had promoted growth had now shifted to socially oriented research, with government sponsorship in the Ministry of Industry in matters which really belonged to the Ministry of Welfare. Professor Oshima said that they were now considering mechanisms to give incentives to private industry to spend money in this socially oriented field. In the past the government and industry had had a large measure of consensus. Now the government was concerned with the environment while industry continued to concentrate on growth. This difference of emphasis was likely to become an important issue in the coming years. Professor Oshima concluded by saying that he was personally very grateful for the opportunity to take part in this discussion. In Japan there were few economists who were interested in the problems that he had discussed in his paper.

13 Technological Change in the Industrial Growth of Hong Kong[1]

Ronald Hsia

UNIVERSITY OF HONG KONG

I. INTRODUCTORY

In two decades, Hong Kong has transformed from an entrepôt to an industrial economy. The record of its industrial growth is the envy of many a developing nation. From 1962–70, for which years the necessary data are obtainable, the gross industrial output of Hong Kong increased at an annual average rate of 16·2 per cent. The corresponding growth rates for industrial employment and labour productivity are 10·3 and 5·5 per cent respectively.

Behind the impressive industrial growth was the rapid technological change. The latter term is used in this paper to refer to output variation not attributable to changes in capital and labour inputs. In Hong Kong where R & D expenditures are extremely limited, the bulk of new technology is imported. Technology imports into Hong Kong are greatly facilitated by the *laissez-faire* policy of its government regarding trade and capital flow, and by its outward-looking entrepreneurs.

In this paper, the rate of technological change in the manufacturing sector for the period 1962–70 is estimated by the Johansen model.[2] This model is particularly suitable for the present inquiry, inasmuch as data on Hong Kong's industrial capital stock are not obtainable. The index thus arrived at reflects the change in disembodied technology, while embodied technological change is implicit in capital accumulation. The paper, in addition, estimates and analyses the rate of technology diffusion in six major industries on the basis of survey findings covering twenty-six innovations. Finally, changes in technology are related to industrial growth in terms of output, employment and productivity.

II. THE RATE OF TECHNOLOGICAL CHANGE

Since the rate of technological change in Hong Kong manufacturing industries is to be measured by the Johansen model, it would be

[1] I am grateful to Edward Chen for his assistance in preparing this paper.

[2] L. Johansen, 'A Method for Separating the Effects of Capital Accumulation and Shifts in Production Functions upon Growth in Labour Productivity', *Economic Journal* (December 1961), pp. 775–82.

useful to examine the model first. It assumes (1) the production function of the Cobb-Douglas type, (2) constant factor shares, (3) neutral technological change, (4) cost minimisation by each producer,[1] and (5) the same relative increase in wage rates in all industries. While the first three assumptions are shared by most other models measuring technological change, assumptions (4) and (5) are not. However, the latter are generally accepted assumptions in economic theory. Moreover, empirically the assumption of the same relative increase in wage rates in all industries appears to be valid on the basis of the inter-industry wage pattern in Hong Kong.[2]

The defining equation of the model is derived as follows:

By assumption (1),

$$X_{it} = A_{it} N_{it}^{\alpha_i} K_{it}^{\beta_i} \tag{A1}$$

where X_{it} is the output of the ith industry at time t;

A_{it}, technological change;

N_{it}, labour input;

K_{it}, capital input;

α_i, labour share; and

β_i, capital share.

Labour productivity (a_{it}) can thus be expressed as

$$a_{it} = \frac{X_{it}}{N_{it}} = A_{it} N_{it}^{\alpha_i - 1} K_{it}^{\beta_i} = A_{it}\left[\frac{K_{it}}{N_{it}}\right]^{\beta_i} \tag{A2}$$

When two periods are compared, equation (A2) gives

$$\frac{a_{i2}}{a_{i1}} = \frac{A_{i2}}{A_{i1}} \left(\frac{K_{i2}/N_{i2}}{K_{i1}/N_{i1}}\right)^{\beta_i} = \frac{A_{i2}}{A_{i1}} \left(\frac{K_{i2}/K_{i1}}{N_{i2}/N_{i1}}\right)^{\beta_i} \tag{A3}$$

By assumption (4),

$$\frac{W_{it}N_{it}}{\alpha_i} = \frac{R_{it}K_{it}}{\beta_i} \tag{A4}$$

where, W_{it} is the wage rate, and R_{it}, the rate of return to capital. When two periods are compared, equation (A4) gives

$$\frac{W_{i2}N_{i2}}{W_{i1}N_{i1}} = \frac{R_{i2}K_{i2}}{R_{i1}K_{i1}} \tag{A5}$$

or

---

[1] In the minimisation, the rates of returns to capital and labour are regarded as given.

[2] The inter-industry wage pattern is shown in the *Annual Reports* of the Commissioner of Labour of the Hong Kong Government for the years 1962–70.

$$\frac{K_{i2}/N_{i2}}{K_{i1}/N_{i1}} = \frac{W_{i2}/W_{i1}}{R_{i2}/R_{i1}} = w_i \qquad (A6)$$

where, w_i is the increase in wage rate relative to the rate of return to capital. By assumption (5),

$$w_1 = w_2 = w$$

Substitute equation (A6) into (A3),

$$\frac{a_{i2}}{a_{i1}} = \frac{A_{i2}}{A_{i1}} \cdot w^{\beta_i} \qquad (A7)$$

Take the $1n$ of equation (A7),

$$1n\left(\frac{a_{i2}}{a_{i1}}\right) = (1n\,w)\beta_i + 1n\left(\frac{A_{i2}}{A_{i1}}\right) \qquad (A8)$$

Let $e_i = 1n\left(\frac{A_{i2}}{A_{i1}}\right)$,

$$1n\left(\frac{a_{i2}}{a_{i1}}\right) = (1n\,w)\beta_i + e_i \qquad (A9)$$

Equation (A9) is the defining equation. Since $1n\,w$ and e_i can be obtained by regression, the effect of capital accumulation on labour productivity can be measured by the anti-$1n$ of $(1n\,w \cdot \beta_i) - 1$, and the disembodied technological change, by the anti-$1n$ of $e_i - 1$.

The two end-point cross-sectional data of the fifteen industries under study[1] are used to fit the defining equation. The major sources of data used in computing the index of technological change are *The Annual Report* and *Quarterly Labour Statistics* published by the Labour Department of the Hong Kong Government and the input-output study undertaken by the Economics Department of the University of Hong Kong.[2] Productivity estimates are based on employment data of the Labour Department and output data of the input-output study. Capital share (β_i) is calculated as $1 - \alpha_i \cdot \alpha_i$ is the ratio of compensation of employees to total value added. Data on wages and fringe benefits (the components of compensation of employees) come from the Labour Department whereas data on the value-added, from the input-output study.

[1] Including (1) textiles, (2) footwear, (3) garments, (4) clothing accessories, (5) wood products, (6) chemicals, (7) non-metallic mineral products, (8) base metals, (9) metal products, (10) non-electrical machinery, (11) transport equipment, (12) electronic and electrical products, (13) plastic dolls and toys, (14) other plastic products, and (15) wigs.
[2] See the forthcoming monograph on *The Structure and Growth of the Hong Kong Economy*.

Regressing $1n\left(\dfrac{a_{i2}}{a_{i1}}\right)$ on β_i, we obtain the following equation:[1]

$$1n\left(\frac{a_{i2}}{a_{i1}}\right) = 0\cdot5534\beta_i + 0\cdot1560 \qquad (A10.1)$$

The coefficient of β_i (0·5534) represents the value of $1n\ w$ and the constant term (0·1560) represents the value of e_i which is $1n\left(\dfrac{A_{i2}}{A_{i1}}\right)$.

[1] For data used, see Appendix A.

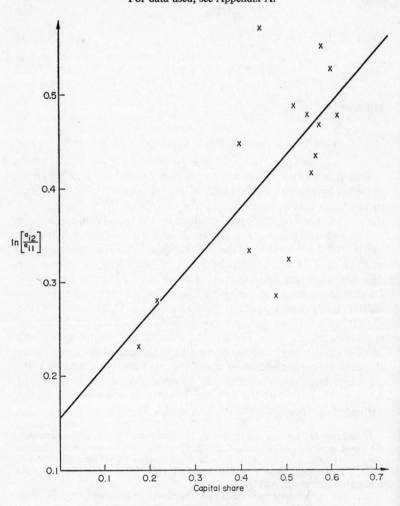

FIG. 13.1   Scatter diagram for equation A.10.1

It should be noted that the anti-$\ln$ of e_i measures essentially the disembodied technological change, inasmuch as the Johansen model is methodologically akin to the group of technological change models treating capital goods as homogeneous. Since all capital equipment regardless of vintage is considered to have the same productivity, any qualitative change of factor input is thus ignored. Consequently, these models measure merely the disembodied technological change.[1]

Our finding is that the disembodied technological change in the manufacturing sector of Hong Kong during the period under study amounted to 16·8 per cent.[2] The annual average rate of technological change thus came to 2·0 per cent. With this, it becomes possible to make a rough comparison of Hong Kong with some other economies on the basis of methodological similarity in measuring technological change. Before attempting such a comparison, a note of caution is necessary. In addition to the difficulties usually involved in comparing different economies,[3] the difference in weighting the technological change index should be noted. The arithmetic index is weighted by factor shares and the geometric index, by factor prices.[4]

[1] Included in this group were models constructed by M. Abramovitz, 'Resource and Output Trends in the United States since 1870', *American Economic Review* (May 1956); J. Kendrick, *Productivity Trends in the United States* (Princeton, 1961); and R. Solow, 'Technical Change and the Aggregate Production Function', *Review of Economics and Statistics* (August 1957). It may be of interest to note that two years before he developed the disembodied model adopted in this paper, L. Johansen had constructed a model measuring embodied technological change, 'Substitution *vs.* Fixed Production Coefficients in the Theory of Economic Growth: A Synthesis', *Econometrica* (April 1959). Among others working on embodied technological change models based on vintage production functions were notably R. Solow, 'Investment and Technical Progress', in *Mathematical Methods in the Social Sciences*, ed. by K. Arrow, S. Karlin and P. Suppes (Stanford, 1960); and E. Phelps, 'The New View of Investment: A Neoclassical Analysis', *Quarterly Journal of Economics* (November 1962).

[2] The anti-$\ln$ of e_i (0·1560) gives 1·168.

[3] These include variations in data quality and coverage, and divergent cost, price and output structures. Specifically, in all cases except Hong Kong the rate of technological change is based on output and capital data at constant prices. The rate for Hong Kong, on the other hand, is based on output data at current prices and capital data implicitly defined in terms of current prices (see equations A4 and A5).

It has been argued that when capital input is measured in constant prices, quality changes of capital input and therefore embodied technological change have been accounted for, to some extent, in the index [B. Massell, 'Is Investment Really Unimportant?' *Metroeconomica* (1962), p. 70]. Following this line of argument, the index for Hong Kong perhaps represents disembodied technological change of a higher degree of purity than the other indices.

[4] For a detailed account of the two indices, see E. Domar, 'On the Measurement of Technological Change', *Economic Journal* (December 1961), pp. 710–26. Whereas Solow and Johansen use the geometric index, Abramovitz, Kendrick,

With this cautionary note, an ordinal comparison of the rates of technological change of selected countries can be attempted. Table 13.1 presents indices of the United States, the United Kingdom, Canada, Japan, the Federal Republic of Germany, Australia, Finland, Argentina and Norway, along with the index of Hong Kong. It also shows the method of measurement used and the time period covered. As can be seen readily from the table, the rate of technological change in Hong Kong manufacturing industries compares favourably with the rates in most other economies. Hong Kong lags only behind Japan, Germany and the United States (1948–60), whose post-World War II economic growth has been substantial. Similarly the array of economies in accordance with the rate of technological change, as shown in the table, appears to make sense in terms of their over-all economic performance.

The relative effects of disembodied technological change and capital accumulation on productivity in Hong Kong can be revealed by assigning a value to β_i in equation (A10.1). If the average (0·48) is the assigned value of β_i, equation (A10.1) gives the following results: from 1962–70 labour productivity increased 52·5 per cent,[1] of which 30·4 per cent[2] is accounted for by capital accumulation and 16·8 per cent by technological change, thus leaving 5·3 per cent to be attributed to the interaction of the two factors. The relatively larger contribution by capital accumulation is expected, in view of the considerable new technological knowledge embodied in the up-to-date capital equipment[3] Hong Kong was importing in large quantities; such imports were necessary to meet the needs of its manufacturing industries which were undergoing rapid mechanisation and automation in the 1960s.

Denison and Domar use the arithmetic index. See R. Solow, 'Technical Change and the Aggregate Production Function', *Review of Economics and Statistics* (August 1957), pp. 312–20; L. Johansen, ' A Method for Separating the Effects of Capital Accumulation ahd Shifts in Production Functions upon Growth in Labour Productivity', *Economic Journal* (December 1961), pp. 775–82; M. Abramovitz, 'Resource and Output Trends in the United States since 1870', *American Economic Review* (May 1956), pp. 5–23; J. Kendrick, *Productivity Trends in the United States* (Princeton, 1961); E. Denison, *The Sources of Economic Growth in the United States and the Alternatives before Us* (New York: Committee for Economic Development, 1962); and E. Domar *et al.*, 'Economic Growth and Productivity in the United States, Canada, the United Kingdom, Germany and Japan in the Post-war Period', *Review of Economics and Statistics* (February 1964), pp. 33–40.

[1] Anti-ln (0·5534 × 0·48 + 0·1560) − 1.

[2] Anti-ln (0·5534 × 0·48) − 1.

[3] The unavailability of data required by the embodied technological change models makes impossible any attempt at measuring embodied technological change in Hong Kong.

Table 13.1

RATES OF TECHNOLOGICAL CHANGE IN THE MANUFACTURING SECTOR OF SELECTED ECONOMIES

Economy	Japan	Germany	United States*	Hong Kong	Norway	Australia	United States*	Canada	Argentina	Finland	United Kingdom	United Kingdom
Period	1951–9	1950–9	1948–60	1962–70	1900–55	1949–60	1909–49	1949–60	1946–61	1925–52	1924–50	1950–9
Method of Measurement	Domar	Domar	Domar	Hsia	Aukrust	Lydall	Solow	Domar	Katz	Niitamo	Johansen	Domar
	(A)	(A)	(A)	(G)	(A)	(A)	(G)	(A)	(G)	(A)	(G)	(A)
Annual rate (%)	4·1	3·4	2·6	2·0	1·8	1·66	1·5	1·4	1·3	1·2	0·7	0·7

(A) Arithmetic index.   (G) Geometric index.

* The arithmetic index of J. Kendrick for the period 1899–1953 is 2·0. It is to be noted that Domar's study of the five countries was based on Kendrick's method of measurement.

Sources: R. Solow, 'Technical Change and the Aggregate Production Function', *Review of Economics and Statistics* (August 1957), p. 315; E. Domar, S. Eddie, B. Herrick, P. Hohenberg, M. Ontriligator and I. Miyanota, 'Economic Growth and Productivity in the United States. Canada, the United Kingdom, Germany and Japan in the Post-war Period', *Review of Economics and Statistics* (February 1964), p. 39; H. F. Lydall, 'Technical Progress in Australian Manufacturing', *Economic Journal* (December 1968), p. 818; O. Aukrust, 'Investment and Economic Growth', *Productivity Measurement Review* (February 1959), pp. 40–50; L. Johansen, 'A Method for Separating the Effects of Capital Accumulation and Shifts in Production Functions upon Growth in Labour Productivity', *Economic Journal* (December 1961), p. 779; J. M. Katz, *Production Functions, Foreign Investment and Growth* (Amsterdam, 1969), p. 27.

It remains to consider the assumption that e's are uncorrelated with β's, an assumption necessary for the regression of $1n\left(\dfrac{a_{i2}}{a_{i1}}\right)$ on β_i in equation (A9). Obviously this assumption would be invalidated if e's were correlated with output increases and the latter with β's. Such a possibility, however, can be minimised by grouping the industries under study according to the rate of output increase. The rates of increase in the output of these industries fall into two distinct groups: seven industries each with an annual average increase above 30 per cent and eight industries each with an increase below 15 per cent. The former industries are designated as Group 1 and the latter Group 2.[1] With industries thus divided but assuming that $1n\ w$ (the coefficient of β_i) is still the same for all the industries, the following equation is computed by using dummy variables:

$$1n\left(\frac{a_{i2}}{a_{i1}}\right) = 0{\cdot}5824\ \beta_i + \begin{cases} 0{\cdot}1611\ \text{(Group 1)} \\ 0{\cdot}1256\ \text{(Group 2)} \end{cases} \qquad \text{(A10.2)}$$

Comparing equation (A10.2) with (A10.1), it can be seen that the coefficients of β_i as well as the constant terms in both equations are reasonably close. This closeness can also be shown by computing from equation (A10.2) the separate effects of capital accumulation and disembodied technological change on the increase in productivity, as has been done with equation (A10.1). The results are

TABLE 13.2

CONTRIBUTING FACTORS TO GROWTH OF
LABOUR PRODUCTIVITY, 1962–70
(Per cent)

	From Capital Accumulation	*From Disembodied Technological Change*	*From Interaction of (1) & (2)*	*Total Increase*
	(1)	*(2)*	*(3)*	*(4) = (1) + (2) + (3)*
All industries	30·4	16·8	5·3	52·5
Group 1	32·4	17·5	5·5	55·4
Group 2	32·4	13·4	4·2	50·0

presented in Table 13.2 along with those from equation (A10.1). Thus output increases do not appear to be significantly correlated with either e's or β's. Inasmuch as the possibility of e's being sig-

[1] Group 1 includes garments, clothing accessories, metal products, electronic and electrical products, plastic dolls and toys, other plastic products, and wigs; Group 2 includes textiles, footwear, wood products, chemicals, non-metallic mineral products, base metals, non-electrical machinery and transport equipment.

nificantly correlated with β's through output changes has been largely ruled out, the conclusions drawn from equation (A10.1) should therefore remain valid.

III. THE RATE OF TECHNOLOGY DIFFUSION

Since the diffusion rate of technology is a determinant of the rate of technological change, it would be of interest to estimate the former. With the estimated diffusion rates, it would be possible to examine the relative capacities of different industries for absorbing new technology. The varying capacities can then be analysed in terms of profitability of, and initial investment in, innovations.

To estimate the rate of technology diffusion in selected manufacturing industries in Hong Kong, the Mansfield model[1] is used. The variables of this model are defined as follows:

$\lambda_{ij}(t)$, the proportion of 'hold-outs' at time t that introduce the innovation by time $t+1$;

n_{ij}, the number of firms in the ith industry supplying information on the jth innovation;

$m_{ij}(t)$, the number of n_{ij} using the innovation at time t;

π_{ij}, the profitability ratio of the innovation to alternative investments;

S_{ij}, the initial investment in the innovation relative to the average assets of n_{ij}.

By definition,

$$\lambda_{ij}(t) = \frac{m_{ij}(t+1) - m_{ij}(t)}{n_{ij} - m_{ij}(t)} \qquad (B1)$$

It is hypothesised,

$$\lambda_{ij}(t) = f_i\left(\frac{m_{ij}(t)}{n_{ij}}, \pi_{ij}, S_{ij}, \ldots\right) \qquad (B2)$$

and

$$\frac{\delta\lambda_{ij}(t)}{\delta\left(\frac{m_{ij}(t)}{n_{ij}}\right)} > 0; \quad \frac{\delta\lambda_{ij}(t)}{\delta\pi_{ij}} > 0; \quad \frac{\delta\lambda_{ij}(t)}{\delta S_{ij}} < 0$$

Expanding equation (B2) by Taylor's series, solving the resulting differential equation and further assuming $\lim_{t \to -\infty} m_{ij}(t) = 0$, we obtain

[1] E. Mansfield, *Industrial Research and Technological Change* (New York, 1968), pp. 136–44.

TABLE 13.3

ESTIMATES AND PARAMETERS*

Innovation	Estimates			Parameters		
	l_{ij}	$\hat{\phi}_{ij}$	r_{ij}	n_{ij}†	π_{ij}†	S_{ij}†
Plastics						
Semi-automatic moulding machine	0·3696	0·4416	0·925	44	2·8846	0·00583
Automatic moulding machine	−0·8025	0·1597	0·892	44	1·7010	0·01382
Blow moulding machine	−1·0497	0·2361	0·875	44	2·3077	0·00755
Rota casting moulding machine	−1·6363	0·2190	0·962	44	1·9341	0·00908
Multi-colour moulding machine	−0·5867	0·2010	0·973	44	1·7885	0·00965
Soft plastic materials	−0·7558	0·2321	0·898	44	2·6161	0·00586
Vacuum-plating	−1·5536	0·1084	0·985	44	1·6310	0·01622
Textiles and Garments						
Staple fibre yarn	−1·7086	0·2197	0·982	37	1·1364	0·02326
Polyester yarn	−1·8061	0·4601	0·933	37	1·2048	0·01775
Automatic winding machine	−1·6643	0·4923	0·987	37	1·5000	0·20245
Blowing machine	−0·6351	0·4166	0·943	37	1·5410	0·10319
Drawing frame	−0·6354	0·3279	0·957	37	1·2308	0·14147
Staple fibre cloth	−3·2672	0·2221	0·964	40	1·0200	0·02833
Polyester cloth	−2·1341	0·4795	0·946	40	1·1000	0·01167
Uni-fill machine	−1·1334	0·2050	0·945	40	1·0526	0·63542
Automatic pirn winder	−1·1319	0·5440	0·930	40	1·2274	0·06004
Electrical cuttor	−1·2375	0·1962	0·964	25	1·0366	0·01409

Automatic soldering machine						0·3456
FM/AM radios	0·2026	0·2293	0·995	23	1·3415	0·04482
Multi-band radios	−0·8292	0·2048	0·991	23	1·4300	0·04310
Silicon transistors	−1·0581	0·1628	0·926	10	1·3333	0·08621
Integrated circuits	−0·7573	0·1997	0·980	10	1·3846	0·06300
Electro-static spraying	−2·0145	0·3251	0·766	31	2·1323	0·00697
Impact extrusion process	−1·7830	0·0599	0·825	31	1·1026	0·60610

* The data on the basis of which the estimates and parameters are computed are obtained from a survey of 199 firms with a responsive rate of 95 per cent. The sample includes firms employing more than 200 workers in plastics, more than 300 workers in textiles and garments, and more than 100 workers in electronics and electrical. Firms of such sizes are considered most likely to introduce innovations. In addition, being larger firms, they are able to supply more reliable data.

† $\pi_{ij} = \dfrac{\text{payoff period to justify investments}}{\text{payoff period for the innovation}}$; $\quad S_{ij} = \dfrac{\text{initial investment in the innovation}}{\text{assets of the firms}}$

Here, the payoff periods, initial investment and assets used in calculating π_{ij} and S_{ij} represent the *average* of the relevant firms. Information regarding the payoff periods is perhaps more satisfactory in Hong Kong than in other economies. This can be attributed to the fact that (1) Hong Kong entrepreneurs are particularly payoff-period conscious because of the prevailing political climate, and (2) Hong Kong entrepreneurs in capital budgeting prefer the payoff-period method to the discounted cash flow method, as the majority of them still adhere to the traditional management practice.

$$m_{ij}(t) = n_{ij}(1 + e^{-(1_{ij} + \phi_{ij}t)})^{-1} \tag{B3}^1$$

where 1_{ij} is a constant of integration;

ϕ_{ij} is the only parameter governing $m_{ij}(t)$ and is therefore a measure of the rate of diffusion;

$$\phi_{ij} = a_{i2} + a_{i5}\pi_{ij} + a_{i6}S_{ij} + \ldots \tag{B4}$$

Taking the $1n$ of equation (B3), we obtain

$$1n\left(\frac{m_{ij}(t)}{n_{ij} - m_{ij}(t)}\right) = 1_{ij} + \phi_{ij}t \tag{B5}$$

Equation (B4) can be written as

$$\phi_{ij} = b_i + a_{i\pi}\pi_{ij} + a_{is}S_{ij} + Z_{ij} \tag{B6}$$

where Z_{ij} is a random variable. Equations (B5) and (B6) are testing equations.

From equation (B5) the estimated value of ϕ_{ij} ($\hat{\phi}_{ij}$) can be obtained. $\hat{\phi}_{ij}$ thus obtained can then be fitted into equation (B6) to find out whether it is linearly correlated with π_{ij} and S_{ij}. The estimates and parameters are summarised in Table 13.3. From these it is possible to obtain the estimated values of b_i, $a_{i\pi}$ and a_{is} in equation (B6). These estimates for the three groups of industries[2] under study are shown below:

Plastics

$$\hat{\phi}_{1j} = -0.1153 + 0.1711\pi_{1j} - 2.0275\,S_{1j} \tag{B7}$$
$$(r = 0.8663;\ r_{\phi\pi} = 0.7502,\ r_{\phi s} = 0.5945)$$

Textiles and Garments

$$\hat{\phi}_{2j} = -0.2837 + 0.5182\pi_{2j} - 0.0915\,S_{2j} \tag{B8}$$
$$(r = 0.6131;\ r_{\phi\pi} = 0.6032,\ r_{\phi s} = 0.0945)$$

Electronics and Electrical

$$\hat{\phi}_{3j} = 0.0071 + 0.1489\pi_{3j} - 0.0187\,S_{3j} \tag{B9}$$
$$(r = 0.9774;\ r_{\phi\pi} = 0.8702,\ r_{\phi s} = 0.7773)$$

The above findings support the hypothesis represented by equation (B2), viz., that the rate of diffusion is positively related to the profita-

[1] This equation shows that the growth of m_{ij} over time conforms to a logistic function or an S-shaped growth curve.

[2] Referring to the list of industries given in note 1, p. 342, the group designated as plastics includes industries 13 and 14; the group designated as textiles and garments includes industries 1, 3 and 4; and electronics and electrical refer to industry 12. The grouping of industries has been necessitated by data deficiency as in the case of garments, and by the fact that some innovations cut across the industry line as in the case of plastic dolls and toys, and other plastic products.

bility of innovations (π_{ij}), but negatively related to the initial investment in innovations (S_{ij}). Not only do the coefficients of π_{ij} and S_{ij} bear the expected signs, they (with the exception of the coefficient of S_{2j}) differ significantly from zero (at the 5 per cent level).[1] As can be seen from the simple correlation coefficients, the correlation is considerably closer between $\hat{\phi}_{ij}$ and π_{ij} than between $\hat{\phi}_{ij}$ and S_{ij}. While the influence of profitability of innovations on the diffusion rate can thus be established, that of initial investment in innovations is less certain.[2]

Our findings also reveal considerable inter-industry differences in the rate of diffusion and in the responsiveness to π_{ij} and S_{ij}. By holding π_{ij} and S_{ij} constant (i.e. assuming that the innovations are equally profitable and require the same amount of initial investment),[3] garments and textiles show the highest rate of diffusion closely followed by electronics and electrical, with plastics falling far behind. Such an inter-industry difference can be attributed to the greater capacity of textiles and electronics to adopt innovations, the former being a well-established industry and the latter with a large component of foreign interests. Furthermore both textiles and electronics are under higher pressure to adopt innovations in view of the highly interdependent structure of firms in these industries, and because of more acute labour and space shortages. The plastics firms, on the other hand, are of smaller prevalent size, less interdependent and less troubled by labour and space problems.

Turning to inter-industry difference in the responsiveness to π_{ij} and S_{ij}, it is found that electronics and electrical are the least responsive to both profitability and initial investment. In π_{ij}, textiles and garments register the highest responsiveness, whereas in S_{ij} plastics is the most responsive. An explanation of the high responsiveness of textiles and garments to profitability lies in the large investment outlays they require for adopting innovations, which make them more sensitive to the profitability of such ventures. In the case of S_{ij}, the small prevalent size of plastics firms accounts, to a large extent, for their high responsiveness to initial investment in innovation.

The foregoing analysis of technology diffusion can be summarised as follows. First, the rate of diffusion varies considerably among

[1] The analysis of variance gives the following F-values: for plastics, $F = 6\cdot02$, significant at the 10 per cent level; for textiles and garments, $F = 2\cdot71$, significant at the 25 per cent level; and for electronics and electrical, $F = 43\cdot03$, significant at the 5 per cent level.

[2] This uncertainty may be attributable partly to measurement error stemming from the reluctance of Hong Kong firms to supply accurate data on investment and assets.

[3] Constant at the average values of π_{ji} and S_{ij}.

innovations. Second, there are evidences supporting the proposition that the diffusion rate is positively related to the profitability of innovations but negatively related to the initial investment in innovations. Finally, there are substantial inter-industry differences in the rate of diffusion and in the responsiveness to profitability and initial investment.

IV. TECHNOLOGICAL CHANGE AND INDUSTRIAL GROWTH

The role of technological change in industrial growth can be better understood if it is related to the other variables of industrial growth. In this section, the rates of technological change of the fifteen industries under study are correlated with their growth in output, employment and labour productivity.[1] In addition, the relative importance of technological change and capital accumulation in industrial growth will be examined.

It is often suggested that industries with high rates of technological change show rapid rates of expansion in output and employment, and conversely that those with rapid rates of growth are likely to have high rates of technological change. To test the validity of this proposition, output and employment are regressed separately on technological change to give the following estimates.

$$\frac{X_{i2}}{X_{i1}} = -294 \cdot 034 + 262 \cdot 677 \frac{A_{i2}}{A_{i1}} \quad R^2 = 0 \cdot 3936 \qquad \text{(C1)}$$
$$(90 \cdot 43)$$

$$\frac{N_{i2}}{N_{i1}} = -116 \cdot 326 + 104 \cdot 360 \frac{A_{i2}}{A_{i1}} \quad R^2 = 0 \cdot 3957 \qquad \text{(C2)}$$
$$(35 \cdot 76)$$

The coefficients of $\frac{A_{i2}}{A_{i1}}$ in both equations are significant (at the 1 per cent level). Correspondingly the coefficients of determination are reasonably high, inasmuch as technological change explains 39 per cent of the growth in output and 40 per cent of the growth in employment. This empirical test thus confirms that the rates of technological change are closely associated with the key variables of industrial growth.

The important role of technological change in industrial growth can be further substantiated by correlating technological change with labour productivity. The regression yields the following equation.

[1] For data used in correlating the variables of industrial growth, see Appendix B.

$$\frac{a_{i2}}{a_{i1}} = 0.010 + 1.298\frac{A_{i2}}{A_{i1}} \quad R^2 = 0.5448 \qquad (C3)$$
$$\qquad\qquad (0.33)$$

The correlation between technological change and labour productivity is positive and strong; the coefficient of $\frac{A_{i2}}{A_{i1}}$ is significant at the 0·5 per cent level. It is interesting to note that the correlation coefficient (0·73) is sufficiently high to show a statistically significant association between the variables, and yet not too high to nullify our effort at computing technological change indices.[1]

In view of the common belief in the economic literature that capital accumulation tends to promote technological change[2] and the recent empirical evidences to the contrary,[3] it would be of interest to examine the relationship between technological change and capital accumulation in the case of Hong Kong. The regression estimates of technological change on capital accumulation for the fifteen industries under study are as follows.

$$\frac{A_{i2}}{A_{i1}} = 1.178 - 0.005 \frac{K'_{i2}}{K'_{i1}} \quad R^2 = 0.00003 \qquad (C4)$$
$$\qquad\qquad (0.27)$$

Equation (C4) shows a negative relationship between technological change and capital accumulation. However, the correlation is extremely weak and the coefficient of $\frac{K'_{i2}}{K'_{i1}}$ is indistinguishable from zero. This points to the probability that capital accumulation may not be closely associated with the inter-industry pattern of growth in output and employment. To test this probability, output and employment are regressed separately on capital accumulation. The following estimates are obtained.

$$\frac{X_{i2}}{X_{i1}} = 41.606 - 21.262 \frac{K'_{i2}}{K'_{i1}} \quad R^2 = 0.0026 \qquad (C5)$$
$$\qquad\qquad (115.60)$$

[1] The correlation coefficient for Hong Kong is lower than that computed by Kendrick for the United States (0·94) and that computed by Domar for the five countries in his study (0·83). Domar felt that his coefficient was too high for comfort since in the extreme case of perfect correlation labour productivity indices could well replace technological change indices. See J. W. Kendrick, *Productivity Trends in the United States* (Princeton, 1961), p. 155; E. Domar *et al.*, 'Economic Growth and Productivity in the United States, Canada, United Kingdom, Germany and Japan in the Post-war period', *Review of Economics and Statistics* (February 1964), p. 40.

[2] Or conversely, rapid technological change in some industries tends to call forth a high rate of capital accumulation.

[3] Some of the evidences will be referred to subsequently.

$$\frac{N_{i2}}{N_{i1}} = 20 \cdot 337 - 10 \cdot 985 \frac{K'_{i2}}{K'_{i1}} \quad R^2 = 0 \cdot 0045 \qquad \text{(C6)}$$
$$\quad\qquad\qquad (45 \cdot 31)$$

Given the results of equations (C1), (C2) and (C4), it is not surprising to find the weak and negative correlation in both equations (C5) and (C6).[1] The latter equations suggest that industries with high rates of capital accumulation generally experience low rates of growth in output and employment. This inference is supported by the findings of other studies. Domar, for example, found a fairly strong positive relationship between output and technological change; but a very weak positive relationship between capital accumulation and technological change, and in some cases (the United Kingdom and Canada) even a negative relationship.[2] Similar results were obtained by Lydall in his study of the Australian manufacturing industries.[3] The weak association between capital accumulation and growth in output, as suggested by Domar, may be attributable to (1) the under-utilisation of capacity caused by unwarrantedly high rates of capital accumulation, (2) diminishing returns to capital, and (3) heavy investment for future use.[4] In the case of Hong Kong, while the third factor can perhaps be ruled out, the first two are worth further investigation.

V. CONCLUDING REMARKS

The rate of disembodied technological change in Hong Kong manufacturing industries for the period 1962–70 is estimated to be

[1] The regression estimates of equations (C5) and (C6) may have been biased by the inclusion of the wigs industry, which shows an unusually high rate of growth in output and employment but only a moderate increase in capital accumulation. In 1962 (the beginning year of the period of our investigation), the wigs industry was relatively new and its output and employment were low. The unusually high rate of growth in output and employment attributable to the small base-year output quantity and volume of employment may thus introduce elements of distortion to our computation. To eliminate this possible bias, we exclude the wigs industry and obtain the following estimates.

$$\frac{X_{i2}}{X_{i1}} = 12 \cdot 952 + 13 \cdot 107 \frac{K'_{i2}}{K'_{i1}} \quad R^2 = 0 \cdot 0433$$
$$\qquad\qquad (17 \cdot 78)$$

$$\frac{N_{i2}}{N_{i1}} = -1 \cdot 375 + 2 \cdot 693 \frac{K'_{i2}}{K'_{i1}} \quad R^2 = 0 \cdot 0233$$
$$\qquad\qquad (5 \cdot 03)$$

The coefficients of $\frac{K'_{i2}}{K'_{i1}}$ in the above equations have become positive. Nevertheless, the correlation remains extremely weak and the coefficients remain statistically non-significant.

[2] E. Domar *et al.*, *Review of Economics and Statistics* (February 1964), p. 40.

[3] H. F. Lydall, 'Technical Progress in Australian Manufacturing', *Economic Journal* (December 1968), pp. 821-2.

[4] E. Domar *et al.*, *Review of Economics and Statistics* (February 1964), p. 39.

2·0 per cent. This rate compares favourably with most other economies for which the rates are known. Hong Kong lags only behind Japan, Germany and the United States.

The rate of diffusion varies considerably from innovation to innovation and from industry to industry. The inter-innovation differences can be attributed preponderantly to variations in the profitability of innovations and the initial investment in innovations. The inter-industry differences are due mainly to the varying responsiveness of different industries to the profitability of, and the initial investment in, innovations.

The association of technological change with changes in output, employment and productivity is found to be positive and strong, i.e. industries with high rates of technological change are at the same time fast-growing industries in terms of output, employment and productivity.

Two observations concerning the relative importance of disembodied technological change and capital accumulation in industrial growth emerge from the present study. First, capital accumulation contributes more to productivity increase than does technological change. Second, the association of capital accumulation with growth in output and employment is statistically non-significant. These two observations though seemingly contradictory can be explained by the fact that the positive association of productivity with output and employment is of limited statistical significance.[1] On the basis of these observations, it is probable that capital has not been used or allocated as efficiently as is feasible in the manufacturing industries of Hong Kong.[2] The policy implications of this inference are (1) that there is a need for redistributing capital resources among industries and (2) that since disembodied technological change can contribute more to industrial growth than a mere increase in capital per worker, promotion efforts should be channelled more in the direction of technological change.

[1] The regression estimates are as follows:

$$\frac{X_{i2}}{X_{i1}} = -140 \cdot 095 + \underset{(59 \cdot 86)}{100 \cdot 509} \frac{a_{i2}}{a_{i1}} \quad R^2 = 0 \cdot 1782$$

$$\frac{N_{i2}}{N_{i1}} = -53 \cdot 699 + \underset{(23 \cdot 82)}{38 \cdot 972} \frac{a_{i2}}{a_{i1}} \quad R^2 = 0 \cdot 1708$$

The coefficients of $\frac{a_{i2}}{a_{i1}}$ are significant only at the 25 per cent level.

[2] The implication is that some of the manufacturing industries are experiencing a rising capital-output ratio.

APPENDIX A

DATA USED IN COMPUTING RATE OF TECHNOLOGICAL CHANGE

Industry	X_{i1} (HK\$ million)	N_{i1} (number)	X_{i2} (HK\$ million)	N_{i2} (number)	a_{i1} HK\$ 0,000)	a_{i2} (HK\$ 0,000)	$\dfrac{a_{i2}}{a_{i1}}$	$\ln \dfrac{a_{i2}}{a_{i1}}$	β_i
1 Textiles	1344·20	70,205	3166·53	125,051	1·9146	2·5321	1·3225	0·2794	0·2180
2 Footwear	164·99	7,621	385·74	13,831	2·1649	2·7890	1·3299	0·2852	0·4778
3 Garments	980·00	38,763	3908·16	91,154	2·5282	4·2874	1·6958	0·5283	0·5997
4 Clothing accessories	163·60	7,390	433·12	14,015	2·2138	3·0904	1·3960	0·3337	0·4189
5 Wood products	126·01	6,076	291·71	8,732	2·0739	3·3407	1·6108	0·4768	0·6143
6 Chemicals	104·00	3,671	190·19	4,210	2·8330	4·5176	1·5946	0·4669	0·5737
7 Non-metallic mineral products	44·23	2,099	98·83	2,908	2·1072	3·3986	1·6129	0·4780	0·5473
8 Base metals	129·00	2,815	191·09	3,013	4·5826	6·3422	1·3840	0·3251	0·5055
9 Metal products	262·00	28,302	667·58	46,182	0·9257	1·4455	1·5615	0·4460	0·3967
10 Non-electrical machinery	115·40	5,676	248·71	7,064	2·0331	3·5298	1·7362	0·5515	0·5779
11 Transport equipment	327·80	15,302	488·71	13,988	2·1422	3·4938	1·6309	0·4892	0·5156
12 Electronics and electrical products	160·62	6,716	1762·88	47,710	2·3916	3·6950	1·5450	0·4350	0·5639
13 Plastic dolls and toys	165·00	8,591	1101·87	37,785	1·9206	2·9162	1·5184	0·4173	0·5557
14 Other plastic products	225·00	17,998	544·02	34,483	1·2501	1·5776	1·2620	0·2327	0·1734
15 Wigs	6·00	484	891·27	28,878	1·7438	3·0863	1·7699	0·5710	0·4451

Sources: Economics Department, University of Hong Kong, *The Structure and Growth of the Hong Kong Economy* (to be published by the Hong Kong Economic Association); Commission of Labour, *Annual Department Reports, 1961/62 and 1969/70*, Hong Kong, 1962 and 1970, *passim*.

APPENDIX B

DATA USED IN CORRELATING VARIABLES OF INDUSTRIAL GROWTH

	Industry	$\dfrac{K'_{i2}}{K'_{i1}}$	$\dfrac{A_{i2}}{A_{i1}}$	$\dfrac{a_{i2}}{a_{i1}}$	$\dfrac{X_{i2}}{X_{i1}}$	$\dfrac{N_{i2}}{N_{i1}}$
1	Textiles	1·129	1·172	1·322	2·3557	1·7812
2	Footwear	1·303	1·021	1·330	2·3380	1·8149
3	Garments	1·394	1·217	1·696	3·9879	2·3516
4	Clothing accessories	1·261	1·107	1·396	2·6474	1·8965
5	Wood products	1·405	1·147	1·611	2·3150	1·4371
6	Chemicals	1·374	1·161	1·595	1·8290	1·1468
7	Non-metallic mineral products	1·352	1·191	1·613	2·2345	1·3854
8	Base metals	1·323	1·046	1·384	1·4813	1·0703
9	Metal products	1·245	1·254	1·562	2·5480	1·6318
10	Non-electrical machinery	1·377	1·261	1·736	2·1552	1·2414
11	Transport equipment	1·330	1·226	1·631	1·4909	0·9141
12	Electronic and electrical products	1·367	1·131	1·545	24·3317	7·1039
13	Plastic dolls and toys	1·360	1·116	1·518	6·6780	4·3982
14	Other plastic products	1·101	1·147	1·262	2·4179	1·9159
15	Wigs	1·279	1·383	1·770	148·5450	59·6653

Discussion of the Paper by Professor Hsia

Professor Robinson said that when he undertook to introduce Professor Hsia's paper he had not yet read it. If he had he would have asked to be excused, because it invoked methods of analysis that he was very ill-equipped technologically to appraise and criticise. He therefore hoped that there would be ample opportunity for Professor Hsia to bring out the inferences that he believed we might derive from his study of Hong Kong, and that other, better equipped critics would comment on the paper. As he saw it, what Professor Hsia's paper had been designed to elicit was the extent to which technological progress in Hong Kong has been attributable to embodied technical change and to disembodied forms of technical change respectively. Of the total increase of 52·5 per cent in labour productivity between 1962 and 1968 (5·4 per cent per year) he attributed (Table 13.2) 30·4 per cent (3·1 per cent per year) to capital accumulation, including embodied technical progress, 16·8 per cent (1·7 per cent per year) to disembodied technical progress, and 5·3 per cent (0·6 per cent per year) to the interaction between the two. Professor Hsia told us comparatively little about local research and development in Hong Kong and its contribution to technical progress embodied or non-embodied. But he made it clear that in Hong Kong, where 'R & D expenditures are extremely limited, the bulk of new technology is imported'. Professor Robinson said this agreed with his own impressions, and he hoped that Professor Hsia would tell him whether he was wrong. Hong Kong was one of the great success stories of growth and development – alongside of Japan, Singapore, Taiwan and more recently Korea. Table 13.1 shows how high it came – ahead of almost all countries shown except Japan and Germany and (on the basis of Domar's estimate) the United States. But it had done this with virtually no research and development as ordinarily defined. How had Hong Kong done this, and what were the lessons for others? He would stick his neck out and suggest some for Professor Hsia and others to demolish. Firstly, along with Singapore, Hong Kong was one of the most open and exposed economies in the world. Almost all its machinery was imported. It could and did benefit by technical progress all over the world and absorbed it in the form of imported embodied technology. This was the principal source of progress. Secondly, Hong Kong business management was immensely vigorous, aggressive, hardheaded, receptive to new ideas. Firms had close contacts with Japan, with mainland China, with all European countries, with the United States. New ideas were quickly and effectively applied. Thirdly, because industry was growing very rapidly, there was a lower proportion of all equipment embodied in obsolescent technologies. Fourthly, Hong Kong had a very adaptable and increasingly well educated labour force, which was more interested in raising productivity and wages than in protecting traditional skills. Fifthly, the main weight of Hong Kong development had been in industries (and especially the textile industries) which in all countries had a low R & D ratio. Sixthly, the industries which Hong Kong was developing were for the moment industries which other countries had already developed and were currently quite willingly abandoning. Japan (to take an example) was getting out of the cotton industry which could no longer carry the new levels of Japanese wages. Hong Kong could acquire ready-made current experience which needed little or no research or development to make it suitable for Hong Kong conditions. This could be compared with what was happening in Singapore with West German camera companies.

Professor Robinson went on to say that in his view almost the most important lesson of Hong Kong was that the conference had been greatly exaggerating the cost of adapting technologies to local conditions in such a case as existed in Hong Kong. He conceded that almost all agricultural research needed adaptation, because climates, diseases and the like differed and were very important. But that was far less true of industrial technologies. Where these had been worked out for one Asiatic country (for example Japan) or even for Europe, they were readily applicable with very little R & D to another, granted certain conditions: a great deal of mobility of top management (there was immense mobility, for example, between Hong Kong, Singapore and Taiwan); a great deal of activity of international companies (Chinese/Japanese as well as American and European); a great deal of similarity of basic conditions of real wage rates, capital availability, and so on.

Finally, Professor Robinson argued that transfer of technology became easier for countries as one went down the income scale. Where the differences of incomes were large, where the products representing the embodiments of the expenditures of the different incomes were very different, where the differences of wage levels, labour productivities, interest rates were great, a great deal of adaptational research was needed. (Even transfer between the United States and Europe and between Europe and India was difficult.) But where the gaps were small and the appropriate technologies of the country under consideration had been very recently improved and developed by another country (as in the case of Hong Kong and Japan) – very little adaptive research was necessary to achieve rapid progress.

In conclusion Professor Robinson apologised for having wandered rather far from Professor Hsia's paper. He did not want to divert the conference from the more technical aspects of an excellent paper, but he thought it was very interesting to ask how Hong Kong has developed so effectively, with relatively little R & D, and he hoped that Professor Hsia would help them to understand this.

Professor Hsia said that he was grateful to Professor Robinson for his comments, particularly as he had called his attention to the bigger issues which were not incorporated in the paper, due mainly to the constraint imposed by the topic, and the specified length. He said it was true Hong Kong had little R & D expenditure and Professor Robinson had already given some answers with which he agreed. Hong Kong has been borrowing available technology with minimum of adaptation expenditures –

minimised by the very way in which entrepreneurs acquired such technology. Hong Kong's entrepreneurs were not merely outward-looking; they actually went out to cultivate friendly relations with firms in other countries. They had been successful in getting contracts with exclusive use of certain processes, and even in entering into partnership with foreign producers. Professor Hsia also agreed with Professor Robinson that people had over-estimated the cost of adopting new technology. Not only did Hong Kong have extremely able entrepreneurs but also a progressive labour force that was highly receptive to new methods of production. On the diffusion model Professor Hsia said he had made use of Professor Mansfield's model and that he was thinking of adding another variable to reflect the interfirm mobility of labour, which he felt was very relevant to the rate of diffusion. Professor Hsia pointed out that the question raised by Professor Robinson regarding the paradox of little R & D and rapid technological change had, in fact, been answered by himself. He would just like to add that the advantage Hong Kong had enjoyed at the beginning of its industrial development was a sudden inflow of capital and entrepreneurship, as well as skilled labour during the political changeover in China which gave Hong Kong a good start.

Professor Tsuru had two questions. Firstly he would like to ask whether Professor Hsia's methodology was adequate to dissociate the effect of disembodied technology from the effects of embodied ones. He noted that he was using Johannson rather than Solow, bringing in the cost minimisation condition which enabled him to skirt around the difficult problem of valuing capital stock, but the algebraic expression was the same thing as that of Solow. His small w was the same as the ratio of capital labour ratios. If this was the case he did not see how Professor Hsia could be certain he was dissociating embodied from disembodied technical change. Secondly, in the latter part of the paper Professor Hsia implied that the capital output ratio was rising. Professor Tsuru asked why? Professor Hsia's capital output ratio, where capital contained embodied technical change, was not shown to be rising. This was puzzling. It might be quite possible that this capital included land, and that land was rising in opportunity cost, so that that was the explanation.

Professor Dupriez said he was very interested by the remarks made by Professor Hsia and also by Professor Robinson, which led him to ask a question. He noted that Hong Kong imported a great part of its technology, and in particular its more traditional technology, and that Hong Kong had a much lower percentage of R & D. He thought this was not surprising because one could not compare the percentages country by country unless one took into account the size of the economies concerned. It might be argued that we were here concerned with products that could be distributed very widely. The great mass of technology was common to mankind and we should give up the idea that we should examine the problem at the national level. It was a common heritage of mankind that we were enlarging, and it was not the nation which did this but the large monopolistic firms. Whether or not major research should be undertaken depended on the nature of the industry concerned. This was less necessary in some industries than in others. For example one could manage without research for textiles in Hong Kong but not in, say, electronics. Professor Dupriez also believed that where a small country was in a less-developed part of the world it was better to use less-developed equipment on the basis of the common fund of knowledge, without having to have recourse to the part of knowledge which was subject to secrecy or patent. He thought Hong Kong could develop on the basis of knowledge common to mankind together with the use of imported machinery.

Professor Oshima said he was most interested in the paper and in Professor Robinson's points, and he saw several things very typical in the Hong Kong industrial growth. The difference between Japan and Hong Kong was in the size of the domestic market. In the experience of Japan most innovations began in the domestic market. In the Hong Kong case he did not know the relative importance of export and domestic markets, but he thought that Hong Kong had a greater dependence on exports, and this had forced them into a position of competition right from the beginning. Professor Oshima said that he was interested in the table given in the Appendix A, and in the fact that the growing sectors were textiles, electronics, and certain other industries. This was quite different from Japan, which was growing chiefly in chemicals and heavy industry. Hong Kong reminded him of the early stage of development of Japan, when they considered whether they ought to put more emphasis on consumer goods and move into direct competition on the international market. Electronics in Hong Kong were much more concerned with consumer goods and thus were not highly R & D oriented. Japanese industry was now more capital-intensive. He thought that Hong Kong was using its international position very cleverly, particularly in the use it made of its free communications and free market position. In Professor Hsia's paper there was some description of the diffusion of technology by sectors. He had mentioned profitability as a measure of diffusion, and found it very different in electronics from the case of plastic dolls and similar products. Professor Oshima thought that in Hong Kong the incentive for diffusion of technology was related to the international competition in the sector. In electronics they had to have a high rate of diffusion of new technology because of competition on the world market using new technology. But this was not the case in, say, toys. The problem was whether this very low level of R & D was a symptom of importing technology or whether it was more specifically related to the special sectors in which Hong Kong is growing very fast.

Professor Mansfield said that he had very much enjoyed the paper, and was impressed by the enormous amount of data that underlay the analysis. He said that in a world where so many papers were almost devoid of data we should be indebted to Professor Hsia for putting this together and doing an excellent job of analysis. He went on to ask whether there was any point in making comparison with the United States to see whether the rates of diffusion did in fact differ. The coefficients were so close to those for the United States that one might suspect an international cartel was maintaining coefficients at a certain level, although he himself pleaded innocence on this.

Professor Robinson said that he had always found the economies of Hong Kong and Singapore fascinating, and suggested that if one was to understand them one must understand the reactions of the United States and of Europe to Japanese competition. There are a number of industries where the aggressive technologies of Japan were making it almost impossible to compete. Some of these were normally regarded as industries of Europe and the United States. For example to take the case of cameras – Voigtlander had just gone out of business but Rolleiflex was going to make cameras for European markets in Singapore because they could buy the necessary skills there at the right price. He said that economists often talked about labour-intensive industries when what they really meant was skill-intensive. Professor Robinson suggested that it was in skill-intensive industries that European countries were finding it increasingly difficult to compete. In such cases the established firms in Europe and the United States would go to countries like Singapore and Hong Kong to buy the skills they needed. But they might in fact continue to do all their R & D at home. Professor Robinson said that he believed this whole process needed to be thought out. In the Singapore case he knew of one big electronics firm which flew out components every week; these were finished and assembled in Singapore and the completed apparatus was flown back to the United States. This had been brought about by a reaction to the Japanese competition. He felt that Hong Kong and Singapore had a great deal to contribute to the world economy by their openness and skills. *Professor Mansfield* intervened to ask whether Professor Hsia could tell us anything about the extent to which multinational firms affected diffusion rates? *Professor Hsia* said this varied between industries. In electronics the greater proportion of output came from multinational firms. But this was not so in textiles. *Professor Mansfield* asked what percentage it was in electronics, and *Professor Hsia* said over two-thirds, whereas in fact it was 15 per cent or less in textiles. *Professor Mansfield* asked what would be the size of these firms, and *Professor Hsia* said in a survey of 200 firms the numbers employed varied considerably; in plastics it tended to be of the order of 100, in electronics 200, and in textiles 300.

Professor Triantis had a small point arising out of Professor Robinson's point that costs of importing technology had been exaggerated. This might be so, since much of that cost might be included in management. He said that to compare the expenditure on R & D as a percentage of income of a country which imported technology with that of a country producing its own did not make much sense. He suggested that there were two separate activities – the positive development of innovation and a great deal of work which had to be done to prevent a firm getting involved in the wrong innovation. The country that imported technology was spared the latter cost. There was some advantage in being able to distinguish between the development and application aspects and the protection aspects of R & D.

Professor Hsia thanked the various contributors and said that he had benefited from their comments. In his earlier intervention he had neglected one question raised by Professor Robinson regarding industries that were well developed in some countries but had since been abandoned. This had, in fact, happened in the case of Hong Kong's electronics industry. Five years ago Japan had voluntarily refrained from shipping to the United States simpler types of electronic products, thus leaving a vacuum in the market. Hong Kong's alert entrepreneurs had quickly filled the vacuum, and this marked the beginning of the electronics industry. On the point made by Professor Dupriez, Professor Hsia said he agreed that the size of the economy, and the type of the industry concerned, deserved special attention. He thought the problem of adopting and adapting technology was simpler for a small economy like Hong Kong than for, say, India or China. Professor Oshima had raised a question about Hong Kong's great dependence upon foreign markets. This was very true, as it exports about 90 per cent of its industrial products. On the question of diffusion in response to profitability, Professor Hsia referred to his paper where he had discussed different responses of different industries. Textiles had shown the highest responsiveness. This was because of the sizeable investment outlays of the large textile firms. Whereas these firms were more sensitive to the profitability angle, the plastics industry was more sensitive to initial investment. On Professor Mansfield's point about a comparative study of the Hong Kong and the United States, Professor Hsia said he agreed wholeheartedly that it would be very interesting, and the difference in size as brought out in Professor Dupriez's comment would make it even more interesting. On the question of why the coefficients differed very little, this would need further investigation. Professor Robinson had suggested a possibility that Hong Kong and Singapore had been in competition with the economy of Japan. Professor Hsia said that Hong Kong believed in competition and had always welcomed it. On Professor Triantis' point about including management in the cost of importing technology, he felt this would be feasible, but there were not much usable data available. Returning to Professor Tsuru's queries on disembodied and embodied technology, Professor Hsia said he would like to refer to Johannson's article in the *Economic Journal*. His own findings showed the capital-output ratio did rise in the period studied. This could possibly be due to diminishing returns to capital and the misallocation of capital. In Hong Kong there was rarely a shortage of capital, even when there had been chronic trade deficits. Finally he was fascinated by Professor Tsuru's remarks about the land values, and he would certainly look into this.

14 The Distribution of Scientific Manpower

J. R. Sargent

I. INTRODUCTORY

As we seek and fail to find convincing reasons why the rate of growth of productive efficiency in the United Kingdom should have lagged behind that of other countries since the war, concern is shifting from whether the resources devoted to economic growth are too small (in terms of investment, R & D, and so on), to whether they are misapplied. This new emphasis emerges clearly in the Brookings study of *Britain's Economic Prospects*, and nowhere more clearly than in the chapter by Merton Peck on Science and Technology. Peck's thesis is the need for a reallocation of technical manpower in the United Kingdom, to be achieved by scaling down 'basic research' and the size of the aircraft industry. The purpose of this paper is to analyse the problem of the allocation of technical manpower in the light of Peck's proposals, and from the standpoint of one who has to say what the government should do.[1] By technical manpower I shall mean men and women with the highest technical qualifications (university degrees or their equivalent), known as Qualified Scientists and Engineers (Q.S.E.s). There are many economies, including the United Kingdom, in which the government can influence, at least in part, the supply and distribution of this type of manpower. What I have to say is offered as an exploration of the tricky ground which separates the power of governments to act in this area from their ability to act correctly.

II. ANALYSIS OF THE PROBLEM

There are three problems requiring a solution. First there is the problem of the optimal allocation of the existing stock of Q.S.E.s between different employments, some of which may be government-determined and some market-determined. Peck asserts the existence of such a problem in the United Kingdom now, in that the produc-

[1] Much of this paper was written while I was Economic Adviser to the (then) Minister of Technology. I am grateful to my assistant Mr. A. J. Buxton for much help in connection with it.

tivity of Q.S.E.s is higher in research-intensive industries other than aircraft than it is in either basic research or aircraft. Theoretically this problem is solved when Q.S.E.s have been reallocated from one employment to another in such a way that the value of their marginal social product is the same everywhere. But there is no guarantee that the optimally-distributed stock is itself of the right size, and this is the second problem. In one sense this is solved when any discrepancy between the value of the marginal social product of Q.S.E.s and their rate of pay is eliminated. For as long as a discrepancy exists, the services of additional Q.S.E.s can be bought (in the long run) for less than society gains in terms of additional social product by employing them, or dispensed with for greater savings than society loses by not employing them. But there is a third problem, which is created by the probability that the rate of pay which will induce people to add themselves to the stock of Q.S.E.s differs from the marginal cost to society of adding them to the stock through higher education whose costs are inaccurately imposed upon its recipients. The theoretical solution to this third problem would be to adjust the scale of higher education producing Q.S.E.s and/or the method of financing it until the marginal social cost and the marginal social product of a Q.S.E. were equated.

This brief summary of the welfare economics of technical manpower is enough to warn the policy-maker what he is up against if he wants to solve the problem systematically. But at least it may help him to take a proper view of symptoms of excess demand for or supply of Q.S.E.s. Much effort has been put into projecting demand and supply in the United Kingdom, with the implication that the government might use its power over both to eliminate projected discrepancies between them. Peck asserts that the research-intensive industries in the United Kingdom have a shortage of Q.S.E.s. If this is so, it must mean that the research-intensive industries would like to employ more Q.S.E.s than they are getting at the going rate of pay, presumably because this is less than the value of what they expect an extra Q.S.E. to add to the product of these industries. But the fact that there is excess demand – or excess supply – does not mean that the government should act to eliminate it. The question must first be asked whether there are other employments in which the marginal product of Q.S.E.s is smaller. If there are none, the excess demand in the research-intensive industries is a symptom of an absolute shortage of Q.S.E.s not of a misallocation of the existing stock. Policy then needs to be directed to removing the absolute shortage of Q.S.E.s by expanding the stock of Q.S.E.s (or expanding it to a larger size than it would otherwise have been); and it is doubtful whether this can be achieved without allowing the excess demand in the research-intensive industries to raise rates of pay and stimulate entry into the channels of higher education from which Q.S.E.s emerge. It would be positively harmful to use an autonomous reduction in the government-controlled demand for Q.S.E.s to eliminate the excess demand in these industries. The excess demand in this case is a symptom of a shortage which will go unresolved in the long-run if it is eliminated in the short-run. In the more general case the excess demand may be a symptom both of misallocation and of long-run shortage. But the objective of policy should still not be the removal of the excess demand as such. It should be relieved as far as it can be by removing from employment outside the research-intensive industries all those Q.S.E.s whose marginal product is less outside than inside the research-intensive industries. But this action should cease at the point when the marginal products of Q.S.E.s have been everywhere equalised. Any excess demand (or excess supply) which then remains must be interpreted as a symptom of a long-run shortage (or excess) of Q.S.E.s, which is unlikely to be eliminated without letting it exert its natural effect on their rate of pay.

Economists can always show off a pretty set of principles, but what about the problem? To apply the principles outlined above calls for a nicety of calculus which is beyond us at present. For the United Kingdom, at least, the present state of knowledge about production functions in individual industries is not enough to disentangle the marginal contribution of one factor input, such as Q.S.E.s, from that of other inputs, such as capital equipment and other types of manpower, to output in any given period. However, it is natural to suppose that the peculiar contribution of those highly qualified in science and engineering when they are recruited to an industry will be felt much less in the current level of production than in the rate at which productive efficiency advances over time through the application of new scientific and engineering knowledge. Some light may be thrown on the factual basis of Peck's argument if we examine the extent to which the growth of productive efficiency in the research-intensive industries (which we have some means of measuring) is associated with their employment of highly qualified technical manpower or Q.S.E.s.

III. TECHNICAL MANPOWER IN INDUSTRY GENERALLY

But there is a prior question to dispose of first. Is it only the research-intensive industries which need more Q.S.E.s? Research-intensity is a matter of degree, and it could be argued that helping the less research-intensive industries to be more so is just as important as helping the more research-intensive to remain so. The argument is important because it greatly strengthens Peck's recommendations for restraint on the use of Q.S.E.s in basic research and aircraft, if the whole of industry, and not simply research-intensive industry, has a priority claim on these. We therefore begin by considering the question: is there evidence that the rate of improvement of productive efficiency in British industry in general is related to its employment of Q.S.E.s?

For lack of anything better – although my current research is designed to evolve something better – I take to measure the rate of improvement of productive efficiency in an industry an index of the growth of its output per unit of its total factor input of the sort used by Denison in the Brookings study. Calculations of this 'residual factor' were made available by Professor R. C. O. Matthews for twelve industries for two periods, 1955–60 and 1960–4. Thus we have twenty-four observations, which the choice of dates should have purged of the effects of changes in the degree of capacity utilisation. Using the manpower surveys,[1] we set the employment of Q.S.E.s as a percentage of the labour force (or 'Q.S.E. density') in 1956 against the 1955–60 residual factor, and the employment of Q.S.E.s in 1959 against the 1960–4 residual factor, for each industry.[2] But no significant correlation between Q.S.E. density and the residuals was found. Two alternative measures of Q.S.E. density were tried: Q.S.E.s on R & D work, and Q.S.E.s on other work, since the latter have been suggested as a possible key element in the recognition and application of opportunities for technical innovations. But in neither case was the correlation significant. Nor was there any tendency for industries which had increased their Q.S.E. density between the two periods to have increased residual factors.

We supplemented this enquiry with a cross-country study of the relation between rates of growth of labour productivity and the density of Q.S.E.s employed on R & D work in manufacturing as a whole, in ten countries of Western Europe and North America. Rates of growth of labour productivity were for 1963–7 and Q.S.E. employment was for various dates between 1963 and 1965 according to country. Growth of labour productivity is, of course, an inadequate measure of efficiency growth, but in any case no significant correlation was found. Nor did any emerge when account was also taken of total R & D expenditure per Q.S.E. employed. As in the

[1] *Scientific and Engineering Manpower in Great Britain, 1969* (Cmnd 902); *Scientific and Technological Manpower in Great Britain, 1962* (Cmnd 2146).

[2] Unfortunately the data now available do not permit us to test the possibility that the Q.S.E. density makes an impact on the residual after a somewhat longer time lag.

national aggregates, so in manufacturing the United States stands first and the United Kingdom relatively high in the Q.S.E. density, but both rank low in the growth of labour productivity.

IV. TECHNICAL MANPOWER IN THE RESEARCH-INTENSIVE INDUSTRIES

Thus the gloss which we put on the Peck thesis – that the availability of Q.S.E.s to industry at large is a significant factor in the general improvement of productive efficiency – stands rejected on the basis of this evidence. We return to the thesis itself, which is concerned with the supply of Q.S.E.s to the research-intensive industries in particular. For this purpose we have supplemented Matthew's data with estimates of our own in order to avoid an extreme shortage of observations. His data contain observations of the residual factor, for each of his two periods, of four industries normally classified as research-intensive: chemicals, electrical engineering, mechanical engineering and vehicles including aircraft. To these eight observations we have added five, for the same industries, but dividing aircraft from motor vehicles, for the period 1962–7. The period overlaps with the later of Matthew's periods, which is 1960–4. To avoid both overlapping and the effects on the residual factor of cyclical fluctuations in utilisation of capacity, we would need 1969 data, which do not yet exist; and there is no necessary reason why the residuals for 1960–4 and 1962–7 should not be statistically independent. To compare with our 1962–7 estimates we have Q.S.E. employment in 1962. Employment, however, can be measured either by the absolute numbers of Q.S.E.s in each industry or by their density. If we choose to test the relationship of Q.S.E. density to the residual factor, we are implicitly theorising that there are no scale economies in the employment of Q.S.E.s: an industry with twice the number of Q.S.E.s and twice the total labour force of another can be expected to have the same residual factor. Since scale economies in the employment of Q.S.E.s have been thought to exist, it seemed best to test for their presence as well as for their absence. Accordingly regression procedures, by simple least squares, were applied to the thirteen observations of the residual factor against, first, the density of Q.S.E.s and, secondly, their numbers. We now found a certain amount of positive correlation, of about the same size in each case. Q.S.E. employment, on either definition, explained about half the variance in the residual factor and the regression coefficients were significantly different from zero.

We next attempted to see whether the residual factor could be more satisfactorily explained if we took into account the distribution of Q.S.E.s between those engaged on R & D work and those engaged on other work. It is often asserted that if the work of Q.S.E.s on R & D is to be recognised and made effective, they need to be complemented by Q.S.E.s in other functions such as production and marketing. Accordingly we tested the relation between the residual factor and two explanatory variables: not only the total number or density of Q.S.E.s employed but also the ratio of Q.S.E.s in other functions to Q.S.E.s in R & D work. When Q.S.E.s were measured by absolute numbers, a fairly high correlation emerged. About two-thirds of the variance of the residual factor was explained, compared with one half before we took in the distribution of Q.S.E.s between R & D and other work as an additional variable. Furthermore, the individual regression coefficients relating the residual factor to Q.S.E.s employed, and to the ratio of those off to those on R & D work, were both statistically significant. The results were better, both from the point of view of the degree of correlation and from that of the significance of the coefficients, than when the input of Q.S.E.s was measured by their density. While this may give us some confidence in asserting that the residual factor is associated with the absolute numbers of Q.S.E.s employed and with their distribution between R & D and other work, it is interesting that the effect of the latter is the opposite of what has often been asserted. For we find that a high ratio of Q.S.E.s on work other than R & D to Q.S.E.s on R & D tends to be associated with a low residual rather than a high one.

A further refinement which can be introduced using the same data is to split Q.S.E.s into scientists and engineers respectively. It is a shortage of engineers which is usually most lamented, and Peck suggests that if this induces scientists to be substituted for them, a loss of efficiency results. The residual factor might be expected, if this is the case, to be positively related to the ratio of engineers to scientists in total Q.S.E. employment. Introducing this additional variable into our regression equation we find the relation to be negative, not positive, implying that the residual factor is larger where there are more scientists relative to engineers. But this inference cannot properly be drawn because the regression coefficient is not statistically significant. In any case, different industries may well achieve the same residual factor on the basis of differing norms for the engineer/scientist ratio. Nevertheless we can say that there is no evidence that a high ratio of engineers to scientists as such has a favourable effect on the residual factor. Nor does the introduction of this additional variable add to our ability to explain variations in it.

A question which now arises is whether the residual factor has been associated with the employment of Q.S.E.s as such or with research intensity in some wider sense, such as might be measured by total R & D expenditure. Ideally one would like to test the influence of not only Q.S.E. employment and its distribution, but also of R & D expenditure other than on the employment of Q.S.E.s, since the effectiveness of the research effort may depend on the extent to which Q.S.E.s are complemented by the availability of non-Q.S.E. labour, materials and equipment. But at present the element of R & D expenditure which consists of salaries of Q.S.E.s cannot be disentangled from the rest.[1] In any case we have no figures for R & D expenditure before 1961/62 to set against the residual factor estimates for the two earlier periods. We are confined to our own estimates of the residual factor for the period 1962–7, i.e. to only five observations for the research-intensive industries. We supplemented these by four more covering Metal Manufacture and the non-research-intensive industries of Food, Textiles and Other Manufacturing. Against these estimates of the residual factor for nine industries for 1962–7 we set 1962 Q.S.E. employment and 1961/62 total R & D expenditure. The positive correlation between the residual factor and Q.S.E. employment again emerged,[2] the latter explaining about 60 per cent of the variance in the former, but practically no correlation was found between the residual factor and R & D expenditure by the industries. When both were included in the regression equation as independent variables, the coefficient of Q.S.E. employment was highly significant, but that of R & D expenditure was not significantly different from zero. This result is consistent with the argument, which Peck bases on some evidence from the United States, that there is no satisfactory substitute for qualified scientists and engineers. Either because these are scarce, or because they are disliked, employers may resort to more non-Q.S.E. technicians in the R & D function, or even to a more lavish expenditure on equipment and materials for use in it; but the contribution of these substitutes for Q.S.E. manpower is insignificant compared with the contribution of Q.S.E. manpower itself. This is perhaps a lesson to be read to the United Kingdom motor industry in particular, which spends quite a lot on research but employs comparatively few Q.S.E.s.

[1] This is an illustration of the disadvantages of the independent data-gathering processes in the United Kingdom for R & D expenditure and for Q.S.E. employment.

[2] This is slightly odd in view of the finding reported on page 365 that when all industries are considered together, and not only the research-intensive, the residual factor does not appear to depend on Q.S.E. employment. But in the present case, although it covers all industries, the majority of the observations are of research-intensive ones.

V. IMPLICATIONS OF THE FOREGOING

These statistical findings allow us to presume that the residual factor in the research-intensive industries would be enhanced if they employed more Q.S.E.s.[1] They also emphasise the prior importance of Q.S.E. manpower over R & D expenditure in general, and cast some doubt on the proposition that importance should be attached to Q.S.E.s on non R & D work as opposed to Q.S.E.s on R & D, and to engineers as opposed to scientists. On the other hand, the association between Q.S.E. employment and efficiency growth is by no means a tight one statistically; two-thirds to a quarter of variance in the residual factor is unexplained by Q.S.E. employment and its distribution. This, and the fact that some non-research-intensive industries have high residual factors, indicates that other forces unknown influence the residual factor; and variations in these could swamp the beneficial effect in research-intensive industries which may be presumed to result from increased employment of Q.S.E.s in them. Nevertheless the presumption stands. The presumptive benefit, however, has to be bought at the cost either of producing additional Q.S.E.s or of transferring them from less productive employment, or of some combination of the two. Our introductory analysis suggests that before committing new resources to the former, it is rational to exhaust the possibilities of the latter. The two main candidates which Peck proposes for the role of releasing Q.S.E. manpower are the aircraft industry and basic research. We consider these in turn.

VI. THE AIRCRAFT INDUSTRY

In the work referred to above we calculated a residual factor for aircraft in the period 1962–7. This was second only to chemicals among the five industries for which estimates were made in 1962–7, and higher than all but one of the eight estimates for the two earlier periods.[2] Taking into account the Q.S.E. manpower which it employed, the aircraft industry performed somewhat better in terms of

[1] Since the data are partly cross-sectional, it may be questioned whether an association between Q.S.E. employment and efficiency growth in different industries can tell us what would be likely to happen to efficiency growth in any given industry if its Q.S.E. employment were raised between two periods. Comparing the same industries over the three periods, however, we find that the residual factor moves in the same direction as Q.S.E. employment for all except vehicles (1960–4 compared with 1955–60: a comparison with 1962–7 is not possible because vehicles is split in the latter period into motors and aircraft).

[2] In 1955–60 and 1960–7, the estimates for vehicles included aircraft, which slightly muddles the comparison.

its residual factor than would be expected from the average relationship between the residual factor and Q.S.E. employment for all the research-intensive industries together. Nevertheless the industry may still be too big, in the sense that all its factors of production, Q.S.E.s included, are contributing less to the social product than they would in some other employment. This could be because of government aid, as a result of which the aircraft industry does not have to exact as high a level of productivity from Q.S.E.s (and other productive factors) at the margin as other research-intensive industries do. Confronting this putative disadvantage is the type of argument which is meant to show that in aircraft production the net benefits commercially assessed understate the net benefits to society. Import-saving, technological 'spin-off', and immobile skills in the industry usually have the largest parts in these arguments; and if true they make a case for aid. We are quite unable at present to do the measurements needed to decide whether the net advantage would lie in reducing, maintaining or increasing the current level of government aid to the industry and the extent to which it consequently absorbs factors of production in general and Q.S.E.s in particular. But the case for government aid has always been powerfully buttressed by a second type of argument, which compared with the first is unsound. This is based upon the proposition (unexceptionable in itself) that even when the most probable outcome is reckoned to be favourable, an aircraft project is hedged about with a very high degree of risk. This may be interpreted to mean that the probability of the most probable outcome is not significantly in excess of the probability of some disastrous outcome. Large amounts of capital have to be committed in advance of any reliable estimate of sales prospects, and these amounts are themselves highly uncertain and subject to escalation. In these circumstances private firms cannot find from private sources the capital required. Because of this government aid is sought.

If aid is given, what the government has to bear is a real economic risk which is in no way lessened by the fact that it happens to be in a financial position to bear it. The government can force people, because it can tax them (or print money), into supporting what they do not freely choose to; but if the project does not pay, society as a whole must still bear the cost in real terms of having diverted resources from other uses to the project in question in excess of the benefits realised by it, and the cost in real terms exists for individuals in tax rates higher (or other benefits from public expenditure lower) than they would otherwise have been. If private firms regard a project as too risky to take on unaided, and they decide not to go ahead, it is presumably because of the disutility which they attach to what might happen in the event of an unfavourable outcome, such as having to make a cut in dividend or to postpone other desirable projects. If the government finances the project, and the outcome is unfavourable, there is an analogous source of disutility: tax rates may have to go up, or other items of expenditure may have to be postponed, to meet the loss. Government may decide to act so that the risk is taken, and can always find the money; but that does not alter the fact that the risk exists, and is a possible source of real economic loss to society at large. The unfavourable outcome involves unpleasant adjustments for some people, whether they be shareholders or taxpayers, and is not conjured away by placing it on taxpayers rather than on shareholders. Government financing does not diminish risk in real terms. Consequently it cannot be supposed that aid for a private project is sufficiently justified by the proposition, which has been widely asserted in the case of aircraft projects, that the initiating firm cannot find the money because the risks are too great. If the proposition is true, it is a sound reason for not undertaking the project, not merely from the firm's private interest, but also from that of society, since the resources which the firm will use could be put to alternative uses by other firms which might more surely yield benefits to the population at large. What is relevant to the government's decision is not the fact that the firm cannot find the risk capital, but whether there is a likelihood that net social benefits will exceed net private benefits. That the government can raise money by means unavailable to private persons is a contingent fact of no relevance to the choice whether the project should be aided or not. But since the aid which the government has given to the aircraft industry has rested heavily on the unsound argument from risk, we must conclude that it ought not to be as large as it is, and is a good candidate for the role of releasing Q.S.E. manpower which other research-intensive industries can benefit by employing. Thus far we underwrite Peck's thesis.

VII. BASIC RESEARCH

What of 'basic research'? This refers mainly to the research which goes on in universities, other institutions of higher education, and government research establishments, although some of the research which is classified as 'basic' goes on in industrial firms. On *a priori* grounds there is a case for subsidising it for 'external economies'. The nature of basic research is that a successful outcome is likely to yield opportunities for the application of new technologies over a range of enterprise considerably exceeding that of any individual firm, which therefore lacks the incentive to finance it on a scale commensurate with its social benefits. Then one can resort to international comparisons. But these do not support the contention that the United Kingdom is devoting a particularly large proportion of her R & D effort to basic research. Criticism of basic research[1] as such often merges into one of R & D which is outside business, and may therefore lack the discipline of a market orientation. But according to O.E.C.D. figures for years around 1963 to 1965, the proportion of total R & D expenditure performed in the business enterprise sector in the United Kingdom is about two-thirds, and this is well in line with the proportions for Germany, Japan, Italy, Belgium and the United States, and above those for France and the Netherlands. A somewhat different pattern emerges, however, when we consider the proportion of total R & D expenditure which is financed by the business enterprise sector. This is relatively low for the United Kingdom; not as low as for France, Canada and the United States but distinctly lower than for Germany, Italy and Japan. When the figures are brought together to show the extent to which R & D performed by business enterprises is financed by them, they offer some support to the contention that success in R & D may be connected with financial responsibility for it, although clearly the generality of this implication cannot be established without further investigation.

But international comparisons are a dubious basis for policy-making; and we are driven back to the fact that we simply do not know how big a loss would be incurred if we were to reduce the number of Q.S.E.s in basic research. All we know is that there would be some positive loss of unknown size; not necessarily large, not necessarily small. But because the loss would probably not show itself in the short run, governments may be tempted, given the presumption that more Q.S.E.s would bring tangible benefits in the medium-term in the research-intensive industries, to raid basic research to supply their needs. Since the demand for Q.S.E.s in the sectors most occupied with basic research is largely government-determined, is there any objection if the government autonomously varies it in order to iron out disequilibrium between the demand and the supply of Q.S.E.s in the research-intensive industries? There certainly is. Excess demand for Q.S.E.s in particular places may be a symptom of misallocation, or of absolute shortage, or of both; and before we begin to act against an absolute shortage we should first redistribute the existing stock of Q.S.E.s until their marginal contribution to the social product is everywhere the same. We can presume that we can work to this end by reducing the numbers employed in the aircraft industry; but beyond this we know nothing.

[1] See Appendix.

If, having exhausted the known possibilities of redistribution, we were left with excess demand, we would be bound to interpret this as the symptom of an absolute shortage. If we were then to yield to the temptation to supply this shortage by raiding basic research, the restoration of market equilibrium would erase the evidence of the shortage and conceal the need to do anything about it. Furthermore, it would remove the possibility of a relative increase in the rates of pay of Q.S.E.s, and remove a remedy for an absolute shortage of them. On past experience in the United Kingdom a problem of this kind is unlikely to be solved by making more places in higher education available for the training of Q.S.E.s without acting on the demand for these places. Nor is the demand for them likely to respond sufficiently to non-monetary expedients meant to raise the 'status' of Q.S.E.s, expedients which in any case would be hard to put into reverse should excess demand turn (as it may now be doing) into excess supply. While different professions, arts- or science-based, may co-exist at different levels of monetary reward, reflecting their different non-monetary attractions, the relative rates at which they recruit are unlikely to be capable of adapting to the changing requirements of the economy without reliance on changing relativities in their rates of pay; and if these are damped out in the interests of short-run market equilibrium, the process of adaptation may fail. Consequently we must dismiss the idea that basic research can serve as a balancing item in the distribution of Q.S.E.s, independently of the marginal productivity of the manpower engaged in it. If there were some means of showing this to be relatively low, then basic research must play its part in releasing Q.S.E.s to the more productive sectors in the interests of the optimal allocation of the existing stock of Q.S.E.s But if it cannot be shown to have any part to play here, it may be positively dangerous to manipulate it simply because it is manipulable. For then, if we reduce the number of Q.S.E.s in basic research to remove excess demand for them elsewhere, not only do we risk whatever loss may turn out to be due to a reduced scale of basic research itself with no presumption that it will be exceeded by gains elsewhere; we also lose the desirable increment to the stock of Q.S.E.s which the excess demand might have brought forth.

Possibly we may need to qualify this argument, because excess demand for Q.S.E.s can only operate effectively on the supply of them after a substantial time-lag. Excess demand is not likely to find a great number of recruits to scientific or engineering professions among those who have already decided against taking science subjects at an earlier stage of education. The moment of choice occurs some six or seven years before the converts to science turn up on the supply side of the Q.S.E. market. However elastic the supply of Q.S.E.s may be in the long run, it takes time to emerge; and meanwhile the excess demand, competing for a supply which is inflexible in the short run, may drive rates of pay to very high levels. This will be mitigated to some extent as older Q.S.E.s are tempted to postpone retirement. But the rate of pay at which the existing supply of Q.S.E.s is rationed out in the short run will inevitably exceed the long-run equilibrium rate, so that the slowly-reacting supply feels a much greater inducement to expand than it should; and the original excess demand will not just disappear gracefully, but become converted into excess supply. An oscillatory or 'cobweb' movement, possibly explosive but in any case unsettling, could then develop in both numbers and in rates of pay. The same thing could occur in the case of excess supply. Because of the time which it takes for the supply of Q.S.E.s to adjust in the face of a surplus, the solution may be continually overshot.

Now unsettling movements of this kind could be prevented if it were possible to remove just enough of the original excess demand for Q.S.E.s as would allow their rates of pay to rise to the long-run equilibrium level only, and not above it. This cut in demand would then be restored as supply expanded in the long run. Similar action would be required in the case of excess supply. This task of manipulating the aggregate demand for Q.S.E.s in such a way as to prevent excess demand (or excess supply) from being excessive without preventing it from being effective, is one which might be performed by the government-determined element of demand, in such a way as to help rather than hinder the adjustments required. It could even be argued that such a policy need not in the long run reduce the scale on which basic research would be carried out; the funds supporting basic research would merely be doled out in a different time-pattern. Nevertheless they would be more variable, and the lack of continuity might well be especially damaging to the long-run effectiveness of research of the basic type. In any case the idea of using compensating variations in basic research to beat the 'cobweb' assumes that the government can recognise not only the existence of excess demand or supply in the market for Q.S.E.s but also its scale. This seems altogether too much to impose upon the current state of the art; and luckily there is something else which probably performs the same function: the rate of migration of Q.S.E.s or 'the brain drain'. For if there is excess demand for Q.S.E.s at home, it will tend to reduce the rate of migration so that the excess demand appears less obtrusive and rates of pay rise less than they would otherwise have done. Similarly some part of an excess supply will be siphoned off by accelerated migration. The effect should be to prevent the adaptation of supply from going to the undesirable extremities which characterise markets in which production has a long gestation period. Since this is the case we can drop the superficially attractive idea that the reservoir of Q.S.E. manpower in basic research might be pumped up and down for this purpose.

VIII. SUMMARY AND CONCLUSIONS

(i) We have tried to define the proper basis for the government policies which mingle with market forces to determine the scale and distribution of our highly-qualified scientific and engineering skill: and to see how far we can sensibly act upon these with the fragments we have of what we need to know. There are two questions: 'can we make better use of our existing stock of Q.S.E.s?'; and 'is the existing stock of Q.S.E.s of the size that we need?' But the number of Q.S.E.s that we need depends upon how well we are using those that we have. So the first question is the one to answer and act on first.

(ii) Accordingly we focussed on this and found:

(a) Some evidence suggesting that the research-intensive industries would benefit from having a higher percentage of Q.S.E.s in their labour force as far as the rate of growth of efficiency was concerned;

(b) That the scale of government support for the aircraft industry had been inflated by unsound arguments, and that therefore a net social benefit was likely if it released resources to other industries;

(c) That since it was quite impossible to say anything about the relative net social benefit contributed (at the margin) by basic research, it was not a candidate which could be rationally selected on economic grounds in the present state of knowledge for action to improve the use of the existing Q.S.E. stock.

(iii) Once the possibilities of redistributing the existing stock of Q.S.E.s are exhausted, we are left with the second question raised above: 'is the existing stock of the size that we need?' It may be too big, or it may be too small. Since we have no reliable evidence for judging which, it seems best to leave it to be identified and resolved by market forces. Indeed it would be against the national interest to swamp these by government action. While it would be tempting to manipulate the government-influenced demand for Q.S.E.s, in basic research for example, to overcome a shortage or a surplus, this would remove the signal that a shortage or a surplus existed and prevent the appropriate movement of rates of pay. While there is a risk that rates of pay and net recruitment might fluctuate too much and overcompensate for a shortage or a surplus, the rate of net migration of Q.S.E.s would help to absorb this.

(iv) As regards the Peck thesis itself as applied to the United Kingdom, we have confirmed, and possibly added to, the strength of what it has to say about the desirability of having more Q.S.E.s in the research-intensive industries. We differ from it, however, in the extent to which we believe that this should be brought about by reallocating rather than by expanding the number of Q.S.E.s available. Reallocation is called for wherever a case is made that some Q.S.E.s could be more productive in employments other than those they are in: but reallocation on a greater scale than this would stifle the incentive provided by excess demand for Q.S.E.s for the stock to expand to the desirable level. A case is made for reallocation out of the aircraft industry (as Peck suggests), but no case is made for reallocation out of basic research (as he also suggests). When we say 'no case is made' we mean it literally: there is no way we know of to measure the marginal productivity of Q.S.E.s in basic research. Until this can be done, it seems best to allow any excess demand for Q.S.E.s (or excess supply) which remains after we have done the reallocation for which a case can be made, to correct itself. A Briton's polemical interpretation of Peck's proposals might be that he wants not only to have our aircraft industry, but also, by removing any short-term excess demand for Q.S.E.s, to prevent us adding to their numbers. Although the former is difficult to gainsay, the latter we ought not to accept.

APPENDIX

	Basic Research as % of Total R & D Expenditure	% of Total R & D Performed in the Business Enterprise Sector	% of R & D Performed in the Business Enterprise Sector Financed by Business Enterprises
Netherlands (1964)	27	56	96
Belgium (1963)	21	69	103
Italy (1963)	19	63	98
France (1963)	17	51	65
United Kingdom (1964/65)	12	67	63
United States (1964)	12	67	48
Germany (1964)	n.a.	66	86
Japan	n.a.	65	100

Source: O.E.C.D.

Discussion of the Paper by Professor Sargent

Professor Tsuru said that the paper by Professor Sargent was a well-packed package of model reasoning and exposition and he could not hope to do full justice to it. He would therefore like to begin his comments with an analogy. Think of a college in the United States Ivy League, for example, trying to achieve an optimum distribution of athletic manpower among such areas as football, track and field, tennis, swimming, and all the other sports. Assume that the objective is to obtain a maximum amount of contributions from alumni who are assumed to respond on the basis of their appreciation of the relative strength of their Alma Mater within the Ivy League in several fields of activities, some valuing the football prowess highest and some the academic standing most. In this example, the athletic manpower corresponds to Q.S.E.s. Note that the question here is that of relative strength. There were three problems requiring a solution: the optimal allocation of the existing stock of athletic manpower between different spheres (we assume that when a boy is recruited into one of the athletic teams, he has to sacrifice his academic pursuits); the question whether the existing stock is itself of the right size – in other words, should we not recruit more into the athletic manpower? The first of these questions was answered by equating the 'marginal social product' (the capability of each athletic activity to elicit alumni contribution at the margin) of different activities; for example, it might be better to shift a boy from the football team to the track and field team for the reason that such a shift would not affect the Ivy League rank of the football team but would raise the rank of the track and field team; the second of the above questions was answered by comparing the gain through expanding the athletic manpower with the loss through the lowering of academic standards; a kind of overall Pareto optimum should thus be reached; there was a third problem of dynamic character – it was possible to recruit a good athletic prospect from high school by giving him financial aid; such an action would result in some gain for one of the athletic teams, but the gain had to be compared with the loss of tuition revenue; in addition, of course, there were questions related to optimal expenditures on athletic equipment, which were analogous to R & D expenditures in business activities.

Professor Tsuru said he had deliberately chosen this analogy in order to point up similarities and differences with, as well as certain difficulties in, Professor Sargent's endeavour to answer his own question.

(i) Most important: 'Boys do enjoy playing in the teams'. Or, put another way, their supply price was not related to any disutility of labour. The same would be the case, more or less, with the supply of Q.S.E.s. Here was the problem of producers' satisfaction which would upset the hedonistic calculation.

(ii) Depending on the type of athletic activities, the degree of indivisibility was too large to permit us to make any marginal calculation. Similarly with the Q.S.E.s.

(iii) Relative ranking was at issue, and not the absolute degree of improvement. Physical productivity improvement was specific to the nature of each industry. That was why we had a shift in relative price structure.

The point made was that 'relative ranking is at issue'. But how was it with Professor Sargent? Let us see how he set out his problem. What the did was to examine, both for industry in general and also for individual industries, the extent to which the growth of productive efficiency was associated with the industry's employment of Q.S.E.s, regarding such growth to be represented by the so-called 'residual factor'. Initial results were not very positive; but with more data and with further refinement, he had obtained some significant results. But prima facie we ought to expect a close correlation between Q.S.E.s and R & D on the one hand and the magnitude of the residual factor on the other. And supposing that the statistical results were most satisfactory, what did we prove? Could we say, as he had implied, that the growth in productive efficiency in the research-intensive industries would be enhanced if they employed more Q.S.E.s? A positive answer to this question would depend greatly upon the significance of the residual factor concept to which he proposed to return later. However, Professor Tsuru pointed out that even if we could obtain a positive answer to the above question, we still needed an answer to a more important question which Professor Sargent himself had raised in connection with the basic research manpower: whether Q.S.E.s marginal contribution to the social product was everywhere the same. This was the crucial question; and in the analogy of athletic manpower it could, in fact, be guessed at *ex ante* and could be definitely answered *ex post*. In the case of the Q.S.E. manpower, their social marginal benefit was impossible to calculate as Professor Sargent himself had admitted, and the question could be answered only by reference to the market, in other words, by reference to the relative performance of each industry in the world market. Such investigation, however, was outside the purview of the paper. There were further complications in the case of Q.S.E.s; that is, the 'ownership externality' character of Q.S.E.s – in other words, the service of Q.S.E.s need not be embraced within a particular firm or industry. From the social point of view, some waste would be involved if so embraced.

Professor Tsuru went on to discuss the methodological problems. Firstly, Professor Sargent's basic methodological framework was that of atomistic calculus (or accountability) with Pigovian modifications. It was highly questionable whether we could be successful in tackling this type of problem with such a framework; for the degree of indivisibility of the contribution of Q.S.E.s was large; marginal social gain was impossible to calculate even when the indivisibility problem was somehow circumvented; and the cost concept had to be qualified in a large measure because of the non-disutility character of Q.S.E. labour. The second methodological problem related itself to the meaning of the residual factor. Before we based ourselves on an analysis of the kind which Professor Sargent performed, we must know clearly what this residual factor really amounted to. The concept started with the well-known production function of the type:

$$R = Ae^{rt} \overset{\alpha}{K}L^{1-\alpha}$$

from which we derive (by letting m stand for Q/L and k for K/L):

$$g(m) = r + \alpha \cdot g(k)$$

where $g(m)$ denotes the rate of growth. r here is the residual factor; and we usually try to find the magnitude of the ratio: $r/g(m)$, which may by written as a, or:

$$a = 1 - \alpha \frac{g(k)}{g(m)}$$

Now, under perfect competition, we must satisfy the equilibrium condition:

$$\frac{cK}{\alpha} = \frac{wL}{1 - \alpha}$$

where c and w stand respectively for the unit cost of capital and labour. If α remains constant, this means:

$$g(k) = g(w) - g(c)$$

and if we can assume $g(c)$ to be zero or negligible, we may say:

$$g(k) \doteq g(w)$$

When α is constant, we can say that $g(w) = g(m)$ since $w/m = 1 - \alpha$. Therefore, we obtain a simple relation:

$$a \doteq 1 - \alpha.$$

This equality, or near equality, states that the contribution of residual factor turns out to be nothing but the expression of the relative share of labour. One need not suspect anything wrong in this relation under perfect competition. But it should be pointed out that a pertains to the real-physical aspect of an economy and α reflects the institutional or societal aspect. In the actual world, the latter was affected by all kinds of things, such as the degree of monopoly, the institutional characteristics as regards the ownership of the means of production, and what we, in fact, calculated was bound to be a figure compounded of these institutional factors and could hardly correspond to any precise measure in the technical aspect of our productive activities. For this reason, it would be extremely dangerous to theorise on the assumption that the calculated measure of the residual factor reflected the degree of contribution of technological progress.

Dr. Teubal referred to Professor Sargent's finding that a high ratio of Q.S.E.s on work other than R & D to Q.S.E.s on R & D tended to be associated with a low residual rather than a high one, and asked whether we could envisage an optimal distribution in Q.S.E.s between R & D and other activities? He said that Q.S.E.s on directly productive activities would enable industry to identify useful new technology arising from R & D in both their own and other sectors in addition to being directly productive. Hence a certain proportion of Q.S.E.s should be allocated to productive activity and if the actual ratio of Q.S.E.s in R & D to Q.S.E.s in productive activities was lower than optimal an increase in the ratio would lead to an increase in productivity, but if higher than optimal, raising it would reduce the growth of productivity. Under certain conditions this would mean that a schedule relating the magnitude of the residual with the ratio of Q.S.E. on R & D to Q.S.E.s on production would look like an inverted U-curve with a maximum level of the residual corresponding to the optimal ratio. Dr. Teubal also asked what Professor Sargent had in mind when he mentioned scale economies in the employment of Q.S.E.s. He wanted to know whether they were in R & D or in directly productive activity or in what combination between the two. Professor Sargent did not differentiate specifically between these categories. Dr. Teubal also referred to the validity of the marginal social productivity relation, as indicators of optimality in the allocation of scientific manpower. Suppose there was a certain stock of Q.S.E.s to allocate between activities at time zero and for simplicity one could not change this allocation in the future. If it was assumed that the objective was to maximise the value of the accumulative output of the sectors, it might very well happen that the accumulated output for a particular industry would be subject to increasing return; for example doubling the total Q.S.E.s might more than double the accumulated output of the sector. This fact did not depend upon the existence of increasing returns to employment of Q.S.E.s in production nor in R & D activity, but might depend upon the length of time horizon considered. Whenever this situation of increasing returns did exist we could not be sure that equating the marginal productivity of Q.S.E.s between sectors reflected a situation of maximum value of the accumulated output of these sectors. Such a position might represent a local maximum or even a minimum, in which case other criteria for optimal allocation should be found.

Professor Sargent replied to Dr. Teubal that he found at first sight his explanation of divergency from the optimum point of the U-shaped curve to be interesting, and it deserved thinking about. On the question of scale economies, when confronting residual factor estimates with Q.S.E.s industry by industry, it was not clear whether one should take the ratio of Q.S.E.s to total employed, that is the density, or the absolute number of Q.S.E.s. If the density was taken, then in setting up econometric regression one would expect the residual to be affected by the proportion of R & D. If one doubled Q.S.E.s and doubled the total employment, then it was probably better to look at absolute total. But both of these were obviously worth testing, as was done in this exercise. On the third point about whether employment of more Q.S.E.s would increase the size of the residual and the benefits would flow over a large number of periods, he agreed that benefits should be calculated as far ahead as one dared. But when they got up to 1985 they felt they were getting into higher flights of fancy. He accepted, however, that the point was right in principle.

Professor Rasmussen said that he had enjoyed the paper very much, but on one particular point he felt more extended discussion was needed. He considered that the conclusions of Professor Sargent in relation to the fundamental research sector of the United Kingdom were interesting but questionable. Professor Sargent had concluded that nothing supported the hypothesis that fundamental research was too great. Professor Rasmussen said that we all knew the United Kingdom was strong in this field and he did not see how one could reconcile this with the fact that in a number of cases the United States had simply taken over the results of the United Kingdom fundamental research and had turned it into useful technology. If this traffic became one-way, as it had been, across the Atlantic, one might argue that given the ability to utilise fundamental research the British had from a national point of view an imbalance between fundamental and applied research. It would probably be difficult and painful to redress this balance by cutting the fundamental research and therefore it might be fruitful to try to redress the balance by trying to change the abilities to utilise fundamental research, and for many reasons this was also a more dynamic approach. This could easily occupy a whole paper, not to mention a whole conference. But he felt the problem needed to be raised.

Professor Matthews said that apropos of Professor Tsuru's comments it was well accepted that the theoretical basis of the residual approach was weak in its assumptions about the determination of the factor prices. But he suggested that the question was whether more acceptable or complicated procedures would make all that difference to the result. He very much doubted it in this particular case. He felt some of Professor Sargent's conclusions were very interesting, but he wondered how far the correlations were due to a few extreme observations. With so few observations correlation coefficients could be misleading and scatter diagrams might be useful. Had Professor Sargent found this to be so? Professor Matthews said that the identification problem appeared throughout the paper. If one obtained a good correlation between a residual and the number of Q.S.E.s which way did the causation go? In certain industries there was a great scope for employing Q.S.E.s. Hence it did not necessarily follow that industries with few Q.S.E.s would get a bigger residual if they employed more, although this is certainly one possible interpretation.

Dr. Sedov said that he had read Professor Sargent's report with great pleasure. It would be very valuable to those responsible for the planning and allocating of scientists and engineers. He suggested that a very skilful diagnosis of the disease had been given, but how to treat the patient was the most important question. Professor Sargent's solution appeared to be a very pessimistic one – that the redistribution of the Q.S.E.s must be left to be identified and resolved by market forces. Dr. Sedov felt that this would lead to a waste of national resources. Market forces generally led not only to waste but to dependence of thousands upon anarchy, and he suggested that the mechanism of the allocation and reallocation of Q.S.E.s must not be done by the market but by planning. This was not just a hypothesis but a fact proved by the experience of many countries, and especially by the experience of the U.S.S.R.

Dr. Nabseth said he would like to raise a few points on the results of the paper. Firstly, Professor Sargent's point that some non-research-intensive industries had a high residual suggested that there were other unknown factors operating. But were they unknown? We had quite a lot of information on such things as the correlation between productivity and the increase in production volume, and Dr. Nabseth suggested that if we removed that factor the results might get us a little further. A second factor about which there were some results in Sweden was the competitive pressures of different industries. Information about this could often be derived from import and export shares. There were obviously other factors, but these were two which might be taken into account, and which were not really unknown. On the question of international comparison, Dr. Nabseth said he was puzzled by some of the figures. He was curious to know whether one should expect a correlation between number of Q.S.E.s and the rate of growth of labour productivity over a number of years. Was it not true that there were wide differences of salary levels and wage levels? Would the high wage level companies have more Q.S.E.s? If so, it would be better if a correction could be made for this in the regression. On the question of private industry and the value of government coming in, Professor Sargent had suggested that if private firms decided it was not worth going ahead it was not worth the government doing it either. Dr. Nabseth suggested this might not be true because the disutilities that the private firms had might not be those we ought to have from the social point of view. For example, if private firms have a risk aversion, they might not undertake work. But from a social point of view it might be well worth doing it. In the Swedish aircraft industry they would not get projects going if the company had to finance them, since they were too big for the company, although they had been successful with government support.

Professor Griliches said he would like to comment on the use of the residual in this type of study. The residual was the result of subtracting a weighted measure of the rate of growth of input from some measure of output. There were several problems in doing this. The more obvious ones were the changes in the skill composition of the labour force. He suggested that one should look at the rate of growth of Q.S.E.s not just at their levels. On the capital side he was not sure how the figures had been manufactured. From the point of view of the model stock figures were not correct. The appropriate measures should be based on service flows with individual items weighted by rents and not capital values. Professor Griliches also said that the depreciation components which were often borrowed from accounting records were of dubious relevance, and it was not clear how quality was evaluated at different points in time. It would be interesting to know more about how the capital figures for different industries had been constructed. As to the regressions themselves, one could usually get hold of only about ten to twenty industries with useful data, and the sample tended to be of R & D intensive industries with results that were probably applicable only to that sector. This was not a very large part of the total economy, and one could not necessarily extrapolate these figures to total G.N.P. growth. He went on to say that one of the more interesting findings by Professor Sargent was the superiority of the absolute versus the relative measure of scientists. In principle the absolute measure made a lot of sense because technology was largely scale-free. However, he would be happier with the conclusions if he understood more about where the industry boundaries had been set. The definition of an industry was often quite arbitrary. Some might be more heterogenous than others and this was a real problem in any context. When one talked about scale it was necessary to have the right units. Returning to the data, Professor Griliches said that he had also found the aircraft industry in the United States an outlier, with a small growth in the residual related to a high level of R & D, but he had interpreted this to mean that the way output was measured in the aircraft industry was not adequate. In the United States 86 per cent of R & D was government financed and sold to government on a cost basis, and he said that he knew of no measure of physical output which would capture such things as the change over from propeller to jet aircraft. He said that he was not sure that the conclusion followed that engineers were less productive in the aircraft industry.

Professor Robinson asked whether Professor Sargent's analysis assumed that the productivity of all individual Q.S.E.s was equal, and that one could measure the input of Q.S.E.s by measuring the numbers. If so, was this a good assumption? Professor Robinson said that when he with others had been involved in allocating government subsidies and grants in the United Kingdom to government research establishments and industrial research, a number of them went round to inspect them, and many of his scientific colleagues were very worried about the quality of the research staff, and particularly of the quality of the leadership. He said that they believed at that time that there was an excessive concentration of the real first-raters in prestige industries, such as atomic research, aeronautics, chemicals and electronics. Equally worrying was the feeling that these first-raters were attracting the younger research workers out of the universities, and that they were getting a polarisation of all first-class workers in prestige industries. This was leading to a greater discrepancy than could be measured by numbers. Professor Robinson wondered whether statistically one could allow for this? He also asked whether one could remedy this – for example by facilities for transferring researchers from one institution to another. He felt that it would often be the case that it would be the poorer scientists who would be transferred rather than the better ones.

Professor Williams remarked that although Professor Sargent set out to test a thesis put forward by Melvyn Peck, he was in fact testing a general thesis expounded in the book on *Industry and Technical Progress* written by Professor Carter and himself about fifteen years ago. Professor Sargent's tests appeared to throw considerable doubt on some of their conclusions. His conclusions on 'the prior importance of Q.S.E. manpower over R & D expenditure in general', were consistent with the Carter-Williams study but not his conclusions on Q.S.E.s outside R & D. Though his tests cast some doubt on previously held theses, the degree

of doubt was not clear. He doubted whether Professor Sargent's tes
could settle the issue. The Carter-Williams conclusions had come out o
detailed case-studies and were intra-industry and not inter-industry tests
Unless there was a common pattern relevant to different industries, he
felt intra-industry tests were what one needed. The need to test in individual
industries rather than across industry led on to a general question of what
it was we should expect to find. What was a plausible type of relationship?
He did not think it was plausible that the optimal distribution of scientific
manpower would be the same between industries – a point made by Dr.
Teubal earlier. If one looked at Professor Sargent's sample there was an
important difference in the way in which R & D could be separated from
other parts of the organisation, for example production and marketing.
In some industries it was possible to segregate research without large
communication losses. In others this was not so. He had no doubt that
in Britain a substantial amount of research concentrated in government
establishments and separated from production and marketing was, in
the final outcome, inefficient. Professor Williams referred to a number of
other important points – he said that neither Peck nor Carter and himself
had argued that the optimal scatter of Q.S.E.s between different areas of
the firm would lead to a particular result. He and Carter argued this as a
necessary, but not a sufficient, condition. One thing that came out clearly
from their case studies was that there was a process of historical growth
in different firms and different industries, and in one period it might be
necessary to have more people concentrated in research than at other
periods. It was also obvious that the way management went about utilising
these people in their different activities was of critical importance. They
had also found in their industry studies that it was very helpful to work
with three concepts: firstly, opportunity to innovate – which might have
to be created within the industry but in other cases might be created outside; secondly, the capacity to use the opportunities – and here the distribution of scientific manpower and the role of scientific management in
decisions in the firm became critical; and thirdly, the pressure to use these
opportunities.

Professor Oshima said that he had found great interest in the paper.
In the early stages the United Kingdom had been the ideal pattern for
Japan, with a concentration in certain fields, and with a number of national,
government-sponsored projects. But Professor Sargent was now saying
that this level of government sponsorship had negative effects on economic
growth. He would therefore like to ask whether from this study one could
establish any quantitative criteria which would enable one to say to what
extent national programmes of development would interfere with the
private sector. Government projects could have a strong effect on high
level technology. But they also obviously had an impact on the allocation
of manpower and of real investment. Was there any possibility that this
sort of study could provide evidence to show how these factors should be
balanced?

Professor Sargent began his reply to the points raised in the discussion
by saying that he was in complete agreement with many of those made on

methodology. He recognised the limitations and accepted that further work was necessary. The criticism of the methodology seemed to hinge on two points, with both of which he agreed, but in different ways. Firstly as regards the application of atomistic calculus, the section in which he had deployed this was put in to set out how ideally an economist trained in the Western tradition might expect a systematic solution to be arrived at. He had used this to demonstrate that there was a considerable gap between the theoretical analysis and technical possibility. The early sections dramatised the contrast between what marginal analysis would suggest and what our limited methods of dealing with them would allow us to do, and in particular what statistics were likely to be available. The paper had attempted to make as many bricks as possible with the limited statistical straw available in the United Kingdom. It was largely written when he was working in a government ministry and was orientated towards the ears of civil servants and its academic limitations were consequently obvious. He said that his experience was that civil servants, who must advise ministers on policy, were prone to pick up simple ideas from the writings of economists, and it was useful to put some quantitative tests on these ideas.

Professor Sargent said that a number of people had criticised the use of the residual factor and it was difficult not to share some of these criticisms. He said that he must express thanks for the help he had had from Professor Matthews who had come out in defence of the use of the residual, although Professor Sargent said he was not entirely swayed by this. He said that his major criticism was that the residual was based on estimates of capital stock which depended on assumptions of life of capital assets, and he believed that this was shortening in the United Kingdom. But having made the criticisms of the residual factor, what else do we have? It was still the best approximation, better than the growth of labour productivity statistics, and we should not refuse to use the material because of criticism but recognise that the results might be overturned when better measures are available. On Professor Matthew's point about the small number of observations in the sample, Professor Sargent said that they had looked at scatter diagrams and some industries did stand out – this did not substantially alter the results, but the tables were available if people were interested. On the identification problem he said the next stage was to construct a model and to put in what determined the number of Q.S.E.s employed. Just before giving the paper he had thought that in research-intensive industries the relationship between the number of Q.S.E.s and size of residual was in fact a definition of a research-intensive industry and no more. On Professor Griliches' remarks made from the vantage point of a statistically highly-developed country Professor Sargent said it would be nice to do what he suggested, particularly with regard to the rate of change of level of Q.S.E. as distinct from the absolute size. However, to take this change into account one would have to reduce the number of observations by half and then the sample would dwindle to almost nothing. Professor Sargent said that Professor Matthews had dealt with the capital stock figures and that he himself was working on the calculation of growth of productive efficiency without the dependence on capital stock figures, but had no success to report as yet. On the output in industry such as aircraft he did not think we did this any better in the United Kingdom than in the United States. On Professor Rasmussen's point about basic research in the United Kingdom and why he had not specifically considered this point in the paper, Professor Sargent said that his purpose was to suggest there was no firm basis for the proposition that the input of resources into basic research was out of line in the United Kingdom. He said we would need to have better support than we yet had for the suggestion that the United Kingdom spends too much on basic research. On Dr. Nabseth's point about the other forces unknown, Professor Sargent agreed that his points were well worth investigating, and on Dr. Sedov's point about his pessimism in using the role of market forces he said he was not an out-and-out proponent of market forces but he did think that if the government used the existence of excess demand or supply for Q.S.E.s to step in and use its power to adjust supply and demand, it might remove some of the material incentives which must go together with planning forces to achieve the optimum results. He thought it was important they should not be removed completely by the government interference. On Professor Robinson's point about the quality of Q.S.E.s Professor Sargent said he again had to fall back upon a major defence of the relative lack of data at the level he was dealing. He said this type of analysis could not get to the bottom of this problem. He suggested that in addition to number of Q.S.E.s one might add measures of salary of Q.S.E.s and this might be a starting point, but at the moment one could not get sufficient data unless one took a much more micro approach, which led to Professor Williams' point about the kind of contributions which this type of aggregated analysis could make to the solution of these problems. Professor Sargent said that this work must be used in combination with studies such as that by Professors Freeman, Carter and Williams at the systematic micro level. He said that neither the aggregate nor the micro approach alone would get to the bottom of the problem. Micro studies needed to be disciplined by what quantitative relationships you would expect to find. He suggested that his analysis needed discipline from the qualitative studies and the two could help each other.

15 Brain Drain: Causes and Effects[1]

Amartya Sen
UNIVERSITY OF DELHI

THE PROBLEM

When in A.D. 1600 the question was asked in Oxford University's Arts Seminar: 'An peregrinatio conducat ad philosophandum?' ('Does migration stimulate philosophical thinking?'), the precaution was taken of requiring the students to answer the question in the affirmative.[2] The problem of 'brain drain' has not been viewed in quite such unequivocal terms in the recent literature on economic development. This is not surprising since a systematic migration of a large part of the skilled and technologically sophisticated labour force from an under-developed country would indeed pose a serious challenge to the economic, technological and scientific development of such a country.

The object of this paper is twofold. In Part I of the paper the *effects* of brain drain are discussed and some models that have been proposed in the literature are examined. The results, I fear, are largely negative. In Part II the *causation* of brain drain is discussed in terms of a cross-country study of emigration of natural scientists, social scientists, engineers, and doctors and dentists, from a large number of under-developed countries into the United States.

I. EFFECTS OF BRAIN DRAIN

There are at least two different ways of measuring the cost of emigration of skill for a country: (i) calculating the cost of doing without the emigrating skilled labourers, and (ii) estimating the cost of replacing the emigrants by similarly trained persons. When both the alternatives are open, the relevant cost is the minimum of the two. This is a reasonable assumption in the long run, unless there are

---

[1] This paper incorporates some of the materials used in an earlier unpublished study, 'The Brain Drain and the Production Function', which I completed during 1968–9 for the Harvard University Project for Quantitative Research in Economic Development. It also uses the statistical findings of a report I prepared for the United Nations during 1968 entitled 'A Quantitative Study of Brain Drain from the Developing Countries to the United States', to be published in the U.N. *Journal of Development Planning*. For helpful discussions I am most grateful to T. C. Chang, Hollis Chenery and Harry Johnson.

[2] See Dedijer (1968), quoting Clark (1887).

non-economic constraints that rule out possibilities of replacement by training. In the short run, however, replacement is not possible, and (i) represents the relevant cost.

It is usual to assume in rather informal studies that there is a significant loss of income per head (of those remaining behind) as a consequence of emigration of skill. As Kenneth Boulding puts it: 'If, however, he migrates, the society which raised him loses a valuable piece of quasi-property without any way of collecting the invisible debt which the migrant has incurred.'[1]

This position has, however, been recently challenged in two closely reasoned articles on brain drain, viz., Grubel and Scott (1966) and Johnson (1967). The position taken in the former paper is qualitative, while the latter paper is concerned primarily with quantitative questions.

The Grubel-Scott approach

While Grubel and Scott (1966) note the loss of 'military and economic power' in skill emigration, they doubt the importance of its welfare impact:

> In a market economy where persons are paid their marginal products, however, such a reduction in per capita income is only a statistical phenomenon which has no influence on the welfare of the remaining people: the emigrant removes both his contribution to national output and the income that gives him a claim to this share, so that other incomes remain unchanged.[2]

If E_i is the number of people of skill i, Q the national output level, and w_i the wage rate of skill i, then we have, by courtesy of the marginal productivity theory of distribution:

$$w_i = \frac{\delta Q}{\delta E_i} \tag{1}$$

The right-hand side of (1) represents the marginal loss and the left-hand side the marginal gain (saved payments) from losing a skilled labourer of type i. Assuming that (a) the marginal productivity theory holds, (b) social marginal product equals the private marginal product, and (c) there are no adjustment costs, clearly a marginal loss of skilled labourer should leave the income of those remaining behind quite unchanged.

Grubel and Scott, therefore, take the position that 'in a market economy any effects that the emigration of a highly skilled person is

[1] Boulding (1968), p. 114. See also Patinkin (1968).
[2] Grubel and Scott (1966), p. 270.

likely to have on the welfare of those remaining behind must be sought either in short-run adjustment costs or in market failures'.[1] Let us leave out, for the moment, market failures (e.g. externalities) and short-run adjustment problems ('unemployment or inefficient employment of factors of production whose effectiveness depends on cooperation with the skills the departing person takes along'). Under these circumstances would skill emigration not affect the welfare of those remaining behind? While that is what the Grubel-Scott argument may appear to imply, this is by no means the case. Strictly speaking, it is the case only for a very special type of production, viz., one with infinite substitutability.

The Grubel-Scott argument holds perfectly well for the emigration of a marginal man. But who is really interested in a one-man brain drain? Would the no-loss argument hold if the emigration is substantial rather than marginal? The answer is that it certainly would not, unless substitution possibilities were infinite. Propositions that are true of marginal changes frequently do not hold at all for substantial changes.[2]

When the availability of skilled resource i is reduced from E_i^0 to E_i^1, given the other factors of production, the loss of income of those remaining behind equals:

$$\lambda = \int_{E_i^1}^{E_i^0} \left(\frac{\delta Q}{\delta E_i}\right) \, dE_i - w_i(E_i^0 - E_i^1). \qquad (2)$$

Assuming that $\frac{\delta Q}{\delta E_i}$ is a monotonic function of E_i, and given marginal productivity equilibrium with $\frac{\delta Q}{\delta E_i^0} = w_i$, the necessary and sufficient condition for λ to be zero, is simply:

$$\frac{\delta Q}{\delta E_i} = \text{constant} \qquad (3)$$

That is, the Grubel-Scott argument will hold for substantial changes if and only if the marginal product of skilled labour of each category is a constant.[3] A production function will have this characteristic if it

[1] *Ibid.*

[2] Compare: 'A monopolist's profit is invariant with respect to changes in his output'. True for marginal changes at the point of profit maximisation (given smoothness), but in general, of course, totally false.

[3] Assumptions (a), (b) and (c) noted earlier are also, of course, profoundly doubtful. Grubel and Scott discuss the extent of possible errors in the last two assumptions, and throw much light on the relevant considerations. See also Weisbrod (1966), Aitken (1968) and Grubel and Scott (1968).

388   *Science and Technology in Economic Growth*

incorporates infinite substitution possibilities, e.g. $Q = \sum_i a_1 E_i$, where a_i is a constant (marginal product) corresponding to resource i. It is, to say the least, a rather special production function.

The Johnson approach

Harry Johnson, in contrast, is concerned explicitly with substantial changes and not merely with purely marginal ones, and specifies his production function fully. His is the first complete statement of the problem in the literature of brain drain.

Johnson (1967) takes a Cobb-Douglas production function, which we can write (denoting Q = output, L = labour, K = capital, and α = a constant, $0 < \alpha < 1$) as:

$$Q = L^\alpha K^{1-\alpha} \qquad (4)$$

Further with n = number of unskilled workers, N = number of skilled workers, k_1 = ratio of material capital to unskilled labour, k_2 = ratio of human capital to the raw labour embodied in skilled labour, he obtains:

$$Q = (n + N)^\alpha (k_1 n + k_2 N)^{1-\alpha} \qquad (5)$$

Originally the unskilled labour force is n_0 and the skilled labour force N_0. Assuming that there is a decline in the skilled labour force by a fraction m, then the proportional loss of the income of those remaining behind will be *approximated* by the following,[1] denoting $k_0 = k_2/k_1$, and $r_0 = \dfrac{N_0}{N_0 + n_0}$:

$$g = \alpha(1 - \alpha)m^2 \left[\frac{r_0(1 - r_0)^2 (k_0 - 1)^2}{(1 + mr_0)(1 - r_0 + (1 + m)k_0 r_0)(1 - r_0 + k_0 r_0)} \right] \qquad (6)$$

Further it is clear that $\alpha(1 - \alpha)$ must reach a maximum for $\alpha = 0.5$, and $\alpha(1 - \alpha)$ must be equal to or less than 0.25. This expression g will generally yield a small fraction as the proportionate loss from skilled immigration. Johnson considers $k_0 = 2$, $r_0 = 0.2$, and $m = 0.1$, which with $\alpha = 0.5$ yields $g = 0.0217$ per cent. A 10 per cent emigration will affect output by only about 0.02 per cent. Even if $m = 0.2$, we have $g = 0.0826$ per cent. The results are striking, and Johnson considers that the loss 'is likely to be a very small fraction in all but exceptional circumstances'.[2]

Is this convincing? I shall argue that it is not. I shall do this first

[1] Strictly, Johnson calculates the proportionate increase in output as a result of skill immigration. For small variations, the proportionate decline figures will be the same except for the sign.
[2] Johnson (1967), p. 409.

in a seemingly perverse way, viz., by strengthening Johnson's results. What if every skilled and educated person leaves the under-developed country? I would surmise that the actual result would be a catastrophe, but what would happen in the Johnson model? After obtaining the answer to this question, the reasons for the results will be examined.

The initial and final (i.e. after total emigration) total outputs are given respectively by Q_0 and Q_1:

$$Q_0 = (n_0 + N_0)^\alpha (n_0 k_1 + N_0 k_2)^{1-\alpha}, \qquad (7)$$

$$Q_1 = (n_0)^\alpha (n_0 k_1)^{1-\alpha} \qquad (8)$$

The income of the non-emigrant part of the nation prior to the departure of their skilled comrades is given by Y_0, denoting $\mu = \dfrac{N_0}{n_0}$:

$$Y_0 = Q_0\alpha\left(\frac{n_0}{n_0 + N_0}\right) + Q_0(1 - \alpha)\left(\frac{n_0 k_1}{n_0 k_1 + N_0 k_2}\right)$$

$$= Q_0\left[\frac{\alpha}{1 + \mu} + \frac{1 - \alpha}{1 + \mu k_0}\right] \qquad (9)$$

Their income, when left in the lurch, is given by:

$$Y_1 = Q_1 = Q_0\left(\frac{1}{1 + \mu}\right)^\alpha \left(\frac{1}{1 + \mu k_0}\right)^{1-\alpha} \qquad (10)$$

The loss of income is given by:

$$\lambda = Y_0 - Y_1 \qquad (11)$$

And the proportionate loss is:

$$\rho = \frac{\lambda}{Y_0} = 1 - \frac{(1 + \mu)^{1-\alpha} (1 + \mu k_0)^\alpha}{\alpha(1 + \mu k_0) + (1 - \alpha) (1 + \mu)} \qquad (12)$$

In the previous example of Johnson we had $k_0 = 2$, $\alpha = 0.5$, and $\mu = 0.25$, the latter corresponding to Johnson's $r_0 = 0.2$. This yields:

$$\rho = 0.0040$$

That is, in this model when every skilled person leaves the under-developed country, leaving behind a land of totally unskilled labour and physical capital, the income of those remaining behind goes down by a mere 0·40 per cent! I would doubt that this is believable.

Johnson uses a Cobb-Douglas production function because it 'can claim traditional authority and empirical support'.[1] This involves a unit elasticity of substitution and he warns that 'this may

[1] Johnson (1967), p. 407.

be too low, overstating the loss from emigration to those remaining behind'.[1] The source of the problem seems to be not so much the unit elasticity of substitution between capital and labour, but the implicit infinite elasticity of substitution between physical and human capital. The capital figure is simply the sum of physical and human capital, and this makes sense only if the two are perfectly substitutable. This is responsible for the striking result.

Consider a variation of the production function. Suppose we take a proper three-factor Cobb-Douglas production function with unit elasticity of substitution between each pair of factors.

$$Q = n^\alpha K^\beta N^{1-\alpha-\beta} \tag{13}$$

Will this make a difference? I should say it would. As all skilled labourers emigrate and N becomes zero, output will be zero too. Rather than a decline of 0·40 per cent, the income of those remaining behind will go down by 100 per cent. The same will be true of other 'well-behaved' production functions, e.g. 'constant elasticity of substitution' production functions, except for the special case of infinite elasticity of substitution.

Johnson's assumption of infinite substitutability between physical and human capital is indeed crucial. It will be particularly open to objection when the emigration of skilled labour force is a substantial proportion of the total skilled labour force, and as such our use of the Johnson model for a 100 per cent emigration, it must be admitted, is more open to objection than Johnson's own use of it. Nevertheless, the assumption is very special in either case and the results are sensitive to it.

The use of the production function in these exercises requires extreme caution. I am not concerned here with general analytical problems in the conception of an aggregate production function, which have been discussed elsewhere and at length, since analytically fuzzy tools might nevertheless yield reasonably good predictions. What one is concerned with here is the much more mundane problem of getting predictions approximately right. It is not at all clear that at the present state of our knowledge of the role of skill in production function that we can really make a profitable use of the tool of an aggregate production function to estimate the impact of skill emigration on those remaining behind. Even in the general family of Cobb-Douglas (I might add that this is a rather restricted family), it makes a difference whether we take a 2-factor Cobb-Douglas like (5) or a 3-factor Cobb-Douglas like (13). The difference in the case discussed may be between a 0·4 per cent reduction in income of those remaining behind and a 100 per cent reduction.

[1] Johnson (1967), p. 407.

Production function and welfare

Suppose, however, we overlook this problem. Imagine that we know what production function to use (and that there is a right production function to use), and suppose we find, like Johnson, that the loss from skill emigration will be small. What does this mean in terms of world welfare? What is the criterion of smallness?

As Johnson (1968) has rightly pointed out, the interesting approach to the question of skill emigration is an 'internationalistic' one, and must take note of the 'welfare of people born in that region who choose to leave it, and the welfare of the outside world in general'.[1] Let N^1, N^2, and N^3 be respectively the number of those emigrating, those remaining behind in the country of emigration, and the inhabitants of the country of immigration. Let the *per capita* income levels of these three groups before and after the migration respectively be Y_0^1 and Y_1^1, Y_0^2 and Y_1^2, and Y_0^3 and Y_1^3. Let $U(Y)$ be a function of *per capita* utility related to *per capita* income. Then the net change in welfare as a consequence of the migration is given by:

$$\Delta W = N^1[U(Y_1^1) - U(Y_0^1)] + N^2[U(Y_1^2) - U(Y_0^2)] \\ + N^3[U(Y_1^3) - U(Y_0^3)] \quad (16)$$

Suppose we know all about production functions and other factors determining income changes and can calculate all Y_j^i. Can we say anything very interesting on the basis of the general properties of a welfare function without knowing its exact shape? The only property that is generally recognised is the concavity of the welfare function, i.e. for any Y and Y', and any $a: 0 < a < 1$,

$$aU(Y) + (1 - a) U(Y') < U(aY + (1 - a)Y') \quad (15)$$

The income levels will have typically the following properties:

$$Y_0^1 > Y_0^2 \quad (16a)$$
$$Y_0^2 > Y_1^2 \quad (16b)$$
$$Y_0^3 > Y_0^2 \quad (16c)$$
$$Y_1^1 > Y_0^1 \quad (16d)$$
$$Y_1^3 > Y_0^3 \quad (16e)$$

The migrants gain (16d), those left behind lose (16b), and those in the country of immigration may gain (16e) from the transfer. Further, immigrants tend to be richer than those left behind, even before emigration (16a), and the people of the immigrating country are richer than the people of the country of emigration (16c), in a typical 'brain drain' case.

Grubel and Scott (1966), Johnson (1967), and others argue that

[1] Johnson (1967), p. 379.

the loss of those left behind (given by (16b)) will be small. Small in what units? The calculations refer to income units. Can we say anything very much about the welfare loss? The answer is that without specifying the welfare function we can never say it is too small, but we can sometimes say that it is too large. The directional asymmetry arises from the concavity of the utility function. We state now these two results more formally.

T.1. For any choice of finite income figures Y_j^i satisfying (16), and any set of finite numbers $N,^1$ N^2 and N^3, there exists a concave welfare function that yields $\Delta W < 0$.

By (16), Y_1^2 is the minimum Y_j^i. It is obvious, therefore, that there exists a concave welfare function such that:

$$U(Y_0^2) - U(Y_1^2) > \left(\frac{N^1}{N^2}\right) [U(Y_1^1) - U(Y_0^1)] + \left(\frac{N^3}{N^2}\right) [U(Y_1^3) - U(Y_0^3)] \quad (17)$$

T.2. For some choice of finite income figures Y_j^i satisfying (16), and some set of finite numbers N^1, N^2 and N^3, there exists no concave utility function that yields $\Delta W > 0$.

PROOF: Consider

$$(Y_0^2 - Y_1^2) > (Y_1^1 - Y_0^1) \left(\frac{N^1}{N^2}\right) + (Y_1^3 - Y_0^3) \left(\frac{N^3}{N^2}\right) \quad (18)$$

Given (15), i.e. concavity, and (16), i.e. the income rankings, Min $Y_j^i = Y_1^2$, and for $(i, j) \neq (2, 1)$, Min $Y_j^i = Y_0^2$, we conclude from (18):

$$U(Y_0^2) - U(Y_1^2) > [U(Y_1^1) - U(Y_0^1)] \left(\frac{N^1}{N^2}\right) + [U(Y_1^3) - U(Y_0^3)] \left(\frac{N^3}{N^2}\right) \quad (19)$$

Hence for all U, $\Delta W < 0$.

Is this an analytical trick of some kind to trap people? The results follow simply from the concavity of the welfare function, i.e. diminishing marginal utility, and the fact that the people of the under-developed country (from which brain drain occurs) are poorer than the skilled migrants as well as poorer than the residents of the advanced country to which immigration occurs. The former, expressed in (15) is a reasonable judgement (and one that is most widely used), and the latter, expressed in (16), is a factual condition that defines precisely one reason why people are concerned about brain drain. There is, I would argue, no trick here. Magnitudes of loss to the poor country can hardly be shown to be small without bringing in the utility function, but it might conceivably be shown to be large even without specifying a utility function as long as we know that there is diminishing marginal utility. The asymmetry is germane to the problem of brain drain, and is a consequence of systematic inequality of income.

The contributions that indicate that the loss to the under-developed countries from brain drain is small are, therefore, open to question not merely for their production function assumptions (infinite elasticity of substitution between skill and all other factors in the Grubel-Scott model and that between physical capital and human capital in the Johnson model), but they also suffer from a lack of a criterion of smallness. Indeed, given the concavity of the welfare function largeness is demonstrable in purely income terms in a manner that smallness is not. Johnson made a major contribution in posing the problem in an 'internationalistic' framework, but this has to be done in terms of social welfare and not merely physical income.

II. IMMIGRATION OF SCIENTISTS, ENGINEERS AND DOCTORS INTO THE UNITED STATES FROM UNDER-DEVELOPED COUNTRIES

After a fairly negative Part I dealing with effects of brain drain, Part II will be concerned with a positive analysis of causation. Immigration of skilled labour into the United States is an interesting field of study for at least two reasons. First, the United States has certainly been the largest recipient of educated and skilled labour from abroad in recent years. Second, the United States data on immigration, while deficient in certain respects, are certainly better than that of any other country, especially in providing a matrix of intercountry-interskill breakdown of immigration.

In Table 15.1 the immigration of four categories of skilled man-power, viz. engineers, natural scientists, doctors and dentists, and social scientists, into the United States are presented for the fiscal years 1966 and 1967. The source is a special report submitted in a hearing before a subcommittee of the Committee on Government Operations, Ninetieth Congress, viz. United States (1968).

The number of countries covered is fifty-one, with eight from Africa, sixteen from Asia and the rest from Latin America, though because of data limitation not all of them could be included for the study of each skill. One of the first things to check is the extent to which this immigration pattern can be explained in terms of the quotas of the developing countries for United States immigration. The quota system was in the process of being liberalised at that time but did exist except for Latin American countries.

In Table 15.2 the actual quotas of the Asian and African countries are presented next to the figures of actual immigration into the

TABLE 15.1

COUNTRY BREAKDOWN OF IMMIGRATION OF
ENGINEERS, NATURAL SCIENTISTS, DOCTORS AND
DENTISTS, SOCIAL SCIENTISTS, FROM UNDER-
DEVELOPED COUNTRIES TO THE UNITED STATES
DURING FISCAL YEARS 1966 AND 1967

Country	Engineers		Natural Scientists		Doctors and Dentists		Social Scientists	
	1966	*1967*	*1966*	*1967*	*1966*	*1967*	*1966*	*1967*
Algeria	0	0	0	1	2	3	0	0
Ethiopia	0	3	0	0	4	1	1	0
Ghana	1	4	0	4	0	8	0	1
Kenya	2	6	0	2	1	1	0	2
Morocco	2	5	0	2	2	2	0	0
Nigeria	4	17	1	4	2	0	0	0
Tunisia	0	0	0	1	2	3	0	0
U.A.R.	9	33	13	20	25	26	2	5
Burma	11	15	1	7	5	5	1	3
Taiwan	172	922	45	348	17	51	8	39
Hong Kong	121	227	47	99	29	49	3	9
India	645	1067	209	269	42	89	21	40
Indonesia	4	11	3	5	0	6	0	3
Iran	86	133	13	28	80	125	3	8
Iraq	16	36	14	12	5	12	1	2
Israel	61	116	23	47	32	43	11	13
Korea (South)	53	108	51	86	35	75	15	37
Lebanon	26	41	13	24	16	41	5	2
Malaysia	9	8	5	9	5	6	0	0
Pakistan	21	57	8	17	11	14	1	3
Philippines	86	348	31	106	280	612	10	15
Syria	9	17	30	3	9	7	2	1
Thailand	4	13	0	1	11	8	1	1
Vietnam (South)	4	6	0	3	1	2	0	2
Argentina	59	92	31	41	126	133	7	10
Bolivia	7	16	5	4	22	20	1	2
Brazil	35	54	19	15	42	26	3	6
Chile	16	20	3	4	12	6	3	0
Colombia	85	73	30	15	90	122	17	8
Costa Rica	8	6	3	3	4	3	0	0
Dominican Rep.	38	17	12	16	65	39	1	2
Ecuador	8	16	6	4	27	19	0	2
El Sadvador	4	2	1	2	4	5	0	2
Guatamala	4	3	2	1	8	8	2	1
Haiti	10	13	4	4	35	41	1	0
Honduras	6	3	2	0	5	6	2	0
Mexico	57	61	36	34	128	93	5	14
Nicaragua	1	1	3	1	6	9	0	0
Panama	4	2	2	3	3	1	1	2
Paraquay	3	1	2	2	6	9	0	0
Peru	13	22	8	3	47	29	1	1
Uruguay	8	3	1	1	9	8	0	0
Venezuela	23	22	6	10	14	17	2	3

United States of engineers, natural scientists, doctors and dentists, and social scientists. It should be quite clear that the pattern does not at all relate closely to the quotas and as a hypothesis the quota has little explanatory power in showing the pattern of immigration of skilled manpower into the United States.

TABLE 15.2

ANNUAL QUOTAS FOR UNITED STATES IMMIGRATION
FOR AFRICAN AND ASIAN COUNTRIES

Country	Annual Quota	Actual Immigration of Engineers Doctors and Dentists, Natural Scientists and Social Scientists in 1966 and 1967
Algeria	574	6
Ethiopia	100	9
Ghana	100	18
Morocco	100	14
Tunisia	100	6
U.A.R.	100	133
Burma	100	48
Taiwan	100	1,602
India	100	2,382
Indonesia	200	32
Iran	100	476
Iraq	100	98
Israel	100	346
Korea (South)	100	460
Lebanon	100	168
Pakistan	100	132
Philippines	100	1,488
Syria	100	78
Thailand	100	39
Vietnam (South)	100	18

Two other hypotheses that are found to be not at all useful relate to: (i) the income difference between the country of origin and the United States, and (ii) cost of transportation involved in the movement. It is not possible to obtain exact salary levels of different skills for many countries, but it can be assumed that the national income of the country is a reasonable proxy. Indeed, some facilities that an educated person will enjoy in an under-developed country will depend not merely on his own salary level but also on the level of prosperity of the country as a whole for which the national income per head is indeed a reasonable indicator. It might be anticipated that a bigger difference in income between the country of origin of the skilled labour and the United States will provide a stronger incentive for moving to the United States, since there is a bigger gain to be made by this movement. Similarly, it might be thought that a higher cost of transportation will discourage movement and it is, therefore, interesting to relate the pattern of immigration to the airfare between the main emigrating point of the country of origin and some spot in the United States, for which Chicago was chosen.

Without reporting in detail on the results, it might be mentioned that the pattern of correlation between national income and immigration and that between airfare and immigration is found to be extremely weak and sometimes even mildly negative. The relations are not significant in terms of any exacting standard and there is little value in the two hypotheses quoted either on their own or combined with other hypotheses. Incidentally, since distance correlates very well with airfare, it also appears that countries far away are no more immune from the pull of America in spite of the usual belief that cultural influences relate to physical closeness.

The finding that larger income differences do not seem to lead to greater brain drain is interesting, but it must be interpreted cautiously. It does not necessarily imply that higher income is not an important motive in brain drain, though it does not rule out that possibility. The United States salary levels being what they are, the movement from any under-developed country to the United States involves a stunning rise in income, and whether this rise is five times or eight times the salary enjoyed at home may not make much difference. The lack of a relation between immigration and income difference may be due to this fact.

It might be thought that the number of students graduating in the country of origin will provide a clue to the pattern of immigration into the United States, and to some extent this expectation is confirmed. But the really remarkable relations that are found relate to the number of students from the country in question studying in the United States in the years preceding the immigration years.

The numbers of students graduating in the country of origin in 1964 were obtained from UNESCO (1967). The numbers of foreign students from different under-developed countries doing graduate studies in the United States in different disciplines during academic years 1963–4, 1964–5, 1965–6 and 1966–7 are obtained from the Institute of International Education (1964), (1965), (1966), (1967). In Tables 15.3, 15.4, 15.5 and 15.6 the regression results for the four skill groups are presented. Immigration is related to the number of students graduating in the country of origin as well as to the number doing graduate studies in the United States during the four preceding academic years.

It would appear that for all cases other than that of doctors and dentists a remarkable amount of explanation is found in relating

TABLE 15.3

REGRESSIONS INVOLVING IMMIGRATION OF
NATURAL SCIENTISTS

	Simple Regression			Multiple Regression			
Independent Variable	*Regression Coefficient*	*R^2*	*F Value*	*Regression Coefficient for Students Graduating in the Country of Origin*	*Regression Coefficient for the Named Variable*	*R^2*	*F Value*
Students graduating in the country of origin	0·112 (4·46)	0·34	19.89				
Graduate students in the United States in 1963–64	0·453 (27·8)	0·95	722·6	0·023 (3·3)	0·425 (25·2)	0·96	486·9
Graduate students in the United States in 1964–65	0·343 25·5	0·94	650·4	0·026 (3·5)	0·320 (23·4)	0·96	423·9
Graduate students in the United States in 1965–66	0·336 (21·5)	0·92	461·6	0·028 (3·1)	0·312 (19·1)	0·94	286·2
Graduate students in the United States in 1966–67	0·261 (16·3)	0·87	264·6	0·038 (3·5)	0·236 (14·9)	0·90	176·1
Sum of United States graduate students for four years, 1963–4 to 1966–7	0·085 (22·6)	0·93	509·2	0·028 (3·4)	0·079 (20·6)	0·95	329·4

Note: (1) Figures in brackets are the respective *t* values.
　　　 (2) Number of observations = 41.
　　　 (3) The variables are all scaled by population.

TABLE 15.4

REGRESSIONS INVOLVING IMMIGRATION OF
SOCIAL SCIENTISTS

Independent Variable	Simple Regression			Multiple Regression			
	Regression Coefficient	R^2	F Value	Regression Coefficient for Students Graduating in the Country of Origin	Regression Coefficient for the Named Variable	R^2	F Value
Students graduating in the country of origin	0·006 (2·2)	0·11	4·7				
Graduate students in the United States in 1963–64	0·153 (20·0)	0·91	400·4	−0·001 (1·1)	0·157 (18·9)	0·91	202·7
Graduate students in the United States in 1964–5	0·134 (18·0)	0·89	324·8	−0·001 (1·2)	0·137 (16·8)	0·89	160·7
Graduate students in the United States in 1965–6	0·149 (17·7	0·89	312·7	−0·001 (0·7)	0·152 (16·5)	0·89	154·8
Graduate students in the United States in 1966–7	0·113 (18·7)	0·90	349·7	−0·000 (1·2)	0·113 (17·4)	0·90	170·9
Sum of United States graduate students for four years 1963–4 to 1966–7	0·034 (19·6)	0·91	384·3	−0·001 (1·4)	0·035 (18·3)	0·91	190·8

Note: (1) Figures in brackets are the respective *t* values.
(2) Number of observations = 41.
(3) The variables are all scaled by population.

REGRESSION INVOLVING IMMIGRATION OF DOCTORS AND DENTISTS

	Simple Regression			*Multiple Regression*			
Independent Variable	*Regression Coefficient*	R^2	*F Value*	*Regression Coefficient for Students Graduating in the Country of Origin*	*Regression Coefficient for the Named Variable*	R^2	*F Value*
Student graduating in the country of origin	0·079 (3·5)	0·24	12·3				
Graduate students in the United States in 1963–4	1·335 (4·5)	0·34	19·8	0·068 (3·7)	1·211 (4·6)	0·51	19·9
Graduate students in the United Statates in 1964–5	1·558 (4·8)	0·37	23·4	0·068 (3·8)	1·427 (5·1)	0·55	23·1
Graduate students in the United States in 1965–6	1·590 (3·9)	0·28	15·5	0·071 (3·7)	1·462 (4·2)	0·48	17·6
Graduate students in the United States in 1966–7	1·258 (4·6)	0·35	21·0	0·068 (3·7)	1·144 (4·8)	0·52	20·9
Sum of United States graduate students for four years 1963–4 to 1966–7	0·365 (4·6)	0·35	20·7	0·068 (3·8)	0·333 (4·8)	0·52	20·9

Note: (1) Figures in brackets are the respective *t* values.
(2) Number of observations = 41.
(3) The variables are all scaled by population.

TABLE 15.6

REGRESSION INVOLVING IMMIGRATION OF ENGINEERS

Independent Variable	Simple Regression			Multiple Regression			
	Regression Coefficient	R^2	F Value	Regression Coefficient for Students Graduating in the Country of Origin	Regression Coefficient for the Named Variable	R^2	F Value
Students graduating in the country of origin	0·196 (3·2)	0·20	10·3				
Graduate students in the United States in 1963–4	1·638 (21·9)	0·92	478·4	−0·034 (1·5)	1·707 (19·7)	0·93	248·0
Graduate students in the United States in 1964–5	1·131 (16·5)	0·87	273·4	−0·034 (1·2)	1·181 (14·6)	0·87	138·9
Graduate students in the United States in 1965–6	1·043 (15·7)	0·86	246·5	−0·035 (1·1)	1·091 (13·8)	0·86	124·7
Graduate students in the United States in 1966–7	0·834 (14·6)	0·84	212·2	−0·028 (·87)	0·866 (12·7)	0·84	105·9
Sum of United States graduate students for four years 1963–4 to 1966–7	0·276 (16·6)	0·87	275·2	−0·034 (1·2)	0·288 (14·7)	0·87	139·7

Note: (1) Figures in brackets are the respective *t* values.
      (2) Number of observations = 43.
      (3) The variables are all scaled by population.

immigration to the number doing graduate work in the United States. Earlier than 1963–4 we do not have data for graduate students broken down country-wise skill-wise, but it appears that within the four-year period the further back we go, by and large the better explanation we get. This is, of course, not surprising in view of the time lag between completing one's education and immigration. Unfortunately the data relate only to the stock of students rather than to flows graduating, but even with the stock figures one does get remarkably good relations. It may be mentioned that the regressions were done after scaling all the variables by the population of the country of origin to avoid spurious relations based on total size variations.

Relating immigration during 1966 and 1967 to the number of students doing graduate studies in the United States in 1963–4, about 95 per cent of the inter-country variation of immigration of natural scientists is explained. The corresponding figures for social scientists is 91 per cent and for engineers is 92 per cent. The only case where it does not work very well is that of doctors and dentists where the amount of explanation that is achieved is only 34 per cent, and while the relationship is significant in the usual sense, the explanatory power of the hypothesis is much less in this case. It should, however, be mentioned that there was a data deficiency in the case of doctors and dentists, since in the tables of Institute of International Education (1964), (1965), (1966), (1967), the number of graduate students in these disciplines are lumped together with graduate students in pharmacy and veterinary medicine. It is therefore possible that having better data of graduate students might have made the explanation more satisfactory. Or else it could be the case that doctors and dentists do not quite fall into this pattern, and indeed there are reasons to believe that as a category of skill, medicine does pose a number of problems not present in others, especially the imposition of arbitrary barriers, which has been commented on by Harry Johnson (1968). Incidentally, the best results for doctors and dentists are obtained by a multiple regression involving both the number of students doing graduate work in the United States and the number graduating at home, which together explain more than half the inter-country variations.

It should be mentioned that while the statistical fits seem to be remarkably good for natural scientists, social scientists and engineers, there are a few problems even in these cases. In particular the regression coefficients seem to be peculiarly high for engineers. In trying to calculate a flow-flow ratio of student entry and immigration from the regression coefficient for the sum of graduate students 1963–7, one would have to multiply the figure by eight to take note of the fact that each of the four years' stock figures represents the sum of approximately four years of entry and the immigration figures relate to two years, so that the adjustment factor will equal four multiplied by four divided by two. Using this adjustment it would appear that if 100 more people came to do graduate work in the natural sciences in the United States, about 68 more natural scientists will immigrate to the United States. The corresponding figure for social scientists is 27. For engineers, however, the figure works out as 221! This suggests at least two different possibilities. Either it can be argued that students studying in America not only tend to immigrate themselves into the United States, but may be responsible for others doing the same in the case of engineers. There might be some point in this since the body of engineers (like doctors and dentists) have the quality of a guild and some times entry is easier for outsiders if they have contacts within the system. A second and perhaps more satisfactory explanation lies in the fact that the pattern of immigration was changing rapidly during 1966 and 1967, since the United States was liberalising its immigration procedures, and many people who had studied in America before and had difficulty in immigrating previously were being able to do so in 1966 and in 1967. Therefore, the results reflect not a normal equilibrium situation but the clearing of an earlier bottleneck. The coefficients are also messed up by the high level of multicollinearity among the variables. In any case it would be wise not to take the exact regression coefficients very seriously in doing predictions.

It is interesting that some of the usual explanations such as income difference, distance, travel cost, or quotas, are of so little use in explaining inter-country variations of immigration of natural scientists, social scientists, engineers and doctors into the United States from the under-developed countries. And it is remarkable that the pattern relates so well to the number doing graduate work in the United States. The reasons for the last are perhaps not far to seek. First, students trained in the United States are exposed to the cultural and social influences of that country which make immigration easier. Second, those who have studied in America are usually more aware of rules and regulations governing immigration and can jump legal hurdles more easily. Third, those who have been educated in the United States may find it easier to obtain jobs there compared with those who were trained at home. Finally, the United States education system, being geared to American needs, may not motivate the foreign student much towards the needs of the under-developed country from which he happens to come and he may find it more fulfilling to stay on in America.

For an under-developed country that wishes to reduce brain drain complex policy issues are raised by the findings. Since income differences do not appear to explain variations of immigration into the United States, it is unlikely the brain drain to America can be reduced significantly just by raising the salary level of scientists, engineers and doctors in the under-developed countries. The changes that can be made (like the inter-country variations in the statistical exercise) are small in comparison with the gap *vis-à-vis* the United States salary levels, and it may matter little whether one's income goes up five times or seven times.

The importance of graduate studies in the United States in inducing immigration into that country raises significant questions of educational policy for the under-developed countries. In particular it outlines the importance of making detailed comparisons of benefits and costs of higher education at home and that abroad taking into account their relative expenses, their relative effectiveness as well as their likely effects on brain drain. If the cross-country results reported here hold for specific countries also, the last may indeed be quantitatively a crucial consideration, especially for engineers, natural scientists and social scientists. This is, evidently, an important aspect of the planning of scientific and technical education in an under-developed country.

REFERENCES

1. Adams, W., *The Brain Drain* (New York: Macmillan, 1968).
2. Aitken, N. D., 'The International Flow of Human Capital', *American Economic Review*, LVIII (June 1968).
3. Boulding, K. E., 'The "National" Importance of Human Capital', in Adams (1968).
4. Clark, A., *Register of the University of Oxford, 1571–1622*, Oxford Historical Society, Vol. II, Part I (Oxford, 1877),
5. Dedijer, B., ' "Early" Migration', in Adams (1968).
6. Grubel, H. G. and Scott, A. D., 'The International Flow of Human Capital', *American Economic Review*, LVI Proceedings (May 1966).
7. Grubel, H. G. and Scott, A. D., 'The International Flow of Human Capital', *American Economic Review*, LVIII, (June 1968).
8. Institute of International Education, *Open Doors 1964*, Report on International Exchange (New York, 1964).
9. Institute of International Education, *Open Doors 1965*, Report on International Exchange (New York, 1965).
10. Institute of International Education, *Open Doors 1966*, Report on International Exchange (New York, 1966).
11. Institute of International Education, *Open Doors 1967*, Report on International Exchange (New York, 1967).
12. Johnson, H. G., 'Some Economic Aspects of Brain Drain', *Pakistan Development Review* (Autumn 1967).
13. Johnson, H. G., 'An "Internationalistic" Model', in Adams (1968).
14. Patinkin, 'A "Nationalist" Model', in Adams (1968).

15. United Nations, *UNESCO Statistical Yearbook 1966* (New York, 1967).
16. United States Congress, *The Brain Drain of Scientists, Engineers and Physicists from the Developing Countries into the United States*. Hearing before a Sub-committee of the Committee on Government Operations, Ninetieth Congress (Washington, D.C., 1968).
17. Weisbrod, B. A., 'Discussion', *American Economic Review*, LVI, Proceedings (May 1966).

Discussion of the Paper by Professor Sen

Professor Machlup, in opening the discussion, said that he was extremely sorry that Professor Sen could not be present, especially because he wanted to criticise slightly one or two points he had made. He regretted that it would not be possible to get the answers from the author at this session of the conference. In any case he would try to be impartial in presenting the paper. When some units of a productive factor, say labour, were moving out of a country, it stood to reason that the total product, or total income, of that country would be reduced. This was not the subject of the paper before us. We were not concerned with the G.N.P., G.D.P. or national income, but only with the income, or welfare, of the people who were left behind. Thus, the relevant product, or income, or welfare before the emigration of the departed units of factors was not that of the whole country, but it was instead the product, income or welfare minus the income that was paid to those ready to move out.

If an Einstein left Germany or a Von Neumann left Hungary, there would be a loss to the remaining Germans or Hungarians only if they had been underpaying Einstein or Von Neumann relatively to their specific contributions to the total products of the country. If those specially skilled emigrants had been paid what they were worth, the remaining compatriots would not be worse off as a result of the emigration. This proposition was applied by Grubel and Scott to larger groups of specially skilled labour, highly qualified human resources. Grubel and Scott concluded that, if their wages had been in accordance with the marginal productivity of that type of factor, then apart from external benefits which their labour had been conferring on the old country and apart from short-term costs of adjustment, the loss of these people would not inflict a loss of income to those they were leaving behind. The conclusion by Grubel and Scott was quite vulnerable on the ground that what holds for the marginal units, or a very small number of units, did not hold for intra-marginal units, or larger numbers, provided the law of variable proportions held and the relevant range of employment of these scarce resources was that of diminishing returns. Professor Machlup said that Professor Sen joined earlier writers, particularly Harry Johnson, in rejecting a conclusion based on constant, rather than diminishing marginal productivity. Harry Johnson specified a traditional production function and looked into the loss of income caused by larger numbers of emigrants. But, unfortunately, he used a Cobb-Douglas function for nothing but labour and capital; this implied unitary elasticities of substitution between these two types of factor regardless of their relative numbers. That was to say, the exponent α went with a mixture of skilled and unskilled labour and the exponent $(1 - \alpha)$ with a mixture of capital invested in machinery and capital invested in human skills. From this production function Johnson derived the conclusion that the loss to the people deprived of a substantial number of their skilled brethren 'is likely to be a very small fraction in all but exceptional circumstances'.

This was the proposition which Professor Sen undertook to demolish. He did it first by a reduction to the absurd. He asked what would happen to the deserted people if all their skilled brothers were to desert them. He showed that Johnson's production function would not indicate the catastrophic loss that most of us would expect to result for the deserted people, but instead only a very small, indeed negligible, loss of income. On the basis of initially rather reasonable factor ratios Johnson's function would tell us, in Professor Sen's words, that when every skilled person is drained out of the country 'leaving behind a land of totally unskilled labour and (only) physical capital, the income of those remaining behind goes down by a mere four-tenths of one per cent'. Professor Sen showed that the use of the Cobb-Douglas function for only two factors was responsible for such an unbelievable conclusion. It was not the unitary elasticity of substitution between labour and capital that was responsible, 'but the implicit infinite elasticity of substitution between physical and human capital'. Professor Sen replaced the production function with only two factors by one with three factors; he stipulated unitary elasticity of substitution between each pair of factors. The three factors were unskilled labour, physical capital and skilled labour. This function, Professor Sen admitted, exaggerated the contribution of skilled labour in the opposite direction. For, since skilled labour was indispensable in this model, its full withdrawal reduced the product of the two remaining factors to zero. This result held for all C.E.S. functions; indeed for all kinds of functions except where the elasticities of substitution between factors were infinite. Perhaps it should be said, for the benefit of any economists who did not believe in aggregate production functions, that the existence of empirical aggregate production functions was not a condition of the relevance of this tool. Professor Machlup, like many others, used it only for analogical reasoning, which he believed was of great importance in economic analysis, despite all neo-positivistic tenets. Professor Sen drew our attention to the fact that Professor Johnson's analogical reasoning led to an income loss of 0·4 per cent for the deserted people, while his own led to a loss of 100 per cent – a sufficiently wide margin for the 'true' answer.

But Professor Sen now struck with a new attack. He held that output and income were not what counted, that only welfare was relevant to the issue – not bundles of goods and services, that is, but the utility of those goods. Accepting Professor Johnson's shift of the focus from the deserted people to all the people of the world, he proceeded to the stipulation of utility functions for the deserted people, the deserters, and the haven people (or host people). But first he stated five basic inequalities:

(a) even before the emigration, the income levels of the skilled were higher than those of the unskilled; and

(b) the income levels in the receiving country were higher than those of the prospective migrants.

The other three comparisons are of income levels before and after the migration:

(c) the unskilled had higher incomes before the desertion than after they were left alone;

(d) the skilled migrants had higher incomes in the new country than they had in the old;

(e) the people in the receiving country had higher incomes after the immigration of the skilled deserters.

Now it was clear that the poorest group got the rawest deal: the poor unskilled left behind without brains suffered a loss of income, of a magnitude between 0·4 and 100 per cent. Professor Sen now presented us with some extremely elegant concave welfare functions which allowed us to compute the utilities of changed income levels of the various groups.

Professor Machlup went on to point out that what Professor Sen said about the changes in utility of the three groups could be put quite simply: If the people in the countries that received the skilful immigrants were not terribly elated by the increase in their already high incomes, but felt only moderately enthusiastic about their greater affluence, and if the migrants were not overwhelmed with joy about their improved standard of life, whereas the people in the countries that lost the skilful emigrants took the loss of their income as a smashing blow, because they had been initially so much poorer than the gainers and were made terribly unhappy by getting still poorer, then it was quite likely that the reduced happiness of the losers exceeded the increased happiness of the gainers, and hence the welfare of the world might be reduced by the migration of brains, even if the bundles of goods and services available to the world as a whole were bigger and better. 'Magnitudes of loss to the poor country can hardly be shown to be small without bringing in the utility function, but it might conceivably be shown to be large even without specifying a utility function as long as we know that there is diminishing marginal utility.'

Before proceeding to the second part of Professor Sen's paper Professor Machlup wished to submit a few critical comments on this result. He believed that the use of utility functions was quite legitimate as long as the same people were involved and interpersonal comparisons of utility were avoided. Such interpersonal comparisons had been made in public finance as early as in the 1880s but he believed that modern welfare analysis had tried to do without them, except in the form of social-welfare functions or world-welfare functions which embodied the utility indices not of the people but of those who made moral judgements about the people and about the distribution of their incomes. Professor Machlup said that he for one did not accept comparisons of utilities of different persons, and he wondered why Professor Sen had not made use of the device widely used by welfare economists – the device of compensation of the losers by the gainers – which could do without interpersonal comparisons of utility. Professor Sen probably agreed that the gain in real income by people who received migrants and by the migrants themselves might be greater than the loss of the people left behind and that there could be actual compensation which would still leave the world better off in goods and services. Of course, this implied actual, and not merely potential, compensation of the losers by the gainers. The compensation could take the form of immigration of unskilled workers. If the rich countries were accepting muscles as well as brains, this would be compensation for losing people; this had been precisely the case in Europe in the last decade.

Professor Machlup continued by saying that Part I of Professor Sen's paper was on the effects of migration. Part II dealt with the causes and these two parts were methodologically different. Part I was an abstract theoretical analysis whereas Part II was an empirical analysis using statistical data. Professor Machlup regarded it as an interesting picture of the state of our science, which allowed such very different methodological approaches to be applied to the same question. Professor Sen used data of United States immigration figures on engineers, natural scientists, social scientists, doctors and dentists, and he tested four hypotheses, namely, that the amount of the brain drain was explained by

(1) immigration quotas set by the United States;
(2) income differentials (using general national-income comparisons);
(3) costs of transportation (using air transport economy class); and
(4) previous studies by the immigrants in the American schools and universities.

Professor Sen found that hypotheses (1), (2) and (3) did not produce good results. The income differentials were so enormous that we could not expect regression analysis to give decent answers. Hypothesis No. (4), however, showed a high significance and explained 95 per cent of the immigration of natural scientists, 92 per cent of engineers, 91 per cent of social scientists though only 34 per cent for doctors and dentists. The low percentage for doctors and dentists was largely due to the fact that they had been lumped with veterinary surgeons and pharmacists. Professor Sen also used regression coefficients to find how many additional immigrants the United States could expect for each 100 foreign students enrolled in American universities. The finding was that 100 natural scientists produce 68 within four years, whereas 100 students of engineering produced 221 immigrant engineers. Foreign students were exposed to cultural and social influences, which operated as attraction to immigration; moreover, they learned to know the rules governing immigration and could jump the usual obstacles. It was easier for them to get jobs and the education they received was geared to United States needs. This had policy implications for developing countries regarding the question whether higher education should be provided abroad or at home; according to Professor Machlup, poor countries that lost so many students to advanced countries should revise their cost-benefit analyses and devise policies to keep their students at home.

Professor Matthews said that it was difficult to follow on after Professor Machlup's brilliant combination of urbanity and tendentiousness. Professor Machlup had shown the proposition to be general that the free flow of all factors would maximise world output. However, although he had said that interpersonal comparison of utility had no place in scientific economics he did in fact revert to the basic Pigovian position that welfare depended on the distribution of income as well as its size, by acknowledging that compensation had actually to be paid, not merely to be possible. Since the amount of 'compensation' paid in practice was not always linked closely to the amount of emigration of skilled manpower, restriction of such emigration was not necessarily an illogical policy.

The most important aspect of the problem was not the short-run one – should skilled manpower be allowed to migrate – but the long-run one: what effect should the fact of the brain drain have on a country's educational policy? Here again considerations about income distribution played a large part. It might well be true that the national income of a poor country (defined to mean the sum of the incomes of all those born there) was maximised by training a very small number of people so that they could earn very large incomes in the United States; but was this the aim of national economic policy?

Professor Matthews suggested that the moral to be drawn from the empirical part of Professor Sen's paper was that the brain drain was to a large extent the consequence not of the free international mobility of factors but of the grants given by governments for study abroad. A further social factor that might be added to Professor Sen's list in this connection was that graduate students were typically of an age highly prone to marriage, and they married American girls. If this was indeed significant, it would provide an argument for developing countries' sending their students for advanced training to the United States in their early thirties rather than in their twenties – a policy for which there was something to be said on other grounds as well.

Professor Rasmussen said we seemed to have this topic on the agenda every year and yet to reach no conclusions. He felt that in a thorough discussion of possible differences between the wage of a skilled man and the respective social return of this man touched upon by Professor Machlup in his introduction, one might argue that the social return exceeded the net wage *inter alia* because these categories usually paid a tax exceeding what they might get free of charge from the state. This might be because the income tax was progressive and/or because he consumed the taxed commodities. This in itself involved a loss from immigration. Further, one very often found that these categories were 'underpaid', for instance the civil servants. His education had usually been paid, up to a certain extent, by the state, and if things worked as they should (though very often they did not) he might be assumed to be rewarded by a wage below his social return, which might explain the 'low' pay of civil servants. Further, one might argue that those leaving were potentially very important contributors to the total, including the cultural, milieu of their country. Again in this sense their absence might involve an economic loss. These were externalities. It was however, at present, almost impossible to evaluate the magnitude of this difference between social marginal product and wages. Thus it was difficult to reach a definite conclusion.

Professor Robinson agreed with Professor Rasmussen that we seemed to be making progress backwards. He was sceptical whether economists were yet capable of measuring the addition to welfare or the subtraction from welfare resulting from the removal of small numbers of people. In his view economists had been less successful than economic historians, when the latter showed the contributions made to British development by the movement of the Huguenots or by the Ismaelis to East African development. He despaired of estimating with any of the apparatus currently available to economists, for example, the damage done to Britain or France by the slaughter of the first world war. As regards the effects of moving the outstanding people from one country to another, could one ever assume that their incomes represented in practice the equivalent of the damage that their removal would cause? Dr. Nurul Islam represented perhaps one-tenth of the whole of applied economic thinking in Pakistan. If he was removed to become one twenty-thousandth of the applied economic thinking of the United States, how did one calculate the net effects on the respective welfares of the two countries? It was quite clear that the personal incomes of these outstanding people bore little or no relationship to their contributions to total welfare and merely reflected conventional rates of pay in the two countries.

Professor Delivanis said that in view of Professor Machlup's brilliant and comprehensive introduction there were very few additional comments which he believed were needed. However he thought that Professor Sen, when trying to estimate the losses of the country which immigrants were leaving, forgot that it was always possible to reduce losses by introducing technology which provided employment for the skilled workers. He also did not calculate the benefits derived from the remittances of immigrants to relatives and to welfare institutions and, even more important, the funds they repatriated when returning home. He wished to suggest that the correlation of brain drain and postgraduate studies in the United States implied that at least the governments of developing countries had not to finance the postgraduate studies of their people in the United States where, provided there was no depression, the possibilities of securing good jobs were greater than anywhere.

Professor Triantis said that he would like to take exception to Professor Sen's use of differences in national income per head as a reasonable proxy for income differential leading to migration. Professor Sen had said that the facilities that an educated person enjoyed in an underdeveloped country would depend not merely on his own salary level but also on the level of prosperity of the country as a whole. Professor Triantis said that welfare was a function of absolute income, relative income, and the rate of growth of income. Whether a person migrated depended on his income at home and in the United States and the comparison between his own income at home and the income of others at home. In the new environment the income of a skilled person would often be a smaller multiple of the unskilled income there; if in his own country he was so much better off than others around him he was less likely to move.

Professor Simai reverting to the question of measurement said that many attempts had been made to quantify the losses and gains of receiving and losing countries, but without much success, finally leading to attempts to measure losses and gains in a very simple way. One possible measure was how much an expert, going to work in a developing country to replace the emigrant, cost the international community. On the basis of United Nations statistics one might expect the cost to be about £8000 per annum (*Professor Williams* intervened to say that this figure was far too low) *Professor Simai* said that this was the figure he was familiar with and referred to a paper by Maddison. On this basis several attempts had been made to examine the exact magnitude of the gains and losses, and this type of measurement was simpler and more evident to the general public than trying to express losses and gains by more complicated functions. On the use of highly skilled manpower in developing countries, Professor Sen's paper did not sufficiently reveal the problems of use of the highly skilled manpower in their native country. Their utilisation was in many cases low and there was either unemployment or under-employment of scientific manpower and Professor Sen's results are not valid without taking this into consideration. The brain drain was partly a consequence of this situation and much more had been said about the brain drain than about the underutilisation of scientific manpower. On the policy of the main receiving countries one could feel a kind of hypocrisy on their side. Immigrant laws encouraged the brain drain and favoured people accord-ing to the level of their skills, and if this policy continued the problem could not easily be solved. We tended to think in most cases only about highly skilled manpower, and in terms of brains. But brains at the technician level often represented equally great losses to the developing countries.

Professor Eastman said that Professor Sen had not taken into account the fact that the host country also was composed of different groups who could be divided as to the brains, the brainless and the property owners. The immigrants in question were more highly qualified than the host country brainless, whose welfare would be affected by influences of opposite direction. Insofar as the labour of the brains and the brainless were complementary, the marginal product and welfare of the latter would be raised by immigration of brains. However the relationship might be one of substitution. More importantly, the pattern of immigration might at times affect the welfare of the host country brainless in a more dynamic context by affecting adversely the rate of return and incentive to them of acquiring skills and so raising their social position and income over time. Furthermore, the ability of the country to import labour trained abroad reduced the need to develop the host educational system to provide particular types of training that open opportunities for upward social mobility for the host country's unskilled workers.

Professor Munthe said that in forming his hypothesis about causation Professor Sen had found no close correlation with the difference in income levels and suggested that this result might be due to the special circumstances of the brain drain to the United States. We might gain more insight if we studied other countries which had a common labour market and had had it for some time. In the Nordic countries there had been a common labour market for fifteen years and in this period there had been one important stream of movement and several small trickles. The strong flow had been of Finnish people into Sweden, both unskilled and skilled workers, and here the income differential was biggest, and many of these people came from places with scarce employment opportunities. However, even though there might be considerable income differentials between the four countries, there were few movements of people with higher degrees. Norwegians, by about five hours travel to Sweden, could increase their incomes by 20 per cent, but they did not do so, and in this case income differential did not seem to matter, which agreed with the hypothesis that there was a diminishing marginal utility if income. At the lower levels of incomes a 20–25 per cent increase meant a lot; but 25 per cent for a well-paid engineer did not really make a great difference. Finally, however, Professor Munthe said that people who had had training in other Nordic countries had a great propensity to stay. That accorded with Professor Sen's results. But in fact the differences between the countries were so small and the knowledge of working conditions so complete that this was not of significance. On the question of calculations, it was very difficult to make welfare calculations on a global basis. So many of these problems were very similar to those which arose in regional problems. There was a specific welfare loss to the country when local communities lost valuable people and they were left with an ageing population when all the young people had migrated. He suggested that the effect of the feeling of backwardness should be included as a welfare function. There were examples in Norway where whole communities had left for the United States. He felt that this was a welfare decreasing effect which had not been brought into the picture when a global analysis was used.

Professor Tsuru said he spoke as someone coming from a country which had depended immensely on returning brains in their early stages of development. He echoed Professor Robinson's plea for common sense. If Professor Sen insisted on using atomistic calculus then he must consider the brain as ownership externality whose social marginal product must be calculated over time. He felt that in the first part of the paper Professor Sen had been a victim of his own brilliance and that he had not contributed to the brain drain in this part. In the second part of the paper, the empirical investigation, the evidence seemed inconclusive. He wanted to suggest that we must look closer at domestic conditions of the country from which the brain has moved, and should look at the statistics and see why the countries with the biggest brain drain had suffered it. Professor Tsuru thought that such a study might be supplemented by taking countries in the middle of the scale, for example Japan, which had sent many graduate students to the United States but most of whom had returned. He believed that if one made such a study as he was suggesting we would find that it was not as simple as Professor Sen had made out.

Professor Khachaturov recognised that the I.E.A. had already discussed this problem on many occasions, but thought the problem was still an acute one and one that was important and unpleasant for developing countries. The brain drain led to an increasing gap between developing and developed countries and the substance of this might be the problem of satisfactory international division of labour. He had the impression that the method of application of the production function by several authors, in particular Professor Johnson, led to the underestimation of the effect of brain drain in developing countries. He would like to suggest to Professor Sen that it was desirable to know the reasons for the brain drain – why people who came to the United States or Britain, or wherever it might be, from Pakistan and other developing countries to complete their education go back. If the reason was not the difference of salaries, as had been suggested, what was it? Perhaps we ought to get information about the scarcity of jobs for skilled people in the United States and Western Europe as compared with the developing countries and see how this was related.

Professor Baumgarten said he would like to make two points. Firstly he felt that what mattered was that training in the United States was more relevant to United States activity and problems than to the country of origin, and that therefore back in his own country the trained man became frustrated. Secondly, he believed that different incomes could be an explanatory variable. He had no exact figures to offer, but recently the Brazilian government had been contacting Brazilian scientists abroad and offering them facilities to come back and had received many applications which suggested that their effort was succeeding.

Mr. Economides said that perhaps the reason why economic analysis did not reach common-sense conclusions about the analysis of the brain drain was because the analysis is a static and not a dynamic one. For example, in looking at the brain drain from Europe to the United States, the value of Einstein, Fermi, Szilard and such people could not be measured by salaries either in the United States or in Europe. In fact had they stayed in Europe their value would have been immense if only they could have solved the technical problems in Europe.

Professor Machlup said that it was not for him to answer the criticisms of the paper but he would like to make a few comments, especially in reply to Professor Robinson on the question whether we should try to measure. He said that neither Professors Johnson nor Sen undertook to measure losses or gains but attempted only to create a theoretical model helpful in obtaining an impression as to whether the losses or gains were large or small. If the losses or gains were negligible, perhaps we should conclude to do nothing about them. But if the losses were serious, then we must consider what could be done and what were the policy implications of our findings. One could play with many ideas, including compensation from gainers to losers. Any kind of policy in this context should be evaluated not only in terms of economic gains and losses but also in terms of personal freedom. He would consider it an enormous loss if a developing country tried to solve the brain drain by refusing exit visas to those who wanted to emigrate. As an old champion of personal liberty, he thought this was more important than economic gains or losses.

Professor Sen, after seeing the record of discussion of his paper, wrote in reply to it that he was very sorry that he could not attend the meeting and was extremely grateful to Professor Machlup for presenting the paper with such clarity and with so many judicious comments. He agreed with all his observations except perhaps with his rejection of interpersonal comparisons of welfare in favour of some compensation criterion. The trouble with compensation criteria, aside from possible inconsistencies and intransitivities, lay in the fact that if compensation was only hypothetical then the criteria were not compelling (What good was it to know that the poor could hypothetically be compensated for their loss if they in fact would not be?), and if it was actual then the compensation criteria were redundant and the Pareto principle was sufficient. In fact, possibilities of actual compensation in this field seemed remote, so that there was no escape from comparing the losses of one group *vis-à-vis* the gains of others. This did involve interpersonal comparisons, but it seemed to him that such judgements were always involved in serious policy issues, even though these judgements were often made implicitly. His purpose was to show that even without specifying a particular welfare function, but by just putting some restrictions on its shape (concavity), results on directional asymmetry between aggregate gains and losses could be deduced.

On the empirical part of the paper, he believed that Professors Triantis and Tsuru might have missed the relevance of one of the main results presented. The fact was not only that inter-country variations of skilled emigration to the United States were not at all explained by the pull of national income differences, but they seemed to be almost totally explainable for engineers, natural scientists and social scientists by the single hypothesis relating immigration to graduate studies in the United States. Certainly, the relative income hypothesis of Professor Triantis and the domestic situations hypothesis of Professor Tsuru were *prima facie* plausible, and in fact had been discussed as such in the literature. But what was interesting and perhaps important was that even without bringing in these factors the one simple hypothesis of graduate studies in the United States seemed to explain so much. Professor Matthews' interesting suggestion about the relevance of the age at which study abroad was undertaken seemed very worth pursuing in the policy context. This might be a fruitful field of study, especially since age-specific data on study and immigration were available for some countries.

Professor Munthe was right in emphasising the psychological impact of skilled emigration which might indeed be demoralising to a community left high and dry. In this context Dr. Simai's suggestion about studying the underutilisation of scientific manpower was particularly relevant. If brain drain took place from a country with much unemployment and underutilisation of the corresponding level of skill, then presumably the psychological and productive impact of emigration might be relatively less. Quantitative studies of brain drain needed to be supplemented by detailed examinations of utilisation of scientific manpower in the developing countries, since this was relevant to assess the effects of brain drain, whether or not it was found to be related to the causation of the phenomenon.

16 The Basis of Science Policy in Market Economics

B. R. Williams
UNIVERSITY OF SYDNEY, AUSTRALIA

I. THE PROBLEMS

The original programme of this conference provided for papers on policy issues in small developed countries and in Western Europe. However, for many purposes this is not a very useful distinction. Developed countries with less than 15 million people are small relative to Britain, France and Germany, but they in turn are small relative to the United States. A special problem in Western Europe is whether economic union will make it possible to bridge the gap between the scale of their research and development expenditure, and that of the United States, and if so whether this would bridge the technology gap.

Policy issues in small developed countries do not differ in kind but only in degree from policy issues in larger developed countries. Analysis of policy issues in, for example, Carter and Williams, *Government, Scientific Policy and Growth of the British Economy* (1964), and in Nelson, Peck and Kalachek, *Technology, Economic Growth and Public Policy* (1967) relating to the United States, are relevant to issues in small developed economies. The main differences are in the range of opportunities open to small and large countries in the creation and application of science and technology, and in the advantageous 'trade off' between home-grown and imported innovations.

Furthermore, the small developed countries are not homogeneous and there is no reason to expect that, even if rationally chosen, they will evolve the same policies on the percentage of resources to be devoted to research and development, on the percentage of the research and development effort to be directed to economic growth, on the importation of technological innovations via licence and know-how agreements and/or direct foreign investments. Thus while as a general rule, small developed countries do not devote a high percentage of their resources to research and development, the Netherlands – a former colonial power with three very large international companies – devotes as high a percentage of its gross national product to research and development as does France and a higher percentage than Italy and West Germany.

In the analysis of its data on research and development, O.E.C.D. has classified expenditure under the headings of 'economically motivated R & D', and 'defence, space and nuclear R & D'. Allocation of expenditure to these different categories is far from precise, and in the O.E.C.D. figures part of the nuclear R & D should be reclassified as 'economically motivated'. As a general rule however – from which Sweden departs – the proportion of 'defence, space and nuclear R & D' is very low in small countries, and the proportion of 'economically motivated R & D' high.

In the O.E.C.D. classification the definition of 'economically motivated' is circular. It is defined as R & D undertaken for economic ends – presumably this means R & D which is expected to add to measured economic growth, or to opulence – in the sense in which Adam Smith used that concept when he wrote that defence may be more important than opulence. To classify some R & D as 'economically motivated' should not carry the implication that other R & D is 'uneconomically motivated' or functionless, though that is how economists' procedures are sometimes interpreted. In his analysis of the complexity of scientific choice, Stephen Toulmin has argued that in the inquiries 'in the United States by economists such as Edward F. Denison; and even more so in the case of British economists, such as Carter and Williams, and Christopher Freeman . . . research and development are seen to be economically functional only in so far as they lead to economic growth in the narrow sense of increases in tangible output. The resulting questions are not easy to answer but they are at any rate *clear*'.[1] Of some of these writers at least Toulmin has misunderstood the approach. He is correct in his final point, but the procedure is not based on the assumption that research and development without impact, or trivial impact, on measured economic growth is economically functionless. However his misunderstanding, which is shared by many other writers on science policy, is not surprising. The O.E.C.D. nomenclature is misleading, and he is able to quote a polemical remark of Harry Johnson – made in the course of a debate on basic science when some distinguished scientists argued that their self-generated and expensive activities were a vital part of 'high civilisation' for which society should be happy to pay – that much of the contemporary 'scientific culture' argument for government support of basic scientific research is such as to put it – intentionally or not – in the class of 'economically functionless' activity. 'The argument that individuals with a talent for such research should be supported by society, for example, differs little from arguments formerly advanced

[1] *Criteria for Scientific Development: Public Policy and National Goals*, Ed. Shils (M.I.T. Press, 1968), pp. 129–30.

in support of the rights of the owners of landed property to a leisured existence, and is accompanied by a similar assumption of superior social worth of the privileged individuals over the common man'.[1]

Misunderstanding by many writers on science policy of the nature of 'the conventional economics of research and development' is also due to something more fundamental. This is an unwillingness to accept that it is legitimate to classify part of R & D as investment and part as consumption. In the treatment of the relation between R & D and economic growth, the investment aspect of R & D is isolated, and it is just this isolation which many scientists dislike. The most articulate scientists, or at least the scientists most articulate in public, work in basic science. They believe that initiative in scientific inquiry must be left to the free decision of the individual scientist, that independence safeguards originality; that expenditure on science should keep on growing; that the decisions on the allocation of funds should be made within the 'scientific community'. However many of them doubt whether society will pay up on a scale which they 'believe to be proper and in the best interests of both society and of the scientists'[2] and are tempted to think that 'basic science viewed as an overhead charge on technology is a more practical way of justifying basic science than is basic science viewed as an analogue of art'.[2] In effect this is an attempt to shift the whole of R & D into the investment category: to treat even the 'purest' of research as an investment in the stock of knowledge which, in the end, brings socially desired increases in goods and services. The obvious difficulty with this approach is that investment is a means to an end, and that projects which basic scientists would rank very highly in terms of intellectual interest and challenge might, when subject to tests applied to investment projects, be rejected in favour of less interesting and challenging problems with greater immediate technological or social merit. The standard attempt to have it both ways – to argue that basic science should be treated as an overhead charge on technology and yet that the scientists should have autonomy in the choice of projects – is based on the assertion that the impact of basic research on technology is large but unpredictable: that it is not possible to predict on which of the trees of knowledge technological fruit will set and ripen.[3] There is something in this line of argument, but not enough to establish the case for treating all

[1] 'Federal Support of Basic Research: Some Economic Issues', in *Basic Research and National Goals* (United States Government Printing Office, 1965).
[2] Alvin Weinberg, *Criteria for Scientific Choice II* in Shils, *Criteria for Scientific Development*, p. 91.
[3] See, for example, Kistiakowsky in *Basic Research and Economic Goals*, pp. 169–88.

basic research that catches the intellectual interests of the scientists as an overhead charge on technology, for there are some areas of science which are very much more likely to produce technological fruit. Nor is there enough in the point concerning unpredictability to establish an objection to treating some research as a consumption activity. (I return to this issue in section III, paragraph 2 below.)

II. WHY R & D RATES ARE NOT A GUIDE TO GROWTH RATES

Research and Development are not sufficiently homogeneous in their nature or effects for the expenditure to be aggregated and treated as a meaningful input to the production process. That is one reason why there is not a correlation which many scientists used to believe (and some still believe) must exist between percentage expenditure on aggregated R & D and (lagged) percentage economic growth. The distinction between research and development corresponds to the distinction between science and technology. Science is the sum total of systematic and formulated knowledge of natural phenomena and research adds to this knowledge. Technology is usually defined as knowledge systematically applied to 'the useful arts', and development adds to this knowledge. In some fields the passage from applied science to technology is simple and inexpensive; in other fields it is very complex and expensive. Although many advances, and probably an increasing proportion of advances, in technology grow out of advances in science, it is important to keep in mind that advances in technology are very often 'empirical' and that practice often outruns theory. Where further advance in technology is checked by this there is an induced demand for more basic knowledge. The recent history of aircraft, atomic reactors, 'defence' systems and space travel, provide many good examples of this interaction. The 'forced feeding' of some of the sciences in the United States has to be explained in these terms. (See Section III, paragraph 2 below.) The distinction between science and technology, between research and development, is also important for the consideration of rates of growth. For a country may decide to do little about adding to science, the knowledge of which flows freely between countries, but a great deal about adding to technology; or even little about adding to either science or technology, but a great deal about the efficient use of existing technology. Many scientists refer to this latter activity as parasitic, though there is no evidence that it kills off the host.

There is not even a correlation between the economically motivated R & D percentages and the economic growth percentages.

The reasons for this lack of correlation – which provide an important basis for the consideration of policy – are to be found in differences in the levels of technology, in the proportions of engineers, scientists and technicians in the work force, in the pressures to innovate, in savings ratios and industrial structures, and in the many ways in which technology is transmitted from country to country.[1]

Investment in industrial R & D is only one factor contributing to economic growth, and even if the social rate of return is substantially greater than the private rate return to other investment, its contribution to economic growth must be relatively small over short periods. The lower a country's level of technology the more it can rely on the existing pool of scientific and technological knowledge as a source of growth.

A country's output per man, and increases in output per man, are very dependent on the skills embodied in the labour force. The skills possessed by 'qualified scientists and engineers' are needed in both the application of existing knowledge and the creation of new knowledge.[2] For two countries with similar levels of technology and industrial structure (see this section, paragraph 5 below) the country with the smaller ratio of qualified scientists and engineers to the total labour force may in the interests of growth use a smaller proportion in economically motivated R & D (or in other forms of R & D). It is possible to reduce the rate of measured growth by drawing too high a proportion of qualified scientists and engineers into R & D. The diffusion of innovations may be impeded; the probability of choosing the most promising R & D projects may be reduced ; the capacity to turn the outputs of R & D departments into inputs to the production process may be weakened.

A major part of the innovation in capital-embodied savings ratios and institutional provisions for the supply of risk capital to potential innovations are therefore important determinants of growth. If the output of R & D outruns the capacity to finance their use, the 'productivity' of R & D will be reduced. Many examples could be given from both rich and poor countries. An interesting example of 'excess R & D' is provided by the British Atomic Energy Authority,

[1] According to Mr. Mark Spade in *How to Run a Bassoon Factory* 'it is not absolutely necessary to have a research department – at least not one of your own. In fact, probably the most efficient and economical type of research is to have an employee whom you can really trust working in the research department of your principal competitor'. In the international sphere licence and know-how agreements and inward investment complement any such trusted employees.

[2] For the significance of the distribution of qualified scientists and engineers between the activities of research, development, production and marketing, see Carter and Williams *Science in Industry* (1959).

which because it underestimated the scarcity of capital invented excessively capital intensive reactors.[1]

The growth effects of R & D are sensitive to pressures to innovate. If the pressure is weak both the first innovatory steps and the diffusion of the innovation may be, and often are, slow. Pressure may be weak because of restrictive practices or it may be weak because of limited capacity to manage innovation. The effects of United States direct investment abroad in increasing the pressures on native firms to innovate have been very obvious and strong. This is a problem that has been highlighted in the discussion of how far 'technology gaps' are due to 'R & D gaps' and how far to differences in management.[2]

Differences in industrial structure also reduce the plausibility of an expectation that R & D rates and growth rates will be correlated. Compare in this respect Britain and Australia. Some years ago I calculated that to match the over-all British expenditure on R & D as a percentage of industrial net output, R & D in Australian industry would need to be $140 million, but that if the sensible procedure of applying the British percentages industry by industry were adopted, this figure would come down to $80 million. Industries with relatively high rates and low rates of R & D expenditure are similar in all industrial countries. In Australia the aircraft, chemical, and electrical industries are relatively less important than in Britain, so that even the same industry by industry research intensiveness in the two countries would leave the Australian industrial R & D percentage well below the British level. Australia like Canada has a lower over-all R & D percentage than other countries with similar levels of income per head. The explanation is not only industrial structure and low percentage expenditures on defence and space. Both have a large foreign ownership of their science-based industries and therefore lean on the industrial R & D of the parent firms. For Australia I calculated that this 'vicarious R & D' practically bridged the gap between actual Australian R & D in industry and what it would be if it matched the British R & D rate (with Australian weights).

Even apart from the foregoing considerations, it was always rather absurd to consider the relation of national R & D and growth rates out of the context of international trade in goods and services, migration and capital flows. There are some very effective procedures for diffusing new technologies throughout the world. Most of the firms with high R & D ratios are in industries producing capital and intermediate goods. The embodied innovations are spread round

[1] See Williams, *Technology, Investment and Growth* (1967), chap. xiv.
[2] See, for example, D. Granick, *Industrial Management in Economically Developed Countries*.

the world through international trade and influence growth in receiving countries, both directly and indirectly. Countries with an ability to copy and adapt the embodied technology may even come to make more effective use of the new technologies than the innovating country or countries – as, for example, Japan in parts of the electronics industry. Where there are impediments to international trade (but not direct investment), or where the new technology has to be adapted to local conditions or could be easily copied, the innovating company may set up subsidiary companies abroad. Historically, direct foreign investment has been a very important way of transmitting new technology[1] and this investment has a considerable impact on the growth rates of the countries concerned. There is the direct effect of the investment, and the indirect effect of the greater pressure on the local firms to increase their innovatory activities.[2] Where the innovating firm has good patent protection (for example, Pilkingtons and float glass) it may choose to license firms in other countries to use the new process. Statistics of royalty and know-how payments are not however restricted to such cases. More than half of the United States receipts from the sale of licences and know-how to Europe comes from charges of parent United States companies to their European subsidiaries. Although international direct investment and royalty and know-how payments have been influential in the international diffusion of technologies, both methods have been questioned in a number of countries – on grounds that economic independence may be more important than opulence, or that by spending more on R & D, advances in technology might be gained more cheaply in the long run than by relying on inward investment and licences. Such arguments do not restore the plausibility of the belief on which many proposals for national policies for science and technology have been based, that higher national R & D rates will bring higher national growth rates. But they do pose important issues concerning the role of government and the proper objectives of policy on science and technology.

III. GOVERNMENT DEMANDS FOR NEW SCIENCE AND TECHNOLOGY

One of the most striking features of R & D is the proportion financed by Governments: in the United States and France – 64 per cent; in

---

[1] Sales by U.S. direct-investment enterprises abroad (some to the United States itself) are approximately four times greater than the value of U.S. exports.

[2] See, for example, J. H. Dunning, *American Investment in British Manufacturing Industry*.

Britain, Canada and Norway – 54 per cent; in Sweden – 48 per cent; in Germany, Austria and the Netherlands – 40 per cent; in Italy, Japan and Belgium between 33 and 28 per cent. Government finance of R & D reflects its interests in the performance of:

  (i) its own public service functions;
  (ii) higher education;
  (iii) sectors of production which operate by the rules of the market – the 'business sector'.

The public service functions of governments include defence, public health (within which control of environmental conditions is likely to play an increasingly important part), standards, roads, bridges and other forms of communication for which reliance on market provision is not possible. Governments of countries which fear violence and invasion from others, devote part of their resources to defence, and they have a direct interest in strengthening their position by advancing the relevant sciences and technologies. Not every country with such an interest has a significant defence R & D programme.[1] For in the last twenty-five years there have been very rapid advances in defence and war technologies, and the R & D costs of making further advances have become very high. This has put defence R & D beyond the resource capacity of most countries (and therefore increased the pressures to align with one of the great powers, or to withdraw from any direct part in the international power system). In so far as defence R & D is intended to provide new or improved technology to enable the government to perform one of its functions it is proper to classify it as an investment activity.[2] Part of the civil R & D to improve the performance of some other public service functions – standards, transport and communications facilities – may also be classified as investment activities, although the expected impact on measured economies may be trivial. As

[1] Of the O.E.C.D. countries only the United States (56 per cent), France (42 per cent), Britain (40 per cent), Sweden (30 per cent), Canada (26 per cent), Italy (20 per cent), Germany (16 per cent) and Norway (12 per cent), devote more than 10 per cent of their R & D effort to defence, nuclear and space. This means that defence, nuclear and space R & D in 1964 was 1·9 per cent of G.N.P. in the United States, 0.9 per cent in Britain, 0·7 per cent in France, 0·5 per cent in Sweden, 0·3 per cent in Canada, 0·2 per cent in Germany, 0·1 per cent in Norway and Italy.

[2] Any attempt to calculate the return to such investment would come up against the problem of measuring quality changes in its most acute form. For improvements in the quality of weapons systems must be measured in terms of known or hypothecated improvements of the weapons sytems of the potential enemy. And fiercely competitive R & D in this field brings a very high rate of obsolescence. The unusual nature of this case is evidenced by the fact that all economists would agree – I think – that agreement to suppress competition would be altogether in the public interest.

investment activities, in principle, investment criteria are applicable to them.

Government finance of university and similar research has increased greatly in recent years. In the United States in particular some of the increase has been due to the expected relevance of the research in government defence, space and nuclear programmes. And in all countries some of the support has been due to the expected effects in improving the supplies and qualities of scientists and engineers. Even so the main explanation is probably the value put on freely chosen university research as an aspect of a civilised way of life.

Interest in the performance of the business sector has led governments in most industrial countries to add to funds available for 'economically motivated' R & D to bring more opportunities and pressures for innovation than the market would otherwise provide. The case for such action exists where there is evidence that although it does not pay individual producers to engage in R & D the potential social return would justify the expenditure, or where the potential marginal social return is high enough to justify a greater R & D effort though the potential marginal private return would not. Examples of the first have come from agriculture and many small firm industries at one end of the scale, and from early R & D in nuclear power in countries where the relevant firms were not large enough to finance the necessarily large and long programmes. Examples of the second – whether good examples is another matter – have come from aircraft, complex machine tools and computers. Many countries have also assumed that there is a generalised tendency for social returns to exceed private returns and so have given general incentives to 'economically motivated' R & D through a variety of bounties or tax remissions.

IV. OBJECTIVES

One implication of the foregoing is that a great deal of R & D in addition to that classified as 'economically motivated' is properly classified as investment, and therefore appraised in terms of whether or not its contribution to production has justified the expenditure or is likely to justify the expenditure. The conclusion is not undermined in principle by the difficulty of measuring even retrospectively the rate of return on cost – a difficulty which is most acute in a field such as defence hardware – though in practice our current inability to measure marginal and average rates of returns to different R & D activities seriously weakens its significance. For it follows from the conclusion that there are ways in which projects should be formulated for initial and continuing appraisal, and proper (sequential) strategies for the finance and administration of projects.

Nor is the conclusion undermined by the argument that too high a proportion of R & D has been devoted to projects which are conveniently and sensibly classified as investment activities: that the public has been short-changed on R & D relevant to public sector wants such as low income public housing, urban services and regulatory problems relating to environment such as control of noise and pollution of the air and water. This objection is relevant to the issue of society's objectives (or 'proper objectives') and is familiar in a more general context of the balance between consumption and investment activities. In all countries R & D as investment is a small part of total investment, but it plays a significant part in maintaining the rate of investment in plant and equipment.

Appraisal of R & D is difficult enough at any time because of the uncertainties involved in both the R & D process and the effects of using the results of apparently successful R & D. Appraisal becomes even more difficult when the R & D, or the R & D proposals, have a mixture of investment and other purposes. For instance many of the proposals to increase West European R & D, to close the technology gap, and end or reduce inward investment and the unfavourable balance of technological payments, are a mixture of two things. The first is a belief that the present degree of reliance on inward investment and licence and know-how agreements for economic growth brings short term advantage at the expense of long term advantage; that the short term advantages of capital inflow will be more than offset by the subsequent payments of interest, dividends, royalties and management fees and the effects of limited export franchises allowed to subsidiary firms by parent companies; that the parent companies in the United States will conduct R & D for their subsidiaries and concentrate an even higher proportion of R & D and (through the brain drain) scientific manpower in the United States; that the competitive strength of the United States companies will be sufficient to prevent the growth or creation of competitive native firms, so reducing West European countries to 'technological colonies'. The second factor in the argument relates to the preservation of national identities. This argument – which may build on the fears of technological colonialism, but is not dependent on such a fear or forecast – is that there are important objectives other than measured economic growth for which a country may need at times to pay a price in terms of measured growth foregone.

It is not possible to put a 'price' on these objectives. In Australia the objective is relatively clear; in Western Europe at the moment, with proposals to strengthen the role of the community and to add to the membership, it is difficult to be certain what is to be protected, whether an emerging European identity or a Gaullist French identity. Nor is it certain what values are being protected, for some of the objections to United States subsidiary firms are objections to change in traditional ways of life which would be equally threatened by a European-generated closure of the technology gap. However it is possible in principle to calculate the price that would have to be paid to preserve certain specified values. If the 'technological colonialism' thesis were valid, some of these values could be protected without long term cost, and the case for government action to remove, directly or indirectly, the underinvestment in R & D due to uncertainties and external economies and problems of scale would be very strong. In *The Research Gap and European Growth*[1] I presented evidence for my conclusion that the research gap would not give the United States a cumulative advantage in technology and economic growth, that there was no evidence, given its scientific manpower position, that Western Europe was underinvesting in R & D, but that there was evidence of managerial deficiencies in the planning, evaluation and the control of R & D, and in the exploitation of opportunities for innovation, which left United States subsidiaries with unduly good opportunities for profit and expansion. I argued also that there were possibilities in the fields of aerospace and electronics which could be realised if appropriately large European companies were created with the support of joint government purchasing arrangements. The difficulty of achieving such an institutional innovation is evident from the history of the European air-bus project, and in any case the key role of co-operation between governments in purchasing, and financing such a development, provides further evidence of the tendency to place excessive weight on the role of R & D. It is important also to keep in mind the many factors other than the size of R & D in the United States which explain the growth of United States direct investment abroad. Apart from the United States skill in integrating plans for R & D with plans for finance and production and marketing, there is the powerful influence of restrictions on trade, of the non-alignment of currencies, of defence alliances, of the outmoded structure of many European industries, of the steps (often competitive steps) taken by European (and other) governments to encourage United States direct investments in the interests of import replacement, higher investment and growth. Unless this sort of thing is appreciated, science and technology policy will be expected to achieve things which are quite outside its range.

[1] Chapter 1 in *Technology, Investment and Growth* (1967).

National circumstances and objectives vary greatly, and appropriate national policies vary with them. Government interest in influencing market conditions and responses in the business sector may now be taken for granted. Perhaps the best answer to the question of how the government should decide where to intervene is to be found by looking at the areas where there appears to be some incongruity between scientific and technological opportunity and the institutions of the industry. The difficulty with policy is not to find areas worth looking at but to ensure that decisions are based on good analysis. The confidence of scientists and engineers that they could transform the technology of the industry is not enough. There must be evidence that if appropriate arrangements were made to apply the forecast new technologies that the prospective rate of return on the R & D and the capital-embodied innovations would pass the appropriate investment test. A considerable part of the nation's activities are of a kind not at the present time ready to be revolutionised by science, and it is a waste of time applying effort at points where it can yield little result. The opportunity must be one which will not be taken, or will not be taken quickly enough or on a sufficient scale, if the industry is left to itself. The term 'institutions of the industry' is used very broadly. The trouble may lie in the size of firms. Thus, in building and farming, many producers are too small to undertake their own research, to realise the potential value of co-operative research, or to make sufficient use of a flow of research results produced elsewhere. Or the trouble may lie in restrictive practices, or in the comfortable ease of a protected market removing the incentive to think carefully of scientific change. This fault can be found in several industries producing specialist machinery. Or the nature of the scientific opportunity may be such as to require very large or very risky expenditure, beyond the resources of firms in the industry or requiring from them longer views than they are accustomed to take; this problem arises in the development of aircraft or of large computers. Again, the industry may have recruited in the past managers who cannot understand what science is now offering them. It is no use simply complaining of a failure which is traceable to recruitment or training policies in past years; positive steps to break down ignorance or to introduce new management may be needed. In these and other ways, an industry (or trade, or profession, or public service) may be found to have a structure, or a command of resources, or a type of manpower or management, or a habit of mind or method of action, which slows down or prevents the application of science. There is not one problem here, but many. The forms of government countervailing action need to be as varied. A different mixture of remedies is required for different circumstances, and almost any form of government action has potentially harmful side-effects if used too vigorously. For instance, subsidies can alter people's actions, but may in time become a form of protection which discourages further effort. The wise policy-maker uses many instruments, and none to excess.[1]

If it is not possible to be certain whether investment in R & D in or for the business sector is optimal either absolutely or relatively to investment in human and physical capital, it is still less possible in the case of defence. Fierce competition to render obsolete the weapons systems of the potential enemy is very costly in resources, but while the fears – real or imaginary – persist the imputed value of improvements in these weapons systems will be high. R & D to prepare better weapons for a type of war that will not be fought, or to produce a deterrent that does not deter, is a waste of resources. The difficulty is to get the information which makes such a judgement possible. For a given objective, the key policy questions are whether R & D is sufficiently well organised to prevent waste, and whether development and production activities are conducted in such a manner that potential technological spin-off to other activities is likely to be identified and communicated effectively. So far it has proved easier to invent cost-effectiveness and sequential procedures to minimise waste than to get these procedures into use. Here the major impediment to innovation is the power of group pressures. (Those of us with direct experience in this field know for instance how often it is argued in effect that 'design capacity' is an output of the system and that skilled workers have no alternative use value, and how influential these arguments can be.) On the diffusion of defence R & D, there is now strong evidence that the value of identifiable spin-off is very small in relation to cost, and that it is most likely to be 'caught' if R & D is not segregated from civil production activities.

I referred earlier to the importance of scientific and technological manpower. Where this is small absolutely and relatively to the labour force sensible policy on science and technology has to be based on acceptance of the obvious – namely, that policy on R & D and on the encouragement of production activities that make heavy demands on scientists and engineers has to be selective. It is an important part of policy to identify and encourage the optimum distribution of such manpower between alternative uses – research, development and application. For application there is a threefold need. Many new or improved technologies require scientists and/or

[1] For a list of these instruments see Carter and Williams, *Government Scientific Policy and the Growth of the British Economy*.

engineers to operate them and achieve the 'learning curve' growth potential; many new technologies create great problems when introduced and resistance to change is less where those involved understand the technology and are confident that they can 'get the bugs out of it'; there may be a serious communication loss, through failure to identify the most promising opportunities for R & D (from whatever source), unless there is a scattering of scientists and engineers throughout firms.

It does not follow that if there is a great shortage of scientists and engineers, say, less than 0·75 per cent of the work force, all research the results of which flow easily and cheaply between countries should be left to the countries better endowed with scientific manpower (or improvident enough to set goals which conflict with measured growth). But this does not follow. Scientists and engineers can be trained, and where there is a great shortage, the prospective social yield from investment in training more will be high. A certain amount of 'pure' research is a condition for getting more scientists and engineers. Such research in universities is in part a consumption good – the community pays for universities to do some research for the same reason that it subsidises opera and music. But university research is in part an investment good (as opera and music in certain cities). Without that research, scientists and engineers could not be properly trained for many of the growth fields in relevant technology; without that research it would not be possible to catch the imagination of many of the able young and attract them to higher education in science and engineering. Such is the variety of national circumstances, it is not possible to make simple generalisations about optimal distributions of scientific manpower. Because pure research and the production of scientists and engineers are in some degree complementary processes, it may be rational for a poor country to have a high proportion of its scientists and engineers engaged in education and research, though the proportion should fall through time as the ratio of new to total scientists and engineers falls. How far in a growth-oriented economy the bulk of the scientists and engineers would shift into trying to make good use of known technologies would then depend on the production structure, on the ownership patterns of that structure, on the number of unique problems presented by agriculture and raw material conditions. In a dominantly agricultural society without great civil engineering problems, scientists and engineers would be needed far less for production work than for applied research and extension activities. In a small country (such as Australia) with a relatively poor scientific manpower supply, a big agricultural and raw material sector with distinctive problems, and a considerable industrial sector owned by foreign companies not making great calls on the supplies of native scientists and engineers, the proportion of scientists and engineers in R & D and the proportion of them engaged in basic and applied research may sensibly be higher than in the United Kingdom. In a country such as Japan with a better relative and absolute scientific manpower supply than Australia, but with a lower level of output *per capita*, a poor raw material position and very little foreign ownership of her wide spread of industries, optimal distribution might entail a smaller proportion of scientific manpower in R & D and a higher proportion of R & D manpower close to the problems of application. However it follows from the general nature of the problem that optimal distribution is not static. In Japan the marked post-war rise in its capacity to improve established technology and to annex a substantial part of learning-curve potentials of capital-embodied innovations has, in various areas, where licence costs have risen significantly, shifted the balance of advantage in the direction of rather more reliance on independent innovation.

Independently of levels of technology and scientific manpower supplies, there are considerable differences between countries in the cultivation of pure or basic science which may have a considerable influence on the respective roles of creation and exploitation of science and technology. Thus even a small country with distinguished researchers in particular fields of science and for which the country has a good endowment in raw materials and relevant technical skills, it may pay handsomely to be original innovators. An example is Sweden and special steels. But even if it has no great natural endowment it may pay to be an original innovator and to maintain a considerable amount of basic research where the quality of that research is high, where both the rate of product obsolescence and the returns from early product innovation are high, where the size structure is appropriate to successful invention, and where market size is such that R & D overhead cost is not too high for unit production cost. Switzerland in certain areas of the chemical industry provides a good example of this. This comment simply gives further emphasis to the point made at the outset of the paper concerning the arbitrary nature of a division between small and large developed economies in policy issues.

No country is concerned only with growth in output, and many of us here I suspect would argue that in the post-war period governments in many rich countries have given an unduly high priority to growth objectives. Non-growth objectives are relevant to policy on science and technology in various ways. Interest in 'defence' provides an outstanding example of the way in which a community may expand knowledge of science and technology in areas that are not very relevant to opulence, and may indeed seriously threaten it. There are other established community goals particularly relevant to science policy which may have only loose or accidental relevance to policies concerned with opulence. Research which is justified only as good science and which is central to the arguments referred to earlier concerning the autonomy of scientific choice is an example. There is no way in which we can derive the optimal rates of expenditure on what the scientists rank as 'good science' from the community's interest in extending its knowledge of man and nature and in nurturing the value systems of the groups within society who are responsible for this extension of knowledge. Doubtless judgement on support for such activities will be influenced by knowledge that some of them may prove to have relevance to material opulence, and that some of them are called forth by puzzling practical problems. A high proportion of medical research, for example, arises from difficulties in the practice of medicine. However in terms of science policy it is still true, I think, that the judgement on the size of the budget must be non-rational. There is a strong case for the view that within that budget, the scientific community should settle the programme, so long as the government or government agency acts to prevent gerontocratic tendencies within the scientific community. We can expect the growth of such research and the related growth of higher education to be one of the fruits of opulence.

The importance of non-growth goals is also emphasised in the claim that citizens (particularly in big cities) have been short-changed on R & D relevant to environment. It is sometimes argued that to cope with this and related problems, policy on science and technology should include a clear statement of national technological goals and involve public agencies to do the R & D. The former should, if supported by good analysis of the relation between purposes and social costs, contribute to the quality of discussion of and decisions on the extent of public support for R & D in various fields. The latter is a matter of procedures. Many of our environmental problems do not call directly for further research, but only for decisions that certain business sector or public sector activities have undesirable social consequences and must cease. The pollution of air and water can be prevented, though it is costly to do so. If new standards are introduced, and enforced, the desire to reduce the cost of conforming to them will induce relevant R & D. The question of how far in the business sector governments can safely leave this to market forces (as changed by the introduction of new constraints) is essentially the same question as that considered earlier in the context of growth R & D.

Discussion of the Paper by Professor Williams

Professor Matthews, in opening the discussion, remarked that Professor Williams' admirable paper on policy touched on a number of issues already discussed in this conference, as well as on some issues which had not been discussed. It might be useful to try to link his arguments to some of the points already raised in earlier sessions. Professor Williams enumerated three reasons why the government was involved in the finance of R & D. The first of these was its public service function. The leading item here was defence, which had been relatively little discussed in our sessions so far. Professor Williams pointed out that, leaving aside possible disputes about objectives, defence R & D was a form of investment, and should be subject to critical appraisal as in other areas. There was a great scope for wasteful R & D in this area, and for the influence of pressure groups. The whole business of international competition in defence R & D resembled oligopoly in its most harmful aspect, and the lessons of economics in handling oligopoly might be of relevance in the handling of underlying policies here.

Public service functions were closely related to the existence of objectives undervalued by the market. Three of these objectives were mentioned in Professor Williams' paper, pure science, national independence discussed in conjunction with the technological gap, and the environment. To these might be added two objectives, possibly subject to adverse influence by R & D – equality, mentioned by Professor Griliches in the discussion of Professor Galbraith's paper, and the formation of consumers' attitudes, Professor Galbraith's own point. With regard to pure science, Professor Williams considered that this was worth doing, but was doubtful about the use of economic principles to establish the scale. He was suspicious of spin-off, but conceded the Haldane principle, that is, the pure science allocation to be left to the scientific community.

The second reason for government involvement was education. This had been touched on many times in our discussions but had not been made the main subject of any of the papers. The third set of reasons concerned the malfunctioning of the business sector in R & D and led to government intervention there. As far as this was concerned, Professor Williams was sceptical of over-all statistical measures as a guide to policy. He was not surprised to find poor correlation between aggregate magnitudes and he declined to answer the question whether more R & D would be a good thing (a question tentatively answered in the affirmative by Professor Griliches and Sargent, answered in the negative by Professor Galbraith, and declined by Professor Mansfield). However, Professor Williams was not at all averse to government action in appropriate cases, most obviously when there was a high degree of non-appropriability, as in agriculture, more doubtfully where high risk was involved. More generally it was clear that Professor Williams did not believe that optimum market structures would necessarily emerge, and when they did not public intervention might be appropriate.

Like others who had undertaken detailed studies of R & D at the micro level Professor Williams had been impressed by the complexity of the issues involved. One consequence of this was that a major part of the contribution to be made by economists lay in debunking oversimple doctrines espoused by scientists or by politicians. Another consequence was that it was difficult to separate policy on science and technology from policy in other areas; for example towards international trade, which could have an effect on the international flow of knowledge as well as on the flow of brains – both subjects we had discussed at this conference. Other important factors were such things as anti-trust policy and anti-cyclical policy, both of which were touched on in the session on Professor Mansfield's paper.

Given the complexity of the problem and the inappropriateness of simple answers, just how positive could economists be in their policy prescriptions? At the micro level where we were concerned as much with management as with economics, there had been considerable agreement in the course of the conference on the paramount importance of articulation. That is to say, the need for research and development to be carried out in close conjunction with production and marketing policies. This came out in the contributions of Professors Freeman, Mansfield and Khachaturov, as well as in Professor Williams' own paper, and of course in the earlier writing by Carter and Williams. An important point raised in discussion had been whether success in this regard was predictable or whether on the other hand it was something which could be diagnosed only after the event. An extreme view in the latter sense had been taken by Professor Griliches in the discussion of Professor Freeman's paper. Professor Griliches had maintained that the people in the business were doing the best they could and economists and all outsiders were unlikely to be able to suggest a better way of setting about things. The impression gained from Professor Williams' paper was that he would not share this view.

At the macro level it appeared that economists had not yet devised a very satisfactory conceptual framework. More needed to be done to build a bridge between the micro work and the macro work. This was brought out clearly by Professor Rasmussen in an earlier session. He pointed out that in the steady state the speed of diffusion affected the level of income rather than its growth rate. While this was clear in theoretical terms, more study was needed on how much and how quickly economic performance could be improved by faster diffusion, given that we were not in fact in a steady state. More generally, as we passed from management problems at the micro level the development of science and technology and policies bearing thereon had social implications. Professor Galbraith's position on these had been made plain. Professor Robinson had shared Professor Galbraith's view of their importance but had pleaded for modesty on the part of economists in pronouncing on what these implications were in the long run. Patient study was needed in this area; in the meanwhile economists' policy prescriptions on these broader questions were bound to be considerably hedged.

Dr. Sedov had found Professor Williams' paper of great interest and particularly the way in which he had underlined the importance of non-growth goals of economic policy. Dr. Sedov suggested that what we wanted to know were the reasons for this change. Were these goals likely to be temporary or permanent? He suggested we should underline first of all the change in the character of the development of science and technology that was due to the scientific and technological revolution which had already changed the structure of the production forces. This had led to a change in the human being's place in the productive processes and in the relationships between humanity and nature. That was why simple economic growth might not be a goal of the nation. And more than that, economic development could not be isolated from the associated social political issues. The scientific/technological revolution was the result of the growth of productive forces in their social setting, and it demanded new forms of organisation, and new forms of social life. But the development of the revolution was very uneven. Different countries had different levels of readiness for it. They entered the revolution at different times but all wanted to be in the van. Economic policy was directed to facilitating these changes. Dr. Sedov did not necessarily think that it was possible to do this by means of the market mechanism on the basis of private property. That was why the experience of planning and regulation of the development of science and technology was of great importance for all countries. The wider the scale of the revolution the more important were the social aspects. The main task of economics was to work out an integrated system of instruments for macro and micro management and the development of science and technology.

Professor Sargent said that he would like to revert to the question picked out by Professor Matthews when he asked why governments had become so much involved in the whole matter of science policy. The answer to this was implied in several parts of Professor Williams' paper but he would like to argue more boldly that one of the main reasons why governments had become involved was that economists had told them a little prematurely that science and technology were important in explaining the rate of economic growth. When it was shown that capital and labour did not account for all the growth and that there was something left over this was called the residual factor, and economists leapt in to label this technological progress, without using their normal caution. Professor Sargent thought that what we had seen at this conference was a beginning by economists to cover up their tracks on this extreme position of science and technology being of principal importance in explaining that part of growth not explicable by capital and labour. He thought this was right. The papers by Professors Mansfield and Freeman and Dr. Nabseth had shown that we must look more seriously at the less technological and scientific aspects of the process. Professor Sargent's own results were not inconsistent with this. They tended to show that the influence of scientific inputs in the United Kingdom had been felt most in the research-intensive industries and not over-all. The three contributions he had mentioned had provided us with the sort of conceptual framework which was needed for a systematic study of the processes of innovation. We had a theory, originally formulated by Professor Mansfield, now applied and tested, and this had led us at this conference to a degree of consensus. This needed to be extended to factors which entered into the growth process other than the rate of diffusion. It did not answer how important the innovations which were thus diffused had been in the growth process or the factors which had led to the emergence of the innovation in the first place. A wider kind of conceptual framework must be developed to deal with those other questions. Professor Sargent said that he put himself firmly in the macro camp in looking for this conceptual model, without wishing to diminish the importance of the micro camp, so well represented by Professor Williams. The researcher who model-builds must contribute detailed information of individual firms as a result of government policy. The importance of the macro approach was in attempting to formulate a conceptual approach within which the more detailed work could go on.

Professor Simai agreed that Professor Sargent had raised a very important issue about the underlying factors of science and technology in the last twenty to twenty-five years. He felt that the role of the economist was secondary in persuading governments. It was like asking whether the cock crows because it is morning or is it morning because the cock crows? The development of science and technology since 1945 had been based on the huge market created by the vast and ramified intervention of governments in the field of science and technology. We could trace almost all important developments in science and technology to government, and often to the defence sector. This was important to us in evaluating the future. He wanted to ask Professor Williams how, on the basis of all that we had discussed, he visualised the further rate of growth of technological innovation. Would it continue on this basis, and were the policy measures of government and private firms sufficient? Professor Simai doubted that this level would continue, and reminded us that Mr. King had pointed out the limiting factors, particularly those on the social side and had suggested that the rate of innovation might slow down in the 1970s.

Professor Oshima said that Professor Williams' paper had raised many basic issues in which he himself was very interested. In framing policy for market economies we were very much concerned about government functions and intervention in the market mechanism. The Government budget for R & D was one such intervention, but the problem was not unique to R & D but concerned the whole budget allocation over many fields. He would like to ask how far in the present state of the art there were any satisfactory economic criteria by which to judge whether the budget allocation was right? Could the policy maker find the impact of government intervention on the policy instrument concerned and feed this information back to the decision maker? As Professor Simai had stressed, most of the important innovations come from a strong effort of manpower and investment in one field. Much more concern was now attached to social problems, as distinct from space or high-performance aircraft. This was very important, because some objectives were not internal to industry but derived from the needs of society. The government budget was very much related to objective purposes, to science and technological needs. This reflected a considerable influence from the market mechanism. He would therefore like to ask Professor Williams how these aspects of optimal allocation of government resources should be treated from an economic point of view.

Professor Lundberg was grateful to Professor Matthews for a fine survey of the central issues. He would like to ask both Professor Williams and Professor Mansfield why we had not been very much concerned with the concept of the technological gap in this conference. It did exist. But did it depend on so many factors that we were afraid of it? When we tried to explain the difference in productivity between the United States and European countries, did we not have the impression that the gap did not in fact exist between the most advanced plants in the different countries? He wanted to suggest that the most advanced firms were on about the same level with respect to knowledge and application of new technology and would like to have Professor Williams' views on this. He believed that most of the gap was represented by the varying lengths of the tails of laggards. Their policies and actions determined the speed of diffusion and the lags often depended on rational profitability calculations. The question was to what extent the technological gap – and average gaps in productivity – was due to the long tails? To what extent should we accept gaps in a branch of industry because of old capital and skills still in use which on cost grounds should not be discarded? Professor Lundberg said that he had missed Professor Matthew's point when referring to Professor Galbraith on sovereignty. Managements as they picked up new techniques tended to be a ruthless bunch of people whose energy to apply new technology was braked by all kinds of social conditions such as attitudes of consumers, trade unions and not least individual workers. We had said little about these social considerations and their effects on applications of new technological ideas. The attitudes of social groups would in various ways influence the rate of diffusion of the adoption of technology. For example, improvement in job environment of employees was an important side of new technology. Workers had their views and he predicted a changed type of management where such attitudes and preferences of employees had an increasing role to play in decisions regarding the adoption of new techniques. The demand of workers for satisfactory conditions of life was important, and the attitude of, say, British trade unions to the introduction of new technology had some rational basis. The government attitude appeared to be divided in these respects. At the same time as government might want to rush ahead and have high rates of productivity gains it was aiming at improvement of natural and human environment. Governments tended to think they knew what was best both for management and labour, and they would play an increasing role in controlling the rate of diffusion of new technology.

Professor Matthews, in reply to Professor Lundberg, emphasised the distinction between two meanings attached to the technological gap. The first was the proposition that productivity in the United States exceeds that in Europe, and that was indisputable. The second was the proposition that this gap was increasing. The latter interpretation was at the centre of much of the discussion of the technological gap. But its empirical foundations were very weak. With regard to the point raised by Professor Lundberg about international differences in productivity arising from the poor performance of laggards in certain countries, he wanted to suggest that while this was certainly possible, it had not yet been established. Calculations by Downie and others had not clearly shown a larger spread in efficiency in the United Kingdom than elsewhere. Regard had also to be paid to the experience of multinational companies which tended to find that identical plants had different labour productivity in different countries.

Professor Robinson confirmed this last point and added that research in Cambridge on the comparative performance of international firms in different centres had shown that productivity differences, using identical equipment, were very similar to the general differences in productivity between the two countries. In trying to explain why governments should support R & D Professor Robinson wanted to add another reason. Almost all economic, social and political problems were more easily soluble with higher income per head and a higher rate of growth of income per head. He also wanted to suggest that one could not get effective basic research out of firms of only 50–100 people; on the other hand the pooling of the resources of small firms with government help could produce organisations that could do effective research and this too would justify a subsidy from the social point of view.

But he wanted to come back to the big issues raised by Professor Williams and Professor Matthews. The I.E.A. had been pursuing the causes of growth for nearly twenty years, partly for intellectual reasons, but also because of a desire to provide guidance to developing countries. We had looked over the years at various possible explanations. In one of our earliest conferences in 1953 on economic progress we had become fully convinced that capital investment was only a small part of the explanation of growth. We had pursued this further and it had brought us naturally enough to the residual factor. We pursued this will-o'-the-wisp in a series of I.E.A. conferences through the effects of capital, of education, of labour quality and productivity, through to this conference on R & D, which had been intended to be the last effort to find an explanation of the residual factor in growth. But each explanation had produced a series of alibis and no explanation of the residual. The experts in each field believed that their particular field was relatively unimportant as an explanation. Professor Robinson went on to ask if there was something large in the explanation of growth which we had missed or had failed to identify in the factors we had studied. Was there something in growth that has defied economists? What had escaped us? If it was not capital, not education, not R & D, and not improved transfer of technology, where lay the explanation of growth if it was not in one of these? We could not continue to have an amorphous residual factor of which there was no explanation, not even a qualitative explanation. And yet this was the state in which he found himself at the end of this conference. Professor Matthews, sitting by his side, was suggesting 'motivation'. Perhaps we ought to have a conference on that in a future year. We could not leave this central problem of economics indefinitely unanswered.

Dr. Teubal agreed with Professor Sargent on the desirability of evolving a normative framework within which it would be possible to evaluate questions of policy and allocation. On the methodological question he said that everybody had agreed on the value of micro economic research but the usefulness of creating a macro framework for analysing R & D allocation was not so clear. There appeared to be basically two views; the one held by Professor Mansfield amongst others was that the processes of R & D and diffusion were so complex that it was not useful to create a normative framework, but this did not relieve us from the need to create such a framework to assist in the practical problems of allocating R & D expenditure and manpower. Professor Sargent's view appeared to be that despite complexity we must have macro models. Dr. Teubal wanted to know the views of other participants on this point. His own opinion was that we were in a situation, similar to that described by Professor Robinson in an earlier session in which a firm had to decide whether or not to invest today in equipment whose technology was changing rapidly. Our decision now was whether to attempt formulation of additional macro models before knowing more about the underlying micro-processes involved. Dr. Teubal believed a lot of research has to be done at the micro level before useful normative frameworks could be evolved, otherwise there was a danger of such frameworks becoming isolated from the empirical work going on in this area, as had happened in other areas of economics.

Professor Tsuru said that the more he reflected on the wisdom contained in Professor Williams' paper, but often hidden between the lines, the more struck he was with the inadequacy of the analytical framework for dealing with government science policy. It was a very practical problem. For example the Japanese nation through its Central Bank had recently acquired three billion dollars which by many was considered a complete loss of such resources. The government would be spending three hundred million dollars to take over the defence establishments on Okinawa Island originally occupied by the United States. At the same time the government was debating whether they should devote thirty-million dollars to scientific research funds to be spent by the whole body of Japanese universities. The Okinawa local government had asked for one hundred thousand dollars to strengthen its university, and this has been refused by the government. These figures give some indication of the orders of magnitude of the sums involved in acquiring dollars, defence of Okinawa, university research funds and the grant required by Okinawa university. How could economists compare the social marginal benefits in these widely different areas? And with such widely different magnitudes? Science policy involved a much broader framework than we had now. Professor Williams was well aware of this. Professor Tsuru said he would like to plead that in tackling science policy we must be aware of the distance between the abstractness of an analytical framework and the concreteness of the framework we had to have in making decisions. He himself had often despaired of making effective use of any economic analytical tools in negotiations with governments. He felt that Professor Williams had combined the principles of economics with practical wisdom.

Professor Williams replied thanking Professor Matthews for introducing his paper at short notice in such an extremely competent way. He would like to put a gloss on one or two of the points made by Professor Matthews in referring to his treatment of defence as oligopolistic competition in its most harmful form. Professor Williams agreed that this was partly true, but he would put it rather differently. He had suggested that it would be a thoroughly good thing of the oligopolists got together and entered into a very restrictive combination to restrict competition completely. This fell outside the economists' usual treatment of imperfect competition. It would be good to get countries together and get rid of defence R & D expenditures altogether. On the problem of the environment Professor Williams considered that this was being much more consciously considered in all countries. In the advanced countries we were now so much richer that there were other values on which we must place a greater emphasis than had previously been done. Also with the growth of industrialisation we were now destroying more of the environment than we did in the past. More research was needed. But he hoped that was not all we should say, because in some cases we already knew what we must do to stop it. The major problem was to generate sufficient political will to stop certain activities, some of which required international control – for example the effects on the seas and on fishing brought about by shipping and pollution more generally. In saying that we already knew what to do we must put the emphasis on the need for action. If the government took action, in many cases this would lead to more research because firms stopped from behaving as they now did would have to find ways to reduce the cost of meeting the new standards.

Professor Williams went on to emphasise that he was not in agreement with Professor Galbraith. He considered that communities had been short-changed due to the lack of legislative action or administrative control. In reply to Professor Matthews' important point on manpower planning, he hoped that more work would be done on this, but did not consider this to be easy. We had talked about the proportion of Q.S.E.s to the work force without looking closely at what sort of Q.S.E.s and what balance within the stock. There were enormous differences here and we needed many more micro studies before we could take the issues further. Professor Matthews had suggested that he disagreed with Professor Griliches on the question of firms doing their best. He thought that all the micro studies had brought out that all firms were capable of making very costly departures from the best rules. We tended to assume that if they were sufficiently profit-orientated this would be ironed out. This was not so. There was intensive competition within society. But human stupidity often came out and there were many extraordinary decisions made even within firms which were progressing well. In making particular studies of individual decisions of considerable magnitude it was easy to identify mistakes. All the firms which had co-operated in their studies insisted that they produce for them checklists of questions which should be asked at various stages, so that they could ensure that their senior managers used them and did not overlook obvious items.

On Dr. Sedov's very important questions about the relations between economics, politics and society, as to whether policy issues were permanent or temporary, and as to the fact that R & D had brought out important changes, Professor Williams said that he was a sufficient scholar of Marx to know what he had in mind – changes in technology, changing appropriate social and political norms – but he had never seen the crystal ball that would give him the answers to the questions raised. He would guess that in Western societies the set-up would become less clear cut than it was at present. There were certain forms of technology where we were likely to have one or two automated plants producing all society needed, for example in flour milling, and there would be no reason for further technical change. Market pressures would not be important but there would be tremendous emphasis on market mechanisms for as long as he could see. We could dissociate competition from private ownership of private property. This had gone a long way and he thought it would go further. Replying to Professor Sargent's suggestion that governments had become involved because economists had sold the idea prematurely that the residual factor was important and that this was to a considerable extent due to R & D, Professor Williams said he was not so sure. He tended to agree more with Professor Simai in questioning this. The first time he became aware of the residual was in Aukrust's writings, where he suggested that the residual seemed to be principally due to people and therefore more investment in people than in plant would be valuable. This statement did not, in fact, differ very much from Pigou's in the *Economics of Welfare*. Professor Williams said he did not think this argument was very persuasive. The single most persuasive example had been the atomic bomb. Here invention had been planned and produced according to rule. It was costly but it was done. And many scientists and engineers said this was to be the pattern for the future. He clearly remembered how strongly this argument had been used after the war in Britain, when the Scientists in Industry Committee was formed. This tended to take the view that we had provided the wherewithall, and if we were not getting tremendous growth it was because of the nature of management or the shortage of risk capital. This argument had been put forward in many countries. An article by an American chemical engineer – Ewell – suggested that if R & D was not increased, growth would not increase. But this became a nonsense formula when it was clear that almost all manpower would be in science and technology before the year 2000. On Professor Sargent's other point, on which Dr. Teubal had also commented, about the extension of the type of work by Mansfield and others on diffusion to cover the generation of new ideas which were worth diffusing, Professor Williams agreed that it was very desirable to get on with this work, and that macro and micro studies ought to go together. They were essentially complementary. He aligned himself more with Dr. Teubal than Professor Sargent and had a strong preference for building macro studies from the micro, to proceed, that is to say by aggregation and not disaggregation. Professor Simai had asked whether the rate of technical change would continue with growing emphasis on social factors. He had touched on this in his earlier remarks on conservation. In the main he thought that it would tend to encourage different types of research and innovation. He did not pretend to know what the future of technological change would be. There was enormous scope for further diffusion and in fact it seemed to be going on at an increasing rate. Whether this rate would continue he did not know. The present under-developed countries provided some very knotty problems which were not beyond the wit of man to handle, if there was the will to develop. There was, of course, often conflict in developed countries. He had in mind, for example, Professor Lundberg's point of the resistance of trade unions to change. This was a continuing problem. On the one hand the unions wanted a great increase in real wages resulting from technical change, but they did not want the consequential changes in social forms. They had the will for the end but unwillingness for the means, and this was probably just as acute in the less-developed countries.

On Professor Oshima's question about whether intervention in market economies would go further, Professor Williams said that he did not regard a market economy as in some sense inevitable. It was a piece of social engineering or control. Whether one had more or less intervention depended upon which forms of social engineering you wanted. There were a variety of measures and we could deal more smoothly with many problems if we did not regard market mechanism as a part of nature. The community had created competition for its own purposes. On the question of allocating resources between science and other activities Professor Williams said that he had no real answers. He thought that concern for special problems might restrict technical change, though this was uncertain. In the end the question hinged on what fraction of our resources we wanted to put into creating resources for the future. There was a paradox here. The most advanced countries were less willing to maintain a high rate of investment in creation of future resources. The paradox was that advanced countries were putting in less in proportion but more in total than the developing countries. As a result of this there was greater interest in capital saving innovations.

Professor Lundberg's question on the technological gap had partly been answered by Professor Matthews, but he wished to add that one of the fears about the technological gap had been that technological leads would be cumulative, leading to a predominance of the United States and the U.S.S.R. There was, however, nothing in past experience to justify this proposition. When Mr. Wilson referred to Europe becoming nations of industrial helots and politicians in France said the same thing, this was good politics but in terms of past experience was nonsense. There was nothing in the nature of new technology which might justify the argument that we were now in an entirely new situation. The experience of the past five to six years ahd confirmed his opinions on this.

Professor Robinson's question whether there remained something in the residual factor which had defied analysis of economists was a fascinating and important one. He could only say that it seemed to be so. And as to his point about the justification of national subsidy where there were economies of scale to research, here we could go back to Professor Robinson's own *Structure of Competitive Industry*. He had pointed out that the optimum in production and in marketing and in other respects might be lower than the optimum in research. In this case one could justify a subsidy. Another case arose when research was so important that the optimum size itself was changed. Professor Tsuru's last point had already been touched on when he said that he despaired of economists using the right theoretical framework. He agreed with this. But now that he had ceased to be an active economist he could only say that it was immoral to despair – that others should get on with it.

Index